Mathematical Lectures from Peking University

For further volumes:
www.springer.com/series/11574

Chengming Bai • Jürgen Fuchs • Yi-Zhi Huang •
Liang Kong • Ingo Runkel • Christoph Schweigert
Editors

Conformal Field Theories and Tensor Categories

Proceedings of a Workshop Held at Beijing International Center for Mathematical Research

Springer

Editors
Chengming Bai
Chern Institute of Mathematics
Nankai University
Tianjin, People's Republic of China

Liang Kong
Institute for Advanced Study
Tsinghua University
Beijing, People's Republic of China

Jürgen Fuchs
Theoretical Physics
Karlstad University
Karlstad, Sweden

Ingo Runkel
Department of Mathematics
University of Hamburg
Hamburg, Germany

Yi-Zhi Huang
Department of Mathematics
Rutgers University
Piscataway, NJ, USA

Christoph Schweigert
Department of Mathematics
University of Hamburg
Hamburg, Germany

ISSN 2197-4209
ISSN 2197-4217 (electronic)
ISBN 978-3-662-52470-1
ISBN 978-3-642-39383-9 (eBook)
DOI 10.1007/978-3-642-39383-9
Springer Heidelberg New York Dordrecht London

Mathematics Subject Classification: 16T05, 17B37, 17B69, 17B81, 18D10, 81T40, 81T45

Softcover reprint of the hardcover 1st edition 2014

Printed on acid-free paper

Springer is part of Springer Science+Business Media (www.springer.com)

Preface

In the past thirty years, (two-dimensional) conformal field theory has been developed into a deep, rich and beautiful mathematical theory and the study of conformal field theories and their applications in mathematics and physics has become an exciting area of mathematics. It has led to new ideas, surprising results and beautiful solutions of various problems in different branches in mathematics and physics, including, but not limited to, algebra, number theory, combinatorics, topology, geometry, critical phenomena, quantum Hall systems, disorder systems, quantum computing and string theory, and is expected to lead to many more.

During June 13 to June 17, 2011, a workshop "Conformal field theories and tensor categories" was held at Beijing International Center for Mathematical Research, Peking University, Beijing, China. This workshop was one of the main activities of a one-semester program on quantum algebra from February to July, 2011 at Beijing International Center for Mathematical Research. It was the aim of the workshop to bring together experts from several different areas of mathematics and physics who are involved in the new developments in conformal field theories, tensor categories and related research directions. Correspondingly the areas covered by the workshop were broad, including conformal field theories, tensor categories, quantum groups and Hopf algebras, representation theory of vertex operator algebras, nets of von Neumann algebras, topological order and lattice models and other related topics. Each of these fields was represented by leading experts.

The simplest class of conformal field theories are rational conformal field theories. In 1988, Moore and Seiberg obtained polynomial equations for the fusing, braiding and modular transformations in rational conformal field theory. They observed that some of these equations are analogous to some properties of tensor categories. Later, a notion of modular tensor category was formulated mathematically and examples of modular tensor categories were constructed from representations of quantum groups. The work of Moore and Seiberg can be interpreted as deriving a modular tensor category structure from a rational conformal field theory. Since then, the theory of various tensor categories has been greatly developed and has been applied to different areas of mathematics and physics. Now the theory of tensor categories not only provides a unifying language for various parts of mathematics and applications of mathematics in physics, but also gives deep results and fundamental structures in different branches of mathematics and physics.

On the other hand, though a number of examples of modular tensor categories were constructed at the time that the notion of modular tensor category was introduced, it took many years and a lot of efforts for mathematicians to directly construct the modular tensor categories conjectured to appear in rational conformal field theories. Many mathematicians, including in particular Kazhdan-Lusztig, Beilinson-Feigin-Mazur, Finkelberg, Huang-Lepowsky, and Bakalov-Kirillov, contributed in 1990's and early 2000's to the construction of the particular class of examples of the modular tensor categories associated with the Wess-Zumino-Novikov-Witten models (the rational conformal field theories associated with suitable representations of affine Lie algebras). However, the construction of even this particular class of examples was not complete until 2005 when Huang gave a general construction of all the modular tensor categories conjectured to be associated with rational conformal field theories. The corresponding chiral rational conformal field theories are thereby largely under control. Indeed, many problems in rational conformal field theories have meanwhile been solved.

In the workshop, new developments beyond rational conformal field theories and modular tensor categories and new applications in mathematics and physics were presented by top experts. Here we would like to mention especially the following:

1. Construction of interesting tensor categories from representation categories of Hopf algebras, as reviewed by Andruskiewitsch in his overview talk and also in the contribution by Andruskiewitsch, Angiono, García Iglesias, Torrecillas and Vay in this volume.
2. New categorical techniques and structures in tensor categories, as reviewed by Ostrik in his overview talk. One also should include here the Witt group as discussed by Nikshych and Davydov and Hopf-monadic techniques as explained by Virelizier. In a sense, Semikhatov's contribution in this volume using Hopf-algebraic structures in representation categories interpolates between this point and the preceding point.
3. Applications to topological phases and gapped systems as reviewed by Wen in his overview talk and also in the contribution by Wen and Wang in this volume. The study of the Levin-Wen model as discussed by Wu is an important example of such applications.
4. Realization of the tensor-categorical structures in lattice models as in Fendley's overview and Gainutdinov's talk.
5. New developments in the representation theory of vertex operator algebras, especially the nonsemisimple theory corresponding to logarithmic conformal field theory, as reviewed by Lepowsky's overview talk and in the contribution by Huang, Lepowsky and Zhang in this volume. Recent results on representations and the structure of the representation category were reported by Adamovic, Arike, Milas and Miyamoto and also in the contributions by Adamovic and Milas and by Miyamoto in this volume. To some extent, Tsuchiya's talk also went in this direction. Connections to logarithmic conformal field theory were discussed

by Runkel and Semikhatov and also in the contributions by Runkel, Gaberdiel and Wood and by Semikhatov in this volume.

In the workshop, there were 21 invited talks by mathematicians and physicists from Argentina, China, Croatia, France, Germany, Japan, Russia and USA. Some of the invited talks were given by young researchers. The participants benefited a lot from communicating results between the various disciplines and from the attempt to understand them in the framework of conformal field theories and tensor categories. These attempts gave rise to further questions during and after the talks, and, maybe even more importantly, also resulted in numerous and lively private discussions among the participants.

The workshop also had important training impact on students. A number of undergraduate and beginning graduate students in the Enhanced Program for Graduate Study in Beijing International Center for Mathematical Research participated in the workshop. They benefited greatly from the talks, especially the five overview talks, and from discussions with active researchers in the workshop.

The present volume is a collection of seven papers that are either based on the talks presented in the workshop or are extensions of the material presented in the talks in the workshop. We believe that the papers in this volume will be useful to everyone who is interested in conformal field theories, tensor categories and related topics. We hope that these papers will also inspire more research activities in these directions.

We are very grateful to Beijing International Center for Mathematical Research and the National Science Foundation in USA for the funding and support of the workshop. We thank the staff at Beijing International Center for Mathematical Research for their help during the workshop. We thank all the participants, the speakers and, especially, the authors whose papers are included in this volume and the anonymous referees for their careful reviews of the papers included in this volume.

Tianjin, People's Republic of China — Chengming Bai
Karlstad, Sweden — Jürgen Fuchs
Piscataway, NJ, USA — Yi-Zhi Huang
Beijing, People's Republic of China — Liang Kong
Hamburg, Germany — Ingo Runkel
Christoph Schweigert

Contents

From Hopf Algebras to Tensor Categories

N. Andruskiewitsch, I. Angiono, A. García Iglesias, B. Torrecillas, and C. Vay

Abstract This is a survey on spherical Hopf algebras. We give criteria to decide when a Hopf algebra is spherical and collect examples. We discuss tilting modules as a mean to obtain a fusion subcategory of the non-degenerate quotient of the category of representations of a suitable Hopf algebra.

Mathematics Subject Classification (2000) 16W30

1 Introduction

It follows from its very definition that the category $\operatorname{Rep} H$ of finite-dimensional representations of a Hopf algebra H is a tensor category. There is a less obvious way to go from Hopf algebras with some extra structure (called spherical Hopf algebras) to tensor categories. Spherical Hopf algebras and the procedure to obtain a tensor category from them were introduced by Barrett and Westbury [19, 20], inspired by previous work by Reshetikhin and Turaev [69, 70], in turn motivated to give a mathematical foundation to the work of Witten [76].

A spherical Hopf algebra has by definition a group-like element that implements the square of the antipode (called a pivot) and satisfies the left-right trace symmetry

N. Andruskiewitsch (✉) · I. Angiono · A. García Iglesias · C. Vay
FaMAF-CIEM (CONICET), Universidad Nacional de Córdoba, Medina Allende s/n, Ciudad Universitaria, 5000 Córdoba, República Argentina
e-mail: andrus@famaf.unc.edu.ar

I. Angiono
e-mail: angiono@famaf.unc.edu.ar

A. García Iglesias
e-mail: aigarcia@famaf.unc.edu.ar

C. Vay
e-mail: vay@famaf.unc.edu.ar

B. Torrecillas
Dpto. Álgebra y Análisis Matemático, Universidad de Almería, 04120 Almería, Spain
e-mail: btorreci@ual.es

C. Bai et al. (eds.), *Conformal Field Theories and Tensor Categories*,
Mathematical Lectures from Peking University, DOI 10.1007/978-3-642-39383-9_1,

condition (3.2). The classification (or even the characterization) of spherical Hopf algebras is far from being understood, but there are two classes to start with. Let us first observe that semisimple spherical Hopf algebras are excluded from our considerations, since the tensor categories arising from the procedure are identical to the categories of representations. Another remark: any Hopf algebra is embedded in a pivotal one, so that the trace condition (3.2) is really the crucial point. Now the two classes we mean are

- Hopf algebras with involutory pivot, or what is more or less the same, with $\mathcal{S}^4 = \mathrm{id}$. Here the trace condition follows for free, and the quantum dimensions will be (positive and negative) integers.
- Ribbon Hopf algebras [69, 70].

It is easy to characterize pointed or copointed Hopf algebras with $\mathcal{S}^4 = \mathrm{id}$; so we have many examples of (pointed or copointed) spherical Hopf algebras with involutory pivot, most of them not even quasi-triangular, see Sect. 3.6. On the other hand, any quasitriangular Hopf algebra is embedded in a ribbon one [69]; combined with the construction of the Drinfeld double, we see that any finite-dimensional Hopf algebra gives rise to a ribbon one. So, we have plenty of examples of spherical Hopf algebras, although of a rather special type.

The procedure to get a tensor category from a spherical Hopf algebra H consists in taking a suitable quotient $\underline{\operatorname{Rep}} H$ of the category $\operatorname{Rep} H$. This appears in [20] but similar ideas can be found elsewhere, see e.g. [35, 55]. The resulting spherical categories are semisimple but seldom have a finite number of irreducibles, that is, they are seldom fusion categories in the sense of [32]. We are interested in describing fusion tensor subcategories of $\underline{\operatorname{Rep}} H$ for suitable H. This turns out to be a tricky problem. First, if the pivot is involutive, then the fusion subcategories of $\underline{\operatorname{Rep}} H$ are integral, see Proposition 3.12. The only way we know is through tilting modules; but it seems to us that there is no general method, just a clever recipe that works. This procedure has a significant outcome in the case of quantum groups at roots of one, where the celebrated Verlinde categories are obtained [3]; see also [72] for a self-contained exposition and [62] for similar results in the setting of algebraic groups over fields of positive characteristic. One should also mention that the Verlinde categories can be also constructed from vertex operator algebras related to affine Kac-moody algebras, see [18, 45, 46, 50–53] and references therein; the comparison of these two approaches is highly non-trivial. Another approach, at least for $SL(n)$, was proposed in [40] via face algebras (a notion predecessor of weak Hopf algebras).

The paper is organized as follows. Section 2 contains some information about the structure of Hopf algebras and notation used later in the paper. Section 3 is devoted to spherical Hopf algebras. In Sect. 4 we discuss tilting modules and how this recipe would work for some finite-dimensional pointed Hopf algebras associated to Nichols algebras of diagonal type, that might be thought of as generalizations of the small quantum groups of Lusztig.

2 Preliminaries

2.1 Notations

Let $\Bbbk$ be an algebraically closed field of characteristic 0 and $\Bbbk^\times$ its multiplicative group of units. All vector spaces, algebras, unadorned Hom and $\otimes$ are over $\Bbbk$. By convention, $\mathbb{N} = \{1, 2, \dots\}$ and $\mathbb{N}_0 = \mathbb{N} \cup \{0\}$.

Let G be a finite group. We denote by $Z(G)$ the center of G and by $\operatorname{Irr} G$ the set of isomorphism classes of irreducible representations of G. If $g \in G$, we denote by $C_G(g)$ the centralizer of g in G. The conjugacy class of g is denoted by $\mathcal{O}_g$ or by $\mathcal{O}_g^G$, when emphasis on the group is needed. The group algebra of G is denoted $\Bbbk G$, while its dual is $\Bbbk^G$; recall that this is the algebra of functions on G.

Let A be an algebra. We denote by $Z(A)$ the center of A. The category of A-modules is denoted A-Mod; the full subcategory of finite-dimensional objects is denoted A-mod. The set of isomorphism classes of irreducible[1] objects in an abelian category $\mathcal{C}$ is denoted $\operatorname{Irr}\mathcal{C}$; we use the abbreviation $\operatorname{Irr} A$ instead of $\operatorname{Irr} A$-mod.

2.2 Tensor Categories

We refer to [18, 31–33, 60, 64] for basic results and terminology on tensor and monoidal categories. A monoidal category is one with tensor product and unit, denoted **1**; thus End(**1**) is a monoid. A monoidal category is rigid when it has right and left dualities. In this article, we understand by tensor category a monoidal rigid abelian $\Bbbk$-linear category, with End(**1**) $\simeq \Bbbk$. A particular important class of tensor categories is that of fusion categories, that is semisimple tensor categories with finite set of isomorphism classes of simple objects, that includes the unit object, and finite-dimensional spaces of morphisms. Another important class of tensor categories is that of braided tensor categories, i.e. those with a commutativity constraint $c_{V,W} : V \otimes W \to W \otimes V$ for every objects V and W. A braided vector space is a pair (V, c) where V is a vector space and $c \in GL(V \otimes V)$ satisfies $(c \otimes \mathrm{id})(\mathrm{id} \otimes c)(c \otimes \mathrm{id}) = (\mathrm{id} \otimes c)(c \otimes \mathrm{id})(\mathrm{id} \otimes c)$; this notion is closely related to that of braided tensor category.

2.3 Hopf Algebras

We use standard notation for Hopf algebras (always assumed with bijective antipode); Δ, ε, $\mathcal{S}$, denote respectively the comultiplication, the counit, and the antipode. For the first, we use the Heyneman-Sweedler notation $\Delta(x) = x_{(1)} \otimes x_{(2)}$. The tensor category of finite-dimensional representations of a Hopf algebra H is

[1] That is, minimal, see p. 10.

denoted Rep H instead of H-mod, to stress the tensor structure. There are two duality endo-functors of Rep H composing the transpose of the action with either the antipode or its inverse: if $M \in \operatorname{Rep} H$, then $M^* = \operatorname{Hom}(M, \Bbbk) = {}^*M$, with actions

$$\langle h \cdot f, m\rangle = \langle f, \mathcal{S}(h) \cdot m\rangle, \qquad \langle h \cdot g, m\rangle = \langle g, \mathcal{S}^{-1}(h) \cdot m\rangle,$$

for $h \in H$, $f \in M^*$, $g \in {}^*M$, $m \in M$.

The tensor category of finite-dimensional corepresentations of a Hopf algebra H, i.e. right comodules, is denoted Corep H. The coaction map of $V \in \operatorname{Corep} H$ is denoted $\rho = \rho_V : V \to V \otimes H$; in Heyneman-Sweedler notation, $\rho(v) = v_{(0)} \otimes v_{(1)}$, $v \in V$. Also, the coaction map of a left comodule W is denoted $\delta = \delta_W : W \to W \otimes H$; that is, $\delta(w) = w_{(-1)} \otimes w_{(0)}$, $w \in W$.

A Yetter-Drinfeld module V over a Hopf algebra H is simultaneously a left H-module and a left H-comodule, subject to the compatibility condition $\delta(h \cdot v) = h_{(1)} v_{(-1)} \mathcal{S}(h_{(3)}) \otimes h_{(2)} \cdot v_{(0)}$ for $v \in V$, $h \in H$. The category ${}^H_H\mathcal{YD}$ of Yetter-Drinfeld modules over H is a braided tensor category with braiding $c(v \otimes w) = v_{(-1)} \cdot w \otimes v_{(0)}$, see e.g. [8, 63]; when $\dim H < \infty$, ${}^H_H\mathcal{YD}$ coincides with the category of representations of the Drinfeld double $D(H)$.

Let H be a Hopf algebra. A basic list of H-invariants is

- The group $G(H)$ of group-like elements of H,
- the coradical $H_0 =$ largest cosemisimple subcoalgebra of H,
- the coradical filtration of H.

Assume that H_0 is a Hopf subalgebra of H. In this case, another fundamental invariant of H is the *infinitesimal braiding*, a Yetter-Drinfeld module V over H_0, see [8]. We shall consider two particular cases:

- The Hopf algebra H is *pointed* if $H_0 = \Bbbk G(H)$.
- The Hopf algebra H is *copointed* if $H_0 = \Bbbk^G$ for a finite group G.

Recall that $x \in H$ is $(g, 1)$ skew-primitive when $\Delta(x) = x \otimes 1 + g \otimes x$; necessarily, $g \in G(H)$. If H is generated as an algebra by group-like and skew-primitive elements, then it is pointed.

We shall need the description of *Yetter-Drinfeld modules* over $\Bbbk G$, G a finite group; these are G-graded vector spaces $M = \bigoplus_{g \in G} M_g$ provided with a G-module structure such that $g \cdot M_t = M_{gtg^{-1}}$ for any $g, t \in G$. The category ${}^{\Bbbk G}_{\Bbbk G}\mathcal{YD}$ of Yetter-Drinfeld modules over G is semisimple and its irreducible objects are parameterized by pairs $(\mathcal{O}, \rho)$, where $\mathcal{O}$ is a conjugacy class of G and $\rho \in \operatorname{Irr} C_G(g)$, $g \in \mathcal{O}$ fixed. We describe the corresponding irreducible Yetter-Drinfeld module $M(\mathcal{O}, \rho)$. Let $g_1 = g, \dots, g_m$ be a numeration of $\mathcal{O}$ and let $x_i \in G$ such that $x_i g x_i^{-1} = g_i$ for all $1 \le i \le m$. Then

$$M(\mathcal{O}, \rho) = \operatorname{Ind}_{C_G(g)}^G V = \bigoplus_{1 \le i \le m} x_i \otimes V.$$

Let $x_i v := x_i \otimes v \in M(\mathcal{O}, \rho)$, $1 \le i \le m$, $v \in V$. The Yetter-Drinfeld module $M(\mathcal{O}, \rho)$ is a braided vector space with braiding given by

$$c(x_i v \otimes x_j w) = g_i \cdot (x_j w) \otimes x_i v = x_h \rho(\gamma)(w) \otimes x_i v \tag{2.1}$$

for any $1 \leq i, j \leq m$, $v, w \in V$, where $g_i x_j = x_h \gamma$ for unique h, $1 \leq h \leq m$, and $\gamma \in C_G(g)$. Now the categories ${}^{\Bbbk G}_{\Bbbk G}\mathcal{YD}$ and ${}_{\Bbbk G}^{\Bbbk^G}\mathcal{YD}$ are tensor equivalent, so that a similar description of the objects of the latter holds, see e.g. [9], or more generally [6, 66].

The following notion is appropriate to describe all braided vector spaces arising as Yetter-Drinfeld modules over some finite abelian group. A braided vector space (V, c) is of *diagonal type* if there exist $q_{ij} \in \Bbbk^\times$ and a basis $\{x_i\}_{i \in I}$ of V such that $c(x_i \otimes x_j) = q_{ij} x_j \otimes x_i$, for each pair $i, j \in I$. In such case, we say that $i, j \in I$ are *connected* if there exist $i_k \in I$, $k = 0, 1, \dots, n$, such that $i_0 = i$, $i_n = j$, and $q_{i_{k-1}, i_k} q_{i_k, i_{k-1}} \neq 1$, $1 \leq k \leq n$. It establishes an equivalence relation on I. The equivalence classes are called connected components, and V is *connected* if it has a unique component.

When $H = \Bbbk \Gamma$, where Γ is a finite abelian group, each $V \in {}^H_H\mathcal{YD}$ is a braided vector space of diagonal type. Indeed, $V = \bigoplus_{g \in \Gamma, \chi \in \widehat{\Gamma}} V_g^\chi$, where $V_g = \{v \in V \mid \delta(v) = g \otimes v\}$, $V^\chi = \{v \in V \mid g \cdot v = \chi(g) v \text{ for all } g \in \Gamma\}$, $V_g^\chi = V^\chi \cap V_g$. Note that $c(x \otimes y) = \chi(g) y \otimes x$, for each $x \in V_g$, $g \in \Gamma$, $y \in V^\chi$, $\chi \in \widehat{\Gamma}$. On the other hand, we can realize every braided vector space of diagonal type as a Yetter-Drinfeld module over the group algebra of an appropriate abelian group.

2.4 Nichols Algebras

Let H be a Hopf algebra. We shall say *braided* Hopf algebra for a Hopf algebra in the braided tensor category ${}^H_H\mathcal{YD}$. Given $V \in {}^H_H\mathcal{YD}$, the *Nichols algebra* of V is the braided graded Hopf algebra $\mathcal{B}(V) = \bigoplus_{n \geq 0} \mathcal{B}^n(V)$ satisfying the following conditions:

- $\mathcal{B}^0(V) \simeq \Bbbk$, $\mathcal{B}^1(V) \simeq V$ as Yetter-Drinfeld modules over H;
- $\mathcal{B}^1(V) = \mathcal{P}(\mathcal{B}(V))$, the set of primitives elements of $\mathcal{B}(V)$;
- $\mathcal{B}(V)$ is generated as an algebra by $\mathcal{B}^1(V)$.

The Nichols algebra of V exists and is unique up to isomorphism. We sketch a way to construct $\mathcal{B}(V)$ and prove its unicity, see [8]. Note that the tensor algebra $T(V)$ admits a unique structure of graded braided Hopf algebra in ${}^H_H\mathcal{YD}$ such that $V \subseteq \mathcal{P}(V)$. Consider the class $\mathfrak{S}$ of all the homogeneous two-sided Hopf ideals $I \subseteq T(V)$ such that I is generated by homogeneous elements of degree ≥ 2 and is a Yetter-Drinfeld submodule of $T(V)$. Then $\mathcal{B}(V)$ is the quotient of $T(V)$ by the maximal element $I(V)$ of $\mathfrak{S}$. Thus, the canonical projection $\pi : T(V) \to \mathcal{B}(V)$ is a Hopf algebra surjection in ${}^H_H\mathcal{YD}$.

Braided vector spaces of diagonal type with finite-dimensional Nichols algebra were classified in [41]; the explicit defining relations were given in [15], using the results of [14]. Presently we understand that the list of braidings of diagonal type with finite-dimensional Nichols algebra given in [41] is divided into three parts:

(1) Standard type [13], comprising Cartan type [7];
(2) Super type [12];
(3) Unidentified type [16].

3 Spherical Hopf Algebras

3.1 Spherical Hopf Algebras

The notion of *spherical Hopf algebra* was introduced in [20]: this is a pair (H, ω), where H is a Hopf algebra and $\omega \in G(H)$ such that

$$\mathcal{S}^2(x) = \omega x \omega^{-1}, \quad x \in H, \tag{3.1}$$

$$\mathrm{tr}_V(\vartheta\omega) = \mathrm{tr}_V\big(\vartheta\omega^{-1}\big), \quad \vartheta \in \mathrm{End}_H(V), \tag{3.2}$$

for all $V \in \operatorname{Rep} H$. We say that $\omega \in G(H)$ is a *pivot* when it satisfies (3.1); pairs (H, ω) with ω a pivot are called *pivotal Hopf algebras*. The pivot is not unique but it is determined up to multiplication by an element in the group $G(H) \cap Z(H)$. A *spherical element* is a pivot that fulfills (3.2).

The implementation of the square of the antipode by conjugation by a group-like, condition (3.1), is easy to verify. For instance, ω should belong to the center of $G(H)$; thus, if this group is centerless, then (3.1) does not hold in H. Further, the failure of (3.1) is not difficult to remedy by adjoining a group-like element (Sect. 2 in [73]). Namely, given a Hopf algebra H, consider a cyclic group Γ of order $\operatorname{ord} \mathcal{S}^2$ with a generator g; let g act on H as $\mathcal{S}^2$. Then the corresponding smash product $E(H) := H\#\Bbbk\Gamma$ is a Hopf algebra where (3.1) holds.[2]

Condition (3.2) is less apparent. If H is a finite-dimensional Hopf algebra and $\omega \in H$ a pivot, then (3.2) holds in the following instances:

- ω is an involution.
- There exists a Hopf subalgebra K of H such that $\omega \in K$ and (K, ω) is spherical—since $\mathrm{End}_H(V) \subset \mathrm{End}_K(V)$.
- H is ribbon, see Sect. 3.7.
- All finite-dimensional H-modules are naturally self-dual.

Proof By hypothesis, there exists a natural isomorphism $F : \mathrm{id} \to {}^*$. Let $V \in \operatorname{Rep} H$ and $\vartheta \in \mathrm{End}_H(V)$. Then

$$\begin{aligned}\mathrm{tr}(\vartheta\omega) &= \sum_i \langle \alpha_i, \vartheta\omega v_i\rangle = \sum_i \big\langle \vartheta^*\alpha_i, \omega v_i\big\rangle = \sum_i \big\langle F_V \vartheta F_V^{-1}\alpha_i, \omega v_i\big\rangle \\ &= \sum_i \big\langle F_V \vartheta\omega^{-1} F_V^{-1}\alpha_i, v_i\big\rangle = \mathrm{tr}\big(F_V \vartheta\omega^{-1} F_V^{-1}\big) = \mathrm{tr}\big(\vartheta\omega^{-1}\big).\end{aligned}$$

Here $\{v_i\}$ and $\{\alpha_i\}$ are dual basis of V and V^* respectively. □

Proposition 3.1 *A pivotal Hopf algebra (H, ω) is spherical if and only if* (3.2) *holds for all $S \in \operatorname{Irr} H$.*

[2] If H has finite dimension $n \in \mathbb{N}$, then the order of $\mathcal{S}^2$ is finite; in fact, it divides $2n$, by Radford's formula on $\mathcal{S}^4$ and the Nichols-Zöller Theorem.

Proof The Proposition follows from the following two claims.

Claim 1 *If* (3.2) *holds for* $M_1, M_2 \in \operatorname{Rep} H$, *then it holds for* $M_1 \oplus M_2$.

Indeed, let $h \in \operatorname{End}_H(M_1 \oplus M_2)$. Let $\pi_j : M \to M_j$, $\iota_i : M_i \to M$ be the projection and the inclusion, for $1 \le i, j \le 2$. Then $h = \sum_{1\le i,j\le 2} h_{ij}$, where $h_{ij} = \pi_i \circ h \circ \iota_j \in \operatorname{Hom}_H(M_j, M_i)$. In particular, $h_{ij}\omega = (h\omega)_{ij}$ as linear maps. Now, we have that $\operatorname{tr}_M(h) = \operatorname{tr}_{M_1}(h_{11}) + \operatorname{tr}_{M_2}(h_{22})$ and thus

$$\begin{aligned}\operatorname{tr}_M(h\omega) &= \operatorname{tr}_{M_1}(h\omega)_{11} + \operatorname{tr}_{M_2}(h\omega)_{22} = \operatorname{tr}_{M_1}(h_{11}\omega) + \operatorname{tr}_{M_2}(h_{22}\omega)\\ &= \operatorname{tr}_{M_1}\left(h_{11}\omega^{-1}\right) + \operatorname{tr}_{M_2}\left(h_{22}\omega^{-1}\right) = \operatorname{tr}_{M_1}\left(h\omega^{-1}\right)_{11} + \operatorname{tr}_{M_2}\left(h\omega^{-1}\right)_{22}\\ &= \operatorname{tr}_M\left(h\omega^{-1}\right).\end{aligned}$$

Claim 2 *If* (3.2) *holds for every semisimple* H*-module, then* H *is spherical.*

Let $M \in \operatorname{Rep} H$ and let $M_0 \subset M_1 \subset \cdots \subset M_k = M$ be the *Loewy filtration* of M, that is $M_0 = \operatorname{Soc} M$, $M_{i+1}/M_i \simeq \operatorname{Soc}(M/M_i)$, $i = 0, \dots, j-1$. In particular, M_{i+1}/M_i is semisimple. We prove the claim by induction on the Loewy length k of M. The case $k = 0$ is the hypothesis. Assume $k > 0$; set $S = \operatorname{Soc} M$ and consider the exact sequence $0 \to S \to M \to M/S \to 0$. Hence the Loewy length of $\widetilde{M} = M/S$ is $k-1$ and thus (3.2) holds for it. Also, (3.2) holds for S by hypothesis. Let $f \in \operatorname{End}_H(M)$, then $f(S) \subseteq S$ and thus f induces $f_1 = f_{|S} \in \operatorname{End}_H(S)$ by restriction and factorizes through $f_2 \in \operatorname{End}_H(\widetilde{M})$. Therefore, we can choose a basis of S and complete it to a basis of M in such a way that f in this new basis is represented by $[f] = \left[\begin{smallmatrix} [f_1] & * \\ 0 & [f_2] \end{smallmatrix}\right]$. Also, as ω preserves S and f is an H-morphism, it follows that $[f\omega] = \left[\begin{smallmatrix} [f_1\omega] & * \\ 0 & [f_2\omega] \end{smallmatrix}\right]$ and thus $\operatorname{tr}_M(f\omega) = \operatorname{tr}_M(f\omega^{-1})$. □

Example 3.2 Let H be a basic Hopf algebra, i.e. all finite-dimensional simple modules have dimension 1; when H itself is finite-dimensional, this amounts to the dual of H being pointed. If $\omega \in G(H)$ is a pivot, then H is spherical if and only if $\chi(\omega) \in \{\pm 1\}$ for all $\chi \in \operatorname{Alg}(H, \Bbbk)$. Assume that H is finite-dimensional; then H is spherical if and only if ω is involutive. For, $\omega - \omega^{-1} \in \bigcap_{\chi \in \operatorname{Alg}(H,\Bbbk)} \operatorname{Ker}\chi = \operatorname{Rad} H$, and $\operatorname{Rad} H \cap \Bbbk[\omega] = 0$.

3.2 Spherical Categories

A monoidal rigid category $\mathcal{C}$ is pivotal when X^{**} is monoidally isomorphic to X [34]; this implies that the left and right dualities coincide. For instance, if H is a Hopf algebra and $\omega \in G(H)$ is a pivot, then $\operatorname{Rep} H$ is pivotal (Proposition 3.6 in [20]). In a pivotal category $\mathcal{C}$, there are left and right traces $\operatorname{tr}_L, \operatorname{tr}_R : \operatorname{End}(X) \to$

End($\mathbf{1}$), for any $X \in \mathcal{C}$. If $\mathcal{C} = \operatorname{Rep} H$ and $V \in \operatorname{Rep} H$, then these traces are defined by

$$\operatorname{tr}_L(\vartheta) = \operatorname{tr}_V(\vartheta\omega), \qquad \operatorname{tr}_R(\vartheta) = \operatorname{tr}_V\left(\vartheta\omega^{-1}\right), \quad \vartheta \in \operatorname{End}_H(V). \tag{3.3}$$

A spherical category is a pivotal one where the left and right traces coincide. Thus, $\operatorname{Rep} H$ is a spherical category, whenever H is a spherical Hopf algebra.

Remark 3.3 If $\mathcal{D}$ is a rigid monoidal (full) subcategory of a spherical category $\mathcal{C}$, then $\mathcal{D}$ is also spherical.

3.3 *Quantum Dimensions*

The *quantum dimension* of an object X in a spherical category $\mathcal{C}$ is given by $\operatorname{qdim} V := \operatorname{tr}_L(\operatorname{id}_X)$. In particular, if H is a spherical Hopf algebra, then

$$\operatorname{qdim} M = \operatorname{tr}_M(\omega) = \operatorname{tr}_M\left(\omega^{-1}\right), \quad M \in \operatorname{Rep} H. \tag{3.4}$$

3.3.1 Some Properties of Quantum Dimension

If $\mathcal{C}$ is a spherical tensor category, then End($\mathbf{1}$) $= \Bbbk$, and the function $V \mapsto \operatorname{qdim} V$ is a character of the Grothendieck ring of $\mathcal{C}$. In fact, this map is additive on exact sequences, as in the proof of Proposition 3.1; also

$$\operatorname{qdim} V \otimes W = \operatorname{qdim} V \operatorname{qdim} W, \quad V, W \in \mathcal{C}. \tag{3.5}$$

In consequence, the quantum dimension of any object in a *finite* spherical tensor category $\mathcal{C}$ is an algebraic integer in $\Bbbk$, see Corollary 1.38.6 in [33].

3.3.2 Computing the Quantum Dimension

Let H be a Hopf algebra. Given $L \in \operatorname{Irr} H$, $M \in \operatorname{Rep} H$, we set $(M : L) =$ multiplicity of L in M (i.e. the number of times that L appears as a Jordan-Hölder factor of M). Assume that (H, ω) is spherical. Let $M \in \operatorname{Rep} H$. Then

$$\operatorname{qdim} M = \sum_{L \in \operatorname{Irr} H} (M : L) \operatorname{qdim} L. \tag{3.6}$$

Here is a way to compute the quantum dimension of M: consider the decomposition $M = \bigoplus_{\rho \in \operatorname{Irr} G(H)} M_\rho$ into isotypical components of the restriction of M to

$G(H)$. Since $\omega \in Z(G(H))$, it acts by a scalar z_ρ on the $G(H)$-module affording $\rho \in \operatorname{Irr} G(H)$. Hence

$$\operatorname{qdim} M = \sum_{\rho \in \operatorname{Irr} G(H)} z_\rho \dim M_\rho. \tag{3.7}$$

See [24, 25] for a Verlinde formula and other information on the computation of the quantum dimension in terms of the Grothendieck ring.

3.3.3 Projective Objects Have Null Quantum Dimension

If H is a non-semisimple spherical Hopf algebra, then $\operatorname{qdim} M = 0$ for any $M \in \operatorname{Rep} H$ projective (Proposition 6.10 in [19]). More generally, the following result appears in the proofs of Theorem 2.16 in [31], Theorem 1.53.1 in [33].

Proposition 3.4 *Let $\mathcal{C}$ be a non-semisimple pivotal tensor category. Then* $\operatorname{qdim} P = 0$ *for any projective object P.*

3.4 *The Non-degenerate Quotient*

Let $\mathcal{C}$ be an additive $\Bbbk$-linear spherical category with $\operatorname{End}(\mathbf{1}) \simeq \Bbbk$. For any two objects X, Y in $\mathcal{C}$ there is a bilinear pairing

$$\Theta : \operatorname{Hom}_{\mathcal{C}}(X, Y) \times \operatorname{Hom}_{\mathcal{C}}(Y, X) \to \Bbbk, \quad \Theta(fg) = \operatorname{tr}_L(fg) = \operatorname{tr}_R(gf);$$

$\mathcal{C}$ is non-degenerate if Θ is, for any X, Y. By Theorem 2.9 in [20], see also [74], any additive spherical category $\mathcal{C}$ gives rise a factor category $\underline{\mathcal{C}}$, with the same objects[3] as $\mathcal{C}$ and morphisms $\operatorname{Hom}_{\underline{\mathcal{C}}}(X, Y) := \operatorname{Hom}_{\mathcal{C}}(X, Y)/\mathcal{J}(X, Y)$, $X, Y \in \mathcal{C}$, where

$$\mathcal{J}(X, Y) = \left\{ f \in \operatorname{Hom}_{\mathcal{C}}(X, Y) : \operatorname{tr}_L(fg) = 0, \forall g \in \operatorname{Hom}_{\mathcal{C}}(Y, X) \right\}. \tag{3.8}$$

The category $\underline{\mathcal{C}}$ is an additive non-degenerate spherical category, but it is not necessarily abelian, even if $\mathcal{C}$ is abelian. Clearly, the quantum dimensions in $\operatorname{Rep} H$ and $\underline{\operatorname{Rep}}\, H$ are the same. See [55] for a general formalism of tensor ideals, that encompasses the construction above.

We now give more information on $\underline{\operatorname{Rep}}\, H$ following Proposition 3.8 in [20]. Let us first point out some precisions on the terminology used in the literature on additive categories, see e.g. [39] and references therein. We shall stick to this terminology in what follows. Let $\mathcal{C}$ be an additive k-linear category, where k is an arbitrary field.

---

[3]This is a bit misleading, as non-isomorphic objects in $\mathcal{C}$ may became isomorphic in $\underline{\mathcal{C}}$.

- $\mathcal{C}$ is *semisimple* if the algebras $\operatorname{End}(X)$ are semisimple for all $X \in \mathcal{C}$.
- An object X is *minimal* if any monomorphism $Y \to X$ is either 0 or an isomorphism; hence $\operatorname{End}(X)$ is a division algebra over k.
- $\mathcal{C}$ is *completely reducible* if every object is a direct sum of minimal ones.

Beware that in more recent literature in topology (where it is assumed that $k = \Bbbk$), an object S is said to be *simple* when $\operatorname{End}(S) \simeq \Bbbk$; and '$\mathcal{C}$ is *semisimple*' means that every object in $\mathcal{C}$ is a direct sum of simple ones. For instance, if $\mathcal{C} = A\text{-mod}$, A an algebra, then a minimal object in $\mathcal{C}$ is just a simple A-module; then $\operatorname{End}(S) \cong \Bbbk$ by the Schur Lemma. But it is well-known that the converse is not true.

Example 3.5 Let $H = \Bbbk\langle x, g | x^2, g^2 - 1, gx + xg\rangle$ be the Sweedler Hopf algebra. H has two simple modules, both one dimensional, namely the trivial V^+ and V^-, where g acts as -1. Consider the non-trivial extension $V^{\pm} \in \operatorname{Ext}^1_H(V^-, V^+)$, that is $V^{\pm} = \Bbbk v \oplus \Bbbk w$ with action

$$g \cdot v = v, \qquad g \cdot w = -w, \qquad x \cdot v = w, \qquad x \cdot w = 0.$$

Then $V^{\pm}$ is not simple and $\operatorname{End}_H(V^{\pm}) \cong \Bbbk$. It is easy to see that the indecomposable H-modules are V^+, V^-, $V^{\pm}$ and $V^{\mp} := (V^{\pm})^*$; hence $\underline{\operatorname{Rep} H} \simeq \operatorname{Rep} \mathbb{Z}/2$ by Step 2 of the proof of Theorem 3.7 below. Also, notice that in $\overline{\operatorname{Rep}} H$ it is not true that all endomorphism algebras are semisimple (just take the regular representation); and, related to this, that $\operatorname{Hom}_H(V^{\mp}, V^{\pm}) \neq 0$.

However, the converse above is true when the category is semisimple.

Remark 3.6 (Lemmas 1.1, 1.3 in [39]) Let $\mathcal{C}$ be a semisimple additive k-linear category, where k is an arbitrary field.

(a) If $\alpha : V \to W$ is not zero, then there exists $\beta, \gamma : W \to V$ such that $\beta\alpha \neq 0$, $\alpha\gamma \neq 0$. If $\operatorname{Hom}_{\mathcal{C}}(V, W) = 0$, then $\operatorname{Hom}_{\mathcal{C}}(W, V) = 0$.
(b) If $V \in \mathcal{C}$ and $\operatorname{End}(V)$ is a division ring, then V is minimal.
(c) If $V, W \in \mathcal{C}$ are minimal and non-isomorphic, then $\operatorname{Hom}_{\mathcal{C}}(V, W) = 0$.

Proof

(a) Assume that $\operatorname{Hom}_{\mathcal{C}}(W, V)\alpha = 0$. Set $U = V \oplus W$; then $\operatorname{End}_{\mathcal{C}} U\alpha$ is a nilpotent left ideal of $\operatorname{End}_{\mathcal{C}} U$, hence it is 0.
(b) Let $W \neq 0$ and $f \in \operatorname{Hom}(W, V)$ be a monomorphism. Since $f \neq 0$, $\operatorname{Hom}(V, W) \neq 0$ by (a). Since $\operatorname{End}(V)$ is a division ring, the map $f \circ - : \operatorname{Hom}(V, W) \to \operatorname{End}(V)$ is surjective. Therefore there exists $g \in \operatorname{Hom}(V, W)$ such that $f \circ g = \mathrm{id}_V$. On the other hand, the map $\operatorname{End}(W) \to \operatorname{Hom}(W, V)$, $\tilde{h} \mapsto f \circ \tilde{h}$ is injective. Hence $g \circ f = \mathrm{id}_W$ since $f \circ (g \circ f) = f \circ \mathrm{id}_W$. Thus $W \simeq V$.
(c) follows from (a) at once. □

If H is a spherical Hopf algebra, then we denote by $\operatorname{Indec}_q H$ the set of isomorphism classes of indecomposable finite-dimensional H-modules with non-zero quantum dimension.

Theorem 3.7 (Proposition 3.8 in [20]) *Let H be a spherical Hopf algebra with pivot ω. Then the non-degenerate quotient $\underline{\operatorname{Rep}} H$ is a completely reducible spherical tensor category, and $\operatorname{Irr}\underline{\operatorname{Rep}} H$ is in bijective correspondence with $\operatorname{Indec}_q H$.*

By Remark 3.6, 'completely reducible' becomes what is called semisimple in the recent literature.

Proof The crucial step is to show that $\underline{\operatorname{Rep}} H$ is semisimple.

Step 1 (Proposition 3.7 in [20]). The algebra $\operatorname{End}_{\underline{\operatorname{Rep}} H}(X)$ is semisimple for any X in $\underline{\operatorname{Rep}} H$.

Indeed, the Jacobson radical J of $\operatorname{End}_H(X)$ is contained in $\mathcal{J}(X, X)$. For, if $\vartheta \in J$, then $\vartheta\omega$ is nilpotent, hence $\operatorname{tr}_L(\vartheta) = \operatorname{tr}_V(\vartheta\omega) = 0$.

As a consequence of Step 1 and Remark 3.6, $X \in \underline{\operatorname{Rep}} H$ is minimal if and only if $\operatorname{End}_{\underline{\operatorname{Rep}} H}(X) \simeq \Bbbk$, that is if and only if it is simple. Also, $\operatorname{Hom}_{\underline{\operatorname{Rep}} H}(S, T) = 0$, for $S, T \in \underline{\operatorname{Rep}} H$ simple non-isomorphic.

Step 2. V is a simple object in $\underline{\operatorname{Rep}} H$ iff there exists $W \in \operatorname{Rep} H$ indecomposable with $\operatorname{qdim} W \neq 0$ which is isomorphic to V in $\underline{\operatorname{Rep}} H$.

Assume that $W \in \operatorname{Rep} H$ is indecomposable. If $f \in \operatorname{End}_H(W)$, then f is either bijective or nilpotent by the Fitting Lemma. If also $\operatorname{qdim} W \neq 0$, then $\operatorname{End}_{\underline{\operatorname{Rep}} H}(W)$ is a finite dimensional division algebra over $\Bbbk$, necessarily isomorphic to $\Bbbk$. Now, assume that V is a simple object in $\underline{\operatorname{Rep}} H$. Let $\pi \in \operatorname{End}_H(V)$ be a lifting of $\operatorname{id}_V = 1 \in \operatorname{End}_{\underline{\operatorname{Rep}} H}(V) \simeq \Bbbk$. We can choose π to be a primitive idempotent. Then the image W of π is indecomposable and $\pi_{|W}$ induces an isomorphism between W and V in $\underline{\operatorname{Rep}} H$. Again by the Fitting Lemma, $\operatorname{End}_H(W) \simeq \Bbbk\pi_{|W} \oplus \operatorname{Rad}\operatorname{End}_H(W)$. Hence $\operatorname{qdim} W \neq 0$ since π is a lifting of $\operatorname{id}_V \in \operatorname{End}_{\underline{\operatorname{Rep}} H}(V)$.

Step 3. Let $V, W \in \operatorname{Rep} H$ indecomposable with $\operatorname{qdim} V \neq 0$, $\operatorname{qdim} W \neq 0$, which are isomorphic in $\underline{\operatorname{Rep}} H$. Then $V \simeq W$ in $\operatorname{Rep} H$.

Let $f \in \operatorname{Hom}_H(V, W)$ and $g \in \operatorname{Hom}_H(W, V)$ such that $gf = \operatorname{id}_V$ in $\underline{\operatorname{Rep}} H$; that is

$$\operatorname{tr}_V\big(\vartheta(gf - \operatorname{id})\omega\big) = 0 \quad \text{for every } \vartheta \in \operatorname{End}_H(V). \tag{3.9}$$

Since V is indecomposable, gf is invertible in $\operatorname{End}(V)$, or otherwise (3.9) would fail for $\vartheta = \operatorname{id}$. Thus g is surjective and f is injective. But W is also indecomposable, hence f is surjective and g is injective, and both are isomorphisms.

To finish the proof of the statement about the irreducibles, observe that an indecomposable $U \in \operatorname{Rep} H$ with $\operatorname{qdim} U = 0$ satisfies $U \simeq 0$ in $\underline{\operatorname{Rep}} H$. Since any $M \in \operatorname{Rep} H$ is a direct sum of indecomposables, we see that M is isomorphic in $\underline{\operatorname{Rep}} H$ to a direct sum of indecomposables with non-zero quantum dimension.

Finally observe that the additive $\Bbbk$-linear category $\underline{\operatorname{Rep}} H$, being isomorphic to a direct sum of copies of $\operatorname{Vec}\Bbbk$, is abelian. □

Here is a consequence of the Theorem: let $\iota : V \hookrightarrow W$ be a proper inclusion of indecomposable H-modules with $\operatorname{qdim} V \neq 0$, $\operatorname{qdim} W \neq 0$. Then $\iota \in \mathcal{J}(V, W)$.

Remark 3.8 From Theorem 3.7 we see the relation between the constructions of [20] and [35]. For, let $\mathcal{C} = \operatorname{Rep} H$ and let $\mathcal{C}^0$, resp. $\mathcal{C}^\perp$, be the full subcategory whose objects are direct sums of indecomposables with quantum dimension $\neq 0$, resp. 0. Then $\underline{\operatorname{Rep}} H$ is the quotient of $\mathcal{C}$ by $\mathcal{C}^\perp$ as described in Sect. 1 in [35].

Even when H is finite-dimensional, $\operatorname{Irr} \underline{\operatorname{Rep}} H$ is not necessarily finite. It is then natural to look at suitable subcategories of $\overline{\operatorname{Rep}} H$ that give rise to finite tensor subcategories of $\underline{\operatorname{Rep}} H$. A possibility is tilting modules, that proved to be very fruitful in the case of quantum groups at roots of one. We shall discuss this matter in Sect. 4.

3.5 Pointed or Copointed Pivotal Hopf Algebras

3.5.1 Pivots in the Pointed Case

Let H be a pointed Hopf algebra and set $G = G(H)$. We assume that H is generated by group-like and skew-primitive elements. For H finite-dimensional, it was conjectured that this is always the case (Conjecture 1.4 in [7]). So far this is true in all known cases, see [15, 17] and references therein. As explained in Sect. 2.3, there exist $g_1, \dots, g_\theta \in G$ and $\rho_i \in \operatorname{Irr} C_G(g)$, $1 \le i \le \theta$, such that the infinitesimal braiding of H is

$$M(\mathcal{O}_{g_1}, \rho_1) \oplus \cdots \oplus M(\mathcal{O}_{g_\theta}, \rho_\theta).$$

Lemma 3.9 *Let $\omega \in G$. Then the following are equivalent:*

(a) *ω is a pivot.*
(b) *$\omega \in Z(G)$ and $\rho_i(\omega) = \rho_i(g_i)^{-1}$, $1 \le i \le \theta$.*

Proof There exist $x_1, \dots, x_\theta \in H$ such that $\Delta(x_i) = x_i \otimes 1 + g_i \otimes x_i$, $1 \le i \le \theta$, and H is generated by $x_1, \dots, x_\theta$ and G as an algebra. Now $\mathcal{S}^2(x_i) = g_i^{-1} x_i g_i = \rho_i(g_i^{-1}) x_i$. □

3.5.2 Pivots in the Copointed Case

Let G be a finite group and δ_g be the characteristic function of the subset $\{g\}$ of G. If $M \in {}_{\Bbbk^G}\mathcal{M}$, then $M = \bigoplus_{g \in \operatorname{Supp} M} M[g]$ where

$$M[g] := \delta_g \cdot M \quad \text{and} \quad \operatorname{Supp} M := \{g \in G : M[g] \neq 0\}.$$

Lemma 3.10 *Let H be a finite-dimensional copointed Hopf algebra over $\Bbbk^G$ and $\omega = \sum_{g \in G} \omega(g) \delta_g \in G(\Bbbk^G)$. The following are equivalent:*

(a) *ω is a pivot.*
(b) *$\mathcal{S}^2(x) = \omega(g) x$ for all $x \in H[g]$, $g \in G$.*

Proof Consider H as a $\Bbbk^G$-module via the adjoint action; then $\delta_t x = x\delta_{g^{-1}t}$ for $x \in H[g]$, $g, t \in G$ (Lemma 3.1(b) in [9]). Hence $\omega x \omega^{-1} = \omega(g)x$. □

3.6 Spherical Hopf Algebras with Involutory Pivot

There are many examples of Hopf algebras with involutory pivot.

3.6.1 The Pointed Abelian Case

Let H be a finite-dimensional pointed Hopf algebra with $G(H)$ abelian. Then its infinitesimal braiding V is a braided vector space of diagonal type with matrix $(q_{ij})_{1\leq i,j\leq\theta}$, $\theta = \dim V$. Assume that $q_{ii} = -1$, $1 \leq i \leq \theta$. The list of all braided vector spaces with this property and such that the associated Nichols algebra is finite-dimensional can be easily extracted from the main result of [41]. We apply Lemma 3.9 because H is generated by group-like and skew-primitive elements [15]. Hence, if V belongs to this list, then H is a spherical Hopf algebra with involutory pivot, eventually adjoining a group-like if necessary. The argument also works when $G(H)$ is not abelian but the infinitesimal braiding is a direct sum of one-dimensional Yetter-Drinfeld modules (one often says that the infinitesimal braiding *comes from the abelian case*).

3.6.2 The Pointed Nonabelian Case

Let H be a finite-dimensional pointed Hopf algebra with $G(H)$ not abelian and such that the infinitesimal braiding does not come from the abelian case. In all the examples of such infinitesimal braidings that are known,[4] we may apply Lemma 3.9 because H is generated by group-like and skew-primitive elements. Also, in all examples except one in [44], the scalar in Lemma 3.9(b) is -1. Thus H is a spherical Hopf algebra with involutory pivot, eventually adjoining a group-like if necessary.

3.6.3 The Copointed Case

Let H be a finite-dimensional copointed Hopf algebra over $\Bbbk^G$, with G not abelian. Lemma 3.10 makes it easy to check whether H has an involutive pivot. For instance, the Hopf algebras $\mathcal{A}_{[\mathbf{a}]}$, for $\mathbf{a} \in \mathfrak{A}_3$, introduced in [9], have an involutory pivot. These Hopf algebras are liftings of $\mathcal{B}(V_3)\#\Bbbk^{\mathbb{S}_3}$, so they have dimension 72; but they are not quasi-triangular [9].

[4]The list of all known examples is in http://mate.dm.uba.ar/~matiasg/zoo.html, see also Table 1 in [37], except for one example discovered later (Proposition 36 in [44]).

Also, $\operatorname{Irr}\underline{\operatorname{Rep}}\,\mathcal{A}_{[0]}$ is infinite. Namely, $\mathcal{A}_{[0]} \simeq \mathcal{B}(V_3)\#\Bbbk^{\mathbb{S}_3}$ and for $\lambda \in \Bbbk$ we define the $\mathcal{A}_{[0]}$-module $M_\lambda = \langle m_g : e \neq g \in \mathbb{S}_3 \rangle$ by

$$\delta_h \cdot m_g = \delta_h(g) m_g \quad \text{and} \quad x_{(ij)} \cdot m_g = \begin{cases} 0 & \text{if } \operatorname{sgn} g = -1, \\ \lambda_{(ij),g} m_{(ij)g} & \text{if } \operatorname{sgn} g = 1, \end{cases}$$

where $\lambda_{(ij),g} = 1$ except to $\lambda_{(12),(123)} = \lambda$. Then $M \in \operatorname{Indec}_q \mathcal{A}_{[0]}$ with $\operatorname{qdim} M_\lambda = -1$ for all $\lambda \in \Bbbk$ and $M_\lambda \simeq M_\mu$ iff $\lambda = \mu$. Analogously, we can define $N_\lambda \in \operatorname{Indec}_q \mathcal{A}_{[0]}$ with basis $\langle n_g : (12) \neq g \in \mathbb{S}_3 \rangle$, $\operatorname{qdim} N_\lambda = 1$ and which are mutually not isomorphic.

Remark 3.11 The dual of $\mathcal{A}_{[0]}$ is $\mathcal{B}(V_3)\#\Bbbk\mathbb{S}_3$, which is not pivotal because $\mathbb{S}_3$ is centerless. Compare with the main result of [67], where it is shown that the dual of a semisimple spherical Hopf algebra is again spherical. In fact, the dual of $\mathcal{A}_{[\mathbf{a}]}$ is not pivotal for any $\mathbf{a} \in \mathfrak{A}_3$. If $\mathbf{a}$ is generic, then $(\mathcal{A}_{[\mathbf{a}]})^*$ has no non-trivial group-likes (Theorem 1 in [10]). If $\mathbf{a}$ is sub-generic, then the unique non-trivial group-like $\zeta_{(12)}$ of $(\mathcal{A}_{[\mathbf{a}]})^*$ is not a pivot, see Lemma 8 in [10] for notations. Namely, if $g \neq (12), e$, then

$$\zeta_{(12)} \rightharpoonup \delta_g \leftharpoonup \zeta_{(12)} = \sum_{t,s \in \mathbb{S}_3} \delta_s\big((12)\big)\delta_{s^{-1}t}\delta_{t^{-1}g}\big((12)\big) = \delta_{(12)g(12)} \neq \mathcal{S}^2(\delta_g).$$

3.6.4 Fusion subcategories of $\underline{\operatorname{Rep}}\, H$, Involutory Pivot

Let H be a spherical Hopf algebra with involutory pivot ω. Then

- The quantum dimensions are integers.
- If χ is a representation of dimension one, then $\operatorname{qdim} \Bbbk_\chi = \chi(\omega)$.
- If H is not semisimple, then at least one module has negative quantum dimension.
- Assume that there exists $L \in \operatorname{Irr} H$ such that $\operatorname{qdim} L' > 0$ for all $L' \in \operatorname{Irr} H$, $L' \neq L$. Then $\operatorname{qdim} L < 0$.

Proposition 3.12 *Let $\mathcal{C}$ be a fusion subcategory of $\underline{\operatorname{Rep}}\, H$, where H is a spherical Hopf algebra with involutory pivot. Then there exists a semisimple quasi-Hopf algebra K such that $\mathcal{C} \simeq \operatorname{Rep} K$ as fusion categories.*

Proof The quantum dimensions are integers, because the pivot is involutory, and positive by Corollary 2.10 in [32]; here we use that $\mathcal{C}$ is spherical, see Remark 3.3. Then Theorem 8.33 in [32] applies. Indeed, the Perron-Frobenius and quantum dimensions here coincide, see e.g. the proof of Proposition 8.23 in [32]. □

We are inclined to believe, because of some computations in examples, that the quasi-Hopf algebra K in the statement is actually a Hopf algebra quotient of H.

3.7 Ribbon Hopf Algebras

This is a distinguished class of spherical Hopf algebras. Let $(H, \mathcal{R})$ be a quasi-triangular Hopf algebra [30]. We denote the universal matrix as $\mathcal{R} = \mathcal{R}^{(1)} \otimes \mathcal{R}^{(2)}$. The *Drinfeld element* is

$$\mathbf{u} = \mathcal{S}\big(\mathcal{R}^{(2)}\big)\mathcal{R}^{(1)}. \tag{3.10}$$

Let $Q = \mathcal{R}_{21}\mathcal{R}$. The Drinfeld element is invertible and satisfies

$$\Delta(\mathbf{u}) = Q^{-1}(\mathbf{u} \otimes \mathbf{u}) = (\mathbf{u} \otimes \mathbf{u})Q^{-1}, \quad \mathbf{u}^{-1} = \mathcal{R}^{(2)}\mathcal{S}^2\big(\mathcal{R}^{(1)}\big), \tag{3.11}$$

$$g := \mathbf{u}\mathcal{S}(\mathbf{u})^{-1} = \mathcal{S}(\mathbf{u})^{-1}\mathbf{u} \in G(H), \quad \mathbf{u}\mathcal{S}(\mathbf{u}) \in Z(H), \tag{3.12}$$

$$\mathcal{S}^2(h) = \mathbf{u}h\mathbf{u}^{-1}, \quad \mathcal{S}^4(h) = ghg^{-1}, \tag{3.13}$$

for any $h \in H$.

Definition 3.13 ([69]) A quasi-triangular Hopf algebra $(H, \mathcal{R})$ is *ribbon* if there exists $\mathbf{v} \in Z(H)$, called the *ribbon element*, such that

$$\mathbf{v}^2 = \mathbf{u}\mathcal{S}(\mathbf{u}), \qquad \mathcal{S}(\mathbf{v}) = \mathbf{v}, \qquad \Delta(\mathbf{v}) = Q^{-1}(\mathbf{v} \otimes \mathbf{v}). \tag{3.14}$$

The ribbon element is not unique but it is determined up to multiplication by an element in $\{g \in G(H) \cap Z(H) : g^2 = 1\}$.

Let H be a ribbon Hopf algebra. It follows easily that $\omega = \mathbf{u}\mathbf{v}^{-1} \in G(H)$ and $\mathcal{S}^2(h) = \omega h \omega^{-1}$ for all $h \in H$; that is, ω is a pivot. Actually, H is spherical (Example 3.2 in [20]). In fact, the concept of *quantum trace* is defined in any ribbon category using the braiding, see for example Definition XIV.4.1 in [48]. By Proposition XIV.6.4 in [48], the *quantum trace* of $\operatorname{Rep} H$ coincides with tr_L, cf. (3.3). Moreover, Theorem XIV.4.2(c) in [48] asserts that $\operatorname{tr}_L = \operatorname{tr}_R$.

There are quasi-triangular Hopf algebras that are not ribbon, but the failure is not difficult to remedy by adjoining a group-like element (Theorem 3.4 in [69]). Namely, given a quasi-triangular Hopf algebra $(H, \mathcal{R})$, let $\widetilde{H} = H \oplus H\mathbf{v}$, where $\mathbf{v}$ is a formal element not in H. Then $\widetilde{H}$ is a Hopf algebra with product, coproduct, antipode and counit defined for $x, x', y, y' \in H$ by

$$(x + y\mathbf{v}) \cdot \big(x' + y'\mathbf{v}\big) = \big(xx' + yy'\mathbf{u}\mathcal{S}(\mathbf{u})\big) + \big(xy' + yx'\big)\mathbf{v}, \tag{3.15}$$

$$\Delta(x + y\mathbf{v}) = \Delta(x) + \Delta(y)Q^{-1}(\mathbf{v} \otimes \mathbf{v}), \tag{3.16}$$

$$\mathcal{S}(x + y\mathbf{v}) = \mathcal{S}(x) + \mathcal{S}(y)\mathbf{v}, \qquad \varepsilon(x + y\mathbf{v}) = \varepsilon(x) + \varepsilon(y). \tag{3.17}$$

Clearly, H becomes a Hopf subalgebra of $\widetilde{H}$; it can be shown then that $\mathcal{R}$ is a universal R-matrix for $\widetilde{H}$ and that $\mathbf{v}$ is a ribbon element for $(\widetilde{H}, \mathcal{R})$. See Theorem 3.4 in [69]. We shall say that $\widetilde{H}$ is the *ribbon extension* of $(H, \mathcal{R})$.

Remark 3.14 (Y. Sommerhäuser, private communication) The ribbon extension fits into an exact sequence of Hopf algebras $H \hookrightarrow \widetilde{H} \twoheadrightarrow \Bbbk[\mathbb{Z}/2]$, which is cleft. Namely, let ξ be the generator of $\mathbb{Z}/2$ and define

$$\xi \rightharpoonup x = \mathcal{S}^2(x), \quad x \in H, \qquad \sigma(\xi^i \otimes \xi^j) = g^{ij}, \quad i, j \in \{0, 1\}.$$

Then the crossed product defined by this action and cocycle together with the tensor product of coalgebras is a Hopf algebra $H\#_\sigma \Bbbk[\mathbb{Z}/2]$, see for instance [5]. Now the map $\psi : H\#_\sigma \Bbbk[\mathbb{Z}/2] \to \widetilde{H}$, $\psi(x\#\xi^i) = x(\mathcal{S}(u^{-1})\mathbf{v})^i$, is an isomorphism of Hopf algebras.

In conclusion, any finite-dimensional Hopf algebra H gives rise to a ribbon Hopf algebra, namely the ribbon extension of its Drinfeld double:

$$H \rightsquigarrow D(H) \rightsquigarrow \widetilde{D(H)}.$$

Remark 3.15 A natural question is whether the Drinfeld double itself is ribbon; this was addressed in [49], where the following results were obtained. Let H be a finite-dimensional Hopf algebra, $g \in G(H)$ and $\alpha \in G(H^*)$ be the distinguished group-likes.[5] The celebrated Radford's formula for the fourth power of the antipode [68] states that

$$\mathcal{S}^4(h) = g(\alpha \rightharpoonup h \leftharpoonup \alpha^{-1})g^{-1}, \quad h \in H. \tag{3.18}$$

Here $\rightharpoonup$, $\leftharpoonup$ are the transposes of the regular actions.

(a) (Theorem 2 in [49]) Suppose that $(H, \mathcal{R})$ is quasi-triangular and that $G(H)$ has odd order. Then $(H, \mathcal{R})$ admits a (necessarily unique) ribbon element if and only if $\mathcal{S}^2$ has odd order.
(b) (Theorem 3 in [49]) $(D(H), \mathcal{R})$ admits a ribbon element if and only if there exist $\ell \in G(H)$ and $\beta \in G(H^*)$ such that

$$\ell^2 = g, \qquad \beta^2 = \alpha, \quad \mathcal{S}^2(h) = \ell(\beta \rightharpoonup h \leftharpoonup \beta^{-1})\ell^{-1}, \quad h \in H. \tag{3.19}$$

3.8 Cospherical Hopf Algebras

It is natural to look at the notions that insure that the category of comodules of a Hopf algebra is pivotal or spherical. This was done in [21, 65].

Definition 3.16 A *cospherical* Hopf algebra is a pair (H, t), where H is a Hopf algebra and $t \in \mathrm{Alg}(H, \Bbbk)$ is such that

$$\mathcal{S}^2(x_{(1)})t(x_{(2)}) = t(x_{(1)})x_{(2)}, \quad x \in H, \tag{3.20}$$

[5]These control the passage from left to right integrals.

$$\mathrm{tr}_V\big((\mathrm{id}_V \otimes t)\rho_V \vartheta\big) = \mathrm{tr}_V\big((\mathrm{id}_V \otimes t^{-1})\rho_V \vartheta\big), \quad \vartheta \in \mathrm{End}_H(V), \tag{3.21}$$

for all $V \in \mathrm{Corep}\, H$. We say that $t \in \mathrm{Alg}(H, \Bbbk)$ is a *copivot* when it satisfies (3.20); pairs (H, t) with t a copivot are called *copivotal Hopf algebras*. A *cospherical element* is a copivot that fulfills (3.21). Let H be a cospherical Hopf algebra. Then the category $\mathrm{Corep}\, H$ is spherical; in fact, the left and right traces are given by the sides of (3.21).

The set $\mathrm{Alg}(H, \Bbbk)$ is a subgroup of the group $\mathrm{Hom}_\star(H, \Bbbk)$ of convolution-invertible linear functionals, which in turn acts on $\mathrm{End}(H)$ on both sides. Hence (3.20) can be written as $\mathcal{S}^2 * t = t * \mathrm{id}_H$ or else as $\mathcal{S}^2 = t * \mathrm{id}_H * t^{-1}$. The copivot is not unique but it is determined up to multiplication by an element in $\mathrm{Alg}(H, \Bbbk)$ that centralizes id_H. The antipode of a copivotal Hopf algebra is bijective, with inverse given by $\mathcal{S}^{-1}(x) = \sum t^{-1}(x_{(1)})\mathcal{S}(x_{(2)})t(x_{(3)})$, $x \in H$.

The following statement is proved exactly as Proposition 3.1.

Proposition 3.17 *A copivotal Hopf algebra (H, t) is cospherical if and only if* (3.21) *holds for all simple H-comodules.*

Example 3.18

(a) Assume that H is finite-dimensional. Then H is copivotal (resp., cospherical) iff H^* is pivotal (resp., spherical).
(b) Any involutory Hopf algebra is cospherical with $t = \varepsilon$.
(c) A copivotal Hopf algebra with involutive copivot is cospherical.
(d) Condition (3.20) is multiplicative on x; also, it holds for $x \in G(H)$.
(e) Let H be a pointed Hopf algebra generated as an algebra by $G(H)$ and a family $(x_i)_{i \in I}$, where x_i is $(g_i, 1)$ skew-primitive. Assume that $g_i x_i g_i^{-1} = q_i x_i$, with $q_i \in \Bbbk^\times \setminus \{1\}$ for all $i \in I$. If $t \in \mathrm{Alg}(H, \Bbbk)$, then $t(x_i) = 0$, $i \in I$. Hence t is a copivot iff $t(g_i) = q_i^{-1}$, for all $i \in I$.
(f) Let H be a pointed Hopf algebra as in item (3.18) and $t \in \mathrm{Alg}(H, \Bbbk)$ a copivot. Then H is cospherical iff $t(g) \in \{\pm 1\}$ for all $g \in G(H)$.
(g) The notion of coribbon Hopf algebra is formally dual to the notion of ribbon Hopf algebra, see [38, 56]. Coribbon Hopf algebras are cospherical. For instance, the quantized function algebra $\mathcal{O}_q(G)$ of a semisimple algebraic group is cosemisimple and coribbon, when q is not a root of 1.

We recall now the construction of universal copivotal Hopf algebras.

Definition 3.19 ([21]) Let $F \in GL_n(k)$. The Hopf algebra $H(F)$ is the universal algebra with generators $(u_{ij})_{1 \le i,j \le n}$, $(v_{ij})_{1 \le i,j \le n}$ and relations

$$uv^t = v^t u = 1, \qquad vFu^tF^{-1} = Fu^tF^{-1}v = 1.$$

The comultiplication is determined by $\Delta(u_{ij}) = \sum_k u_{ik} \otimes u_{kj}$, $\Delta(v_{ij}) = \sum_k v_{ik} \otimes v_{kj}$ and the antipode by $\mathcal{S}(u) = v^t$, $\mathcal{S}(v) = Fu^tF^{-1}$. The Hopf algebra $H(F)$ is copivotal, the copivot being $t_F(u) = (F^{-1})^t$ and $t_F(v) = F$.

Let H be a Hopf algebra provided with $V \in \operatorname{Corep} H$ of dimension n such that $V \cong V^{**}$. Then there exist a matrix $F \in GL_n(k)$, a coaction $\beta_V : V \to V \otimes H(F)$ and a Hopf algebra morphism $\pi : H(F) \to H$ such that $(\mathrm{id}_V \otimes \pi)\beta_V = \rho_V$.

In conclusion, we would like to explore semisimple tensor categories arising from cospherical, not cosemisimple, Hopf algebras. The first step is the dual version of Theorem 3.7, which is proved exactly in the same way.

Theorem 3.20 *Let H be a cospherical Hopf algebra. Then the non-degenerate quotient* $\underline{\operatorname{Corep} H}$ *of* $\operatorname{Corep} H$ *is a completely reducible spherical tensor category, and* $\operatorname{Irr}\underline{\operatorname{Corep} H}$ *is in bijective correspondence with the set of isomorphism classes of indecomposable finite-dimensional H-comodules with non-zero quantum dimension.*

4 Tilting Modules

The concept of tilting modules appeared in [22] and was extended to quasi-hereditary algebras in [71]. Observe that finite-dimensional quasi-hereditary Hopf algebras are semisimple. Indeed, quasi-hereditary algebras have finite global dimension, but a finite-dimensional Hopf algebra is a Frobenius algebra, hence it has global dimension 0 or infinite. Instead, the context where the recipe of tilting modules works is a suitable category of modules, or comodules, of an infinite dimensional Hopf algebra. The relevant examples are: algebraic semisimple groups over an algebraically closed field of positive characteristic (the representations are comodules over the Hopf algebra of rational functions), quantum groups at roots of one and the category $\mathcal{O}$ over a semisimple Lie algebra [1, 3, 28, 36, 62]. The main features are:

- The suitable category of representations is not artinian, and the simple modules are parameterized by dominant weights; the set of dominant weights admits a total order that refines the usual partial order. To fit into the framework of quasi-hereditary algebras, subcategories of modules with weights in suitable subsets are considered; this allows to define Weyl modules $\Delta(\lambda)$, dual Weyl modules $\nabla(\lambda)$, and eventually tilting modules $T(\lambda)$, for λ a dominant weight. Usually these constructions are performed in an ad-hoc manner, not through quasi-hereditary algebras, albeit those corresponding to this situation are studied in the literature under the name of Schur algebras.
- The tensor product of two tilting modules and the dual of a tilting module are again tilting. The later statement is trivial, the former requires a delicate proof.
- There is an *alcove* inside the chamber defined by the positive roots and bounded by an affine hyperplane. If λ is in the alcove, then the simple module satisfies $L(\lambda) = \Delta(\lambda)$, hence it is the tilting $T(\lambda)$. The tilting modules $T(\lambda)$ outside the alcove are projective, hence have zero quantum dimension. Thus, the fusion category looked for is spanned by the tilting modules in the alcove.

- The fusion rules between the tilting modules is given by the celebrated Verlinde formula [75] or a modular version, see [3, 62].

We would like to adapt these arguments to categories of representations of certain Hopf algebras H arising from finite-dimensional Nichols of diagonal type. The Hopf algebra H would be the Drinfeld double, or a variation thereof, of the bosonization of the corresponding Nichols algebra with a suitable abelian group. We would like to solve the following points:

- The set of irreducible objects in $\operatorname{Rep} H$ (or some appropriate variation) should split as a filtered union $\operatorname{Irr} H = \bigcup_{A \in \mathcal{A}} A$; each A spans an artinian subcategory where tilting modules $\mathcal{T}_A$ can be computed.
- Define the category $\mathcal{T}_H$ of tilting modules over H as the union of the various $\mathcal{T}_A$; this should be a semisimple category.
- The category $\mathcal{T}_H$ of tilting modules is stable by tensor products and duals.
- It is possible to determine which irreducible tilting modules have non-zero quantum dimension; there are a finite number of them.
- The fusion rules are expressed through a variation of the Verlinde formula.

Provided that these considerations are correct, the full subcategory of $\underline{\operatorname{Rep}} H$ generated by the indecomposable tilting modules with non-zero quantum dimension, is a fusion category. In this way, we hope to obtain new examples of non-integral fusion categories.

4.1 Quasi-hereditary Algebras and Tilting Modules

Tilting modules work for our purpose because they span a completely reducible category already in $\operatorname{Rep} H$. We think it is worthwhile to recall the main definitions of the theory of (partial) tilting modules over quasi-hereditary algebras, due to Ringel [71]. A full exposition is available in [29].

Let A be an artin algebra. Consider a family $\Theta = (\Theta(1), \dots, \Theta(n))$ of A-modules such that

$$\operatorname{Ext}^1_A\big(\Theta(j), \Theta(i)\big) = 0, \quad j \geq i. \tag{4.1}$$

We denote by $\mathcal{F}(\Theta)$ the full subcategory of A-Mod with objects M that admit a filtration with sub-factors in Θ. We fix a numbering (that is, a total order) of $\operatorname{Irr} A$: $L(1), \dots, L(n)$. We set

$$\begin{aligned}
P(i) &= \text{projective cover of } L(i), \\
Q(i) &= \text{injective hull of } L(i), \\
\Delta(i) &= P(i)/U(i), \quad \text{where } U(i) = \sum_{j>i} \sum_{\alpha \in \operatorname{Hom}(P(j), P(i))} \operatorname{Im} \alpha,
\end{aligned}$$

$$\nabla(i) = \bigcap_{j>i} \bigcap_{\beta \in \mathrm{Hom}(Q(i),Q(j))} \ker \beta,$$

$1 \leq i \leq n$. Let $\Delta = (\Delta(1), \ldots, \Delta(n))$, $\nabla = (\nabla(1), \ldots, \nabla(n))$; these satisfy (4.1) and then Theorem 1 in [71] applies to them.

Definition 4.1 The artin algebra A is *quasi-hereditary* provided that ${}_A A \in \mathcal{F}(\Delta)$, and $L(i)$ has multiplicity one in $\Delta(i)$, $1 \leq i \leq n$.

Remark 4.2 Quasi-hereditary algebras were introduced by Cline, Parshall and Scott, see e.g. [23]. There are some alternative definitions.

(a) An ideal J of an artin algebra A is *hereditary* provided that

- $J \in A$-Mod is projective,
- $\mathrm{Hom}_A(J, A/J) = 0$ (Ringel assumes $J^2 = J$ instead of this),
- $JNJ = 0$, where N is the radical of A.

It can be shown that A is quasi-hereditary iff there exists a chain of ideals $A = J_0 > J_1 > \cdots > J_m = 0$ with J_i/J_{i+1} hereditary in A/J_{i+1}.

(b) Also, A is quasi-hereditary iff A-Mod is a highest weight category, that is the following holds for all i:

- $Q(i)/\nabla(i) \in \mathcal{F}(\nabla)$.
- If $(Q(i)/\nabla(i) : \nabla(j)) \neq 0$, then $j > i$.

For completeness, we include the definitions of tilting, cotilting and basic modules, see e.g. [71] and its bibliography. First, a module T is tilting provided that

- it has finite projective dimension;
- $\mathrm{Ext}^i(T, T) = 0$ for all $i \geq 1$;
- for any projective module P, there should exist an exact sequence $0 \to P \to T_0 \to \cdots \to T_m \to 0$, with all T_j in the additive subcategory generated by T, denoted $\mathrm{add}\, T$.

Second, a cotilting module should have

- finite injective dimension;
- $\mathrm{Ext}^i(T, T) = 0$ for all $i \geq 1$;
- for any injective module I, there should exist an exact sequence $0 \to T_m \to \cdots \to T_0 \to I \to 0$, with all T_j in $\mathrm{add}\, T$.

Lastly, a basic module is one with no direct summands of the form $N \oplus N$, with $N \neq 0$.

Assume that A is a quasi-hereditary algebra and consider the full subcategory $\mathcal{T} = \mathcal{T}_A = \mathcal{F}(\Delta) \cap \mathcal{F}(\nabla)$ of (partial) tilting modules. It was shown in Theorem 5 in [71] that—for a quasi-hereditary algebra—there is a unique basic module T, that is tilting and cotilting (in the sense just above), and such that $\mathcal{T}$ coincides with $\mathrm{add}\, T$. The relation between this T and the partial tilting modules is clarified by the following result.

Theorem 4.3 (Corollary 5, Proposition 2 in [71]) *Let A be a quasi-hereditary algebra. There exist indecomposable (partial) tilting modules $T(i) \in \mathcal{T}$, for $1 \le i \le n$, with the following properties:*

- *Any indecomposable tilting module is isomorphic to one of them.*
- *$T = T(1) \oplus \cdots \oplus T(n)$ is the tilting module mentioned above.*
- *There are exact sequences*

$$0 \longrightarrow \Delta(i) \xrightarrow{\beta(i)} T(i) \longrightarrow X(i) \longrightarrow 0,$$

$$0 \longrightarrow Y(i) \longrightarrow T(i) \xrightarrow{\gamma(i)} \nabla(i) \longrightarrow 0,$$

where $X(i) \in \mathcal{F}(\{\Delta(j) : j < i\})$, $Y(i) \in \mathcal{F}(\{\nabla(j) : j < i\})$, $\beta(i)$ is a left $\mathcal{F}(\nabla)$-approximation and $\gamma(i)$ is a right $\mathcal{F}(\Delta)$-approximation, $1 \le i \le n$ (see [71] for undefined notions).

4.2 Induced and Produced

Let $B \hookrightarrow A$ be an inclusion of algebras. We denote by Res_B^A the restriction functor from the category A-Mod to B-Mod.

4.2.1 Definition and General Properties

The induced and produced modules of $T \in B$-Mod are

$$\operatorname{Ind}_B^A T = A \otimes_B T, \qquad \operatorname{Pro}_B^A T = \operatorname{Hom}_B(A, T). \tag{4.2}$$

These are equipped with morphisms of B-modules $\iota : T \hookrightarrow \operatorname{Ind}_B^A T$, given by $\iota(t) = 1 \otimes t$ for $t \in T$, and $\pi : \operatorname{Pro}_B^A T \twoheadrightarrow T$, given by $\pi(f) = f(1)$ for $f \in \operatorname{Hom}_B(A, T)$. The following properties are well-known.

(a) $\operatorname{Hom}_B(T, \operatorname{Res}_B^A M) \simeq \operatorname{Hom}_A(\operatorname{Ind}_B^A T, M)$; that is, induction is left adjoint to restriction.
(b) For every $S \in \operatorname{Irr} A$ there exists $T \in \operatorname{Irr} B$ such that S is a quotient of $\operatorname{Ind}_B^A T$.
(c) $\operatorname{Hom}_B(\operatorname{Res}_B^A N, T) \simeq \operatorname{Hom}_A(N, \operatorname{Pro}_B^A T)$; that is, production (also called coinduction) is right adjoint to restriction.
(d) For every $S \in \operatorname{Irr} A$ there exists $T \in \operatorname{Irr} B$ such that S is a submodule of $\operatorname{Pro}_B^A T$.

4.2.2 The Finite Case

Let $B \hookrightarrow A$ still be an inclusion of algebras. Assume that A is a finite B-module. Then Res_B^A, Ind_B^A and Pro_B^A restrict to functors (denoted by the same name) between the categories A-mod and B-mod of finite-dimensional modules; mutatis mutandis, the preceding points hold in this context. Assume also that there exists a *contravariant* $\Bbbk$-linear functor $\mathcal{D}: A\text{-mod} \to A\text{-mod}$ such that $\mathcal{D}(B\text{-mod}) \subseteq B\text{-mod}$ and admits a quasi-inverse $\mathcal{E}: A\text{-mod} \to A\text{-mod}$, so that $\mathcal{D}$ is an equivalence of categories. It follows at once from the universal properties that

$$\operatorname{Ind}_B^A T \simeq \mathcal{D}\big(\operatorname{Pro}_B^A \mathcal{E}T\big) \simeq \mathcal{E}\big(\operatorname{Pro}_B^A \mathcal{D}T\big), \tag{4.3}$$

$$\operatorname{Pro}_B^A T \simeq \mathcal{D}\big(\operatorname{Ind}_B^A \mathcal{E}T\big) \simeq \mathcal{E}\big(\operatorname{Ind}_B^A \mathcal{D}T\big). \tag{4.4}$$

Hence $\mathcal{D}(\operatorname{Ind}_B^A T) \simeq \operatorname{Pro}_B^A \mathcal{D}T$, $\mathcal{D}(\operatorname{Pro}_B^A T) \simeq \operatorname{Ind}_B^A \mathcal{D}T$, and so on.

In this setting, consider the following conditions:

$$\text{For every } S \in \operatorname{Irr} A, \exists \text{ a } \textit{unique } T \in \operatorname{Irr} B \text{ such that } \operatorname{Ind}_B^A T \twoheadrightarrow S. \tag{4.5}$$

$$\text{For every } S \in \operatorname{Irr} A, \exists \text{ a } \textit{unique } U \in \operatorname{Irr} B \text{ such that } S \hookrightarrow \operatorname{Pro}_B^A U. \tag{4.6}$$

$$\text{The head of } \operatorname{Ind}_B^A T \text{ is simple for every } T \in \operatorname{Irr} B. \tag{4.7}$$

$$\text{The socle of } \operatorname{Pro}_B^A U \text{ is simple for every } U \in \operatorname{Irr} B. \tag{4.8}$$

Then (4.5) $\iff$ (4.6) and (4.7) $\iff$ (4.8). If all these conditions hold, then for any $T \in \operatorname{Irr} B$, there exists a unique $U \in \operatorname{Irr} B$ such that

$$\operatorname{Ind}_B^A T \twoheadrightarrow S \hookrightarrow \operatorname{Pro}_B^A U,$$

where S is the head of $\operatorname{Ind}_B^A T$ and the socle of $\operatorname{Pro}_B^A U$. We set $U = w_0(T)$.

4.2.3 Further Properties for Inclusions of Hopf Algebras

Let K be a Hopf subalgebra of a Hopf algebra H, with H finite over K. Then

$$\operatorname{Ind}_K^H T \simeq \big(\operatorname{Pro}_K^H {}^*T\big)^* \simeq {}^*\big(\operatorname{Pro}_K^H T^*\big), \qquad \operatorname{Pro}_K^H T \simeq \big(\operatorname{Ind}_K^H {}^*T\big)^* \simeq {}^*\big(\operatorname{Ind}_K^H T^*\big).$$

If H is pivotal, then these formulae are simpler because the left and right duals coincide.

4.2.4 Quantum Groups

Let $\mathfrak{g}$ be a simple Lie algebra, $\mathfrak{b}$ a Borel subalgebra and q a root of unity of odd order, relatively prime to 3 when $\mathfrak{g}$ is of type G_2. Let $H = \mathcal{U}_q(\mathfrak{g})$ be the Lusztig's q-divided power quantized enveloping algebra and $K = \mathcal{U}_q(\mathfrak{b})$. Let $\mathcal{C}$ be the category of finite-dimensional H-modules of type 1, see [4], and $\mathcal{C}_{\mathfrak{b}}$ the analogous category of K-modules. Then there are induced and produced functors $\operatorname{Ind}_K^H, \operatorname{Pro}_K^H : \mathcal{C}_{\mathfrak{b}} \to \mathcal{C}$. Then the Weyl and dual Weyl modules are the produced and induced modules of the simple objects in $\mathcal{C}_{\mathfrak{b}}$, parameterized conveniently by highest weights. This allows to define Weyl and dual Weyl filtrations and tilting modules; instead of appealing to Theorem 4.3, one establishes the semisimplicity of the category of tilting modules by establishing crucial cohomological results, see [3] for details.

4.3 Finite-Dimensional Nichols Algebras of Diagonal Type

We continue the analysis started in Sect. 2.4.

4.3.1 Weights

Let $\theta \in \mathbb{N}$ and $\mathbb{I} = \{1, \dots, \theta\}$. Let Λ be a free abelian group with basis $\alpha_1, \dots, \alpha_\theta$. Let $\leq$ be the partial order in Λ defined by

$$\lambda \leq \mu \quad \Longleftrightarrow \quad \mu - \lambda \in \Lambda^+ := \sum_{i \in \mathbb{I}} \mathbb{N}_0 \alpha_i. \tag{4.9}$$

Let us fix a $\mathbb{Z}$-linear injective map $E : \Lambda \to \mathbb{R}$ such that $E(\alpha_i) > 0$ for all $i \in \mathbb{I}$. This induces a total order on Λ by $\lambda \preceq \mu \iff E(\lambda) \leq E(\mu)$; clearly, $\lambda \leq \mu$ implies $\lambda \preceq \mu$.

Given a Λ-graded vector space $M = \bigoplus_{\lambda \in \Lambda} M_\lambda$, the λ's such that $M_\lambda \neq 0$ are called the weights of M; the set of all its weights is denoted $\Pi(M)$.

4.3.2 Nichols Algebras of Diagonal Type

Let $(q_{ij})_{i,j \in \mathbb{I}}$ be a symmetric matrix with entries in $\Bbbk^\times$. Let (V, c) be a braided vector space of diagonal type with matrix $(q_{ij})_{i,j \in \mathbb{I}}$, with respect to a basis $(v_i)_{i \in \mathbb{I}}$. The Nichols algebra $\mathcal{B}(V)$ has a Λ-grading determined by $\deg v_i = \alpha_i$, $i \in \mathbb{I}$. By [54], there exists an ordered set $\bar{S}$ of homogeneous elements of $T(V)$ and a function $h : \bar{S} \to \mathbb{N} \cup \{\infty\}$ such that:

- The elements of $\bar{S}$ are hyperletters in $(v_i)_{i \in \mathbb{I}}$.
- The projection $T(V) \to \mathcal{B}(V)$ induces a bijection of $\bar{S}$ with its image S. Denote also by $h : S \to \mathbb{N} \cup \{\infty\}$ the induced function.
- The following elements form a basis of $\mathcal{B}(V)$:

$$\left\{s_1^{e_1} \cdots s_t^{e_t} : t \in \mathbb{N}_0, s_1 > \cdots > s_t, s_i \in S, 0 < e_i < h(s_i)\right\}. \tag{4.10}$$

When S is finite, two distinct elements in S have different degree, and we can label the elements in S with a finite subset Δ_+^V of Λ_+; this is instrumental to define the root system $\mathcal{R}$ of (V, c) [43].

Let W be another braided vector space of diagonal type with matrix $(q_{ij}^{-1})_{i,j \in \mathbb{I}}$, with respect to a basis $w_1, \dots, w_\theta$; we shall consider the Λ-grading on the Nichols algebra $\mathcal{B}(W)$ determined by $\deg w_i = -\alpha_i$, $i \in \mathbb{I}$.

We assume from now on that $\dim \mathcal{B}(V) < \infty$; hence $\dim \mathcal{B}(W) < \infty$. Under this assumption, the connected components of (q_{ij}) belong to the list given in [41]. An easy consequence is that $q_{ii} \neq 1$ and q_{ij} is a root of 1 for all $i, j \in \mathbb{I}$; this last claim follows because the matrix (q_{ij}) is assumed to be symmetric. Also, S is finite and h takes values in $\mathbb{N}$. Thus (4.10) says that

$$\Pi\big(\mathcal{B}(V)\big) = \Big\{\sum_{s\in S} e_s \deg s, 0 \le e_s < h(s)\Big\}.$$

Note that $0 \le \alpha \le \varrho$ for all $\alpha \in \Pi(\mathcal{B}(V))$, where

$$\varrho = \sum_{s\in S}\big(h(s) - 1\big) \deg s = \deg \mathcal{B}^{\mathrm{top}}(V) \in \Pi\big(\mathcal{B}(V)\big). \tag{4.11}$$

4.3.3 Pre-Nichols Algebras

A *pre-Nichols algebra* of V is any graded braided Hopf algebra $\mathfrak{T}$ intermediate between $\mathcal{B}(V)$ and $T(V)$: $T(V) \twoheadrightarrow \mathfrak{T} \twoheadrightarrow \mathcal{B}(V)$ (Masuoka). The defining relations of the Nichols algebra $\mathcal{B}(V) = T(V)/\mathcal{J}(V)$ are listed as (40), …, (68) in Theorem 3.1 in [15]. Now we observe that, since $\dim \mathcal{B}(V) < \infty$, the following integers exist:

$$-a_{ij} := \min\big\{n \in \mathbb{N}_0 : (n+1)_{q_{ii}}\big(1 - q_{ii}^n q_{ij}^2\big) = 0\big\} \tag{4.12}$$

for all $j \in \mathbb{I} - \{i\}$. Set also $a_{ii} = 2$. The *distinguished pre-Nichols algebra* of V is $\widehat{\mathcal{B}}(V) = T(V)/\widehat{\mathcal{J}}(V) = \bigoplus_{n\in\mathbb{N}_0} \widehat{\mathcal{B}}^n(V)$, where $\widehat{\mathcal{J}}(V)$ is the ideal of $T(V)$ generated by

- relations (41), …, (68) in Theorem 3.1 in [15],
- the quantum Serre relations $(\operatorname{ad}_c x_i)^{1-a_{ij}} x_j$ for those vertices such that $q_{ii}^{a_{ij}} = q_{ij}q_{ji}$.

The ideal $\widehat{\mathcal{J}}(V)$ was introduced in [15], see the paragraph after Theorem 3.1; $\widehat{\mathcal{J}}(V)$ is a braided bi-ideal of $T(V)$, so that there is a projection of braided Hopf algebras $\widehat{\mathcal{B}}(V) \twoheadrightarrow \mathcal{B}(V)$ (Proposition 3.3 in [15]).

Definition 4.4 We say that $p \in \{1, \dots, \theta\}$ is a *Cartan vertex* if, for every $j \neq p$, $q_{pp}^{a_{pj}} = q_{pj}q_{jp}$. In such case, $\operatorname{ord} q_{pp} \ge 1 - a_{pj}$ by hypothesis.

Clearly the projection $T(V) \to \widehat{\mathcal{B}}(V)$ induces a bijection of $\bar{S}$ with its image $\widehat{S}$. Denote again by h the induced function. Let $\widehat{h} : \widehat{S} \to \mathbb{N} \cup \{\infty\}$ be the function given by

$$\widehat{h}(s) = \begin{cases} \infty, & \text{if } s \text{ is conjugated to a Cartan vertex} \\ h(s), & \text{otherwise.} \end{cases}$$

Then the following set is a basis of $\widehat{\mathcal{B}}(V)$, see the end of the proof of Theorem 3.1 in [15]:

$$\left\{s_1^{e_1}\cdots s_t^{e_t} : t\in\mathbb{N}_0, s_1>\cdots>s_t, s_i\in\widehat{S}, 0<e_i<\widehat{h}(s_i)\right\}. \tag{4.13}$$

4.3.4 Lusztig Algebras

The *Lusztig algebra* $\mathcal{L}(V)$ of V is the graded dual of $\widehat{\mathcal{B}}(V)$, that is $\mathcal{L}(V) = \bigoplus_{n\in\mathbb{N}_0}\mathcal{L}^n(V)$, where $\mathcal{L}^n(V)=\widehat{\mathcal{B}}^n(V)^*$. The Lusztig algebra $\mathcal{L}(V)$ of V is the analogue of the q-divided powers algebra introduced in [57, 58].

4.4 The Small Quantum Groups

We consider finite-dimensional pointed Hopf algebras attached to the matrix $(q_{ij})_{i,j\in\mathbb{I}}$, analogues of the small quantum groups or Frobenius-Lusztig kernels. We need the following additional data: A finite abelian group Γ, provided with elements $g_1,\dots,g_\theta\in\Gamma$ and characters $\chi_1,\dots,\chi_\theta\in\operatorname{Hom}_{\mathbb{Z}}(\Gamma,\Bbbk^\times)$ such that

$$\chi_j(g_i)=q_{ij},\quad i,j\in\mathbb{I}. \tag{4.14}$$

We define a structure of Yetter-Drinfeld module over $\Bbbk\Gamma$ on $W\oplus V$ by

$$v_i\in V_{g_i}^{\chi_i},\qquad w_i\in W_{g_i}^{\chi_i^{-1}},\quad i\in\mathbb{I}. \tag{4.15}$$

Let $\mathfrak{u}$ be the Hopf algebra $T(W\oplus V)\#\Bbbk\Gamma/\mathfrak{J}$, where $\mathfrak{J}$ is the ideal generated by $\mathcal{J}(V)$, $\mathcal{J}(W)$ and the relations

$$v_iw_j-\chi_j^{-1}(g_i)w_jv_i-\delta_{ij}\left(g_i^2-1\right)\quad i,j\in\mathbb{I}. \tag{4.16}$$

This is a pointed quasi-triangular Hopf algebra with $\dim\mathfrak{u}=|\Gamma|\dim\mathcal{B}(V)^2$. The freedom to choose the abelian group Γ allows more flexibility, but otherwise this is very close to the small quantum groups (with more general Nichols algebras). By choosing Γ appropriately, $\mathfrak{u}$ is a spherical Hopf algebra. Let $\mathfrak{u}^{\mathfrak{b}}$ (resp. $\mathfrak{u}^-$) be the subalgebra of $\mathfrak{u}$ generated by $v_1,\dots,v_\theta$ and Γ (resp. $w_1,\dots,w_\theta$). Consider the morphisms of algebras $\rho_V:\mathcal{B}(V)\to\mathfrak{u}$, $\rho_W:\mathcal{B}(W)\to\mathfrak{u}$ and $\rho_\Gamma:\Bbbk\Gamma\to\mathfrak{u}$, given by $\rho_V(v_i)=v_i$, $\rho_W(w_i)=w_i$, $\rho_\Gamma(g_i)=g_i$, $i\in\mathbb{I}$. Then

(a) $\rho_V,\rho_W,\rho_\Gamma$ give rise to isomorphisms $\mathcal{B}(W)\simeq\mathfrak{u}^-$, $\mathcal{B}(V)\#\Bbbk\Gamma\simeq\mathfrak{u}^{\mathfrak{b}}$.
(b) The map $\mathcal{B}(V)\otimes\mathcal{B}(W)\otimes\Bbbk\Gamma\to\mathfrak{u}$, $v\otimes w\otimes g\mapsto\rho_V(v)\rho_W(w)\rho_\Gamma(g)$ is a coalgebra isomorphism.
(c) The multiplication maps $\mathfrak{u}^-\otimes\mathfrak{u}^{\mathfrak{b}}\to\mathfrak{u}$, $\mathfrak{u}^{\mathfrak{b}}\otimes\mathfrak{u}^-\to\mathfrak{u}$ are linear isomorphisms.

See Theorem 5.2 in [61], Corollary 3.8 in [11]. Now suppose that one would like to define tilting modules over $\mathfrak{u}$, ignoring that this is not a quasi-hereditary algebra. Inducing from $\mathfrak{u}^{\mathfrak{b}}$, we see that simple modules correspond to characters of Γ; but the set of simple modules could not be totally ordered and (4.6) and (4.7) do not necessarily hold. A first approach to remedy this that might come to the mind is to assume the following extra hypothesis: *There exists a* $\mathbb{Z}$*-bilinear form* $\langle\, ,\, \rangle : \Gamma \times \Lambda \to \Bbbk^{\times}$ *such that*

$$\langle g_i, \alpha_j \rangle = q_{ij}, \quad i, j \in \mathbb{I}. \tag{4.17}$$

Then we may consider a category that is an analogue of category of representations of the algebraic group G_1T in positive characteristic, or else of its quantum analogue in the literature of quantum groups. Let $\mathcal{C}_{\mathfrak{u}}$ be the category of finite-dimensional $\mathfrak{u}$-modules M with a Λ-grading $M = \bigoplus_{\lambda \in \Lambda} M_\lambda$, compatible with the action of $\mathfrak{u}$ in the sense

$$M_\lambda = \{m \in M : g \cdot m = \langle g, \lambda \rangle m, g \in \Gamma\}, \quad \lambda \in \Lambda, \tag{4.18}$$

$$v_i \cdot M_\lambda = M_{\lambda+\alpha_i}, \qquad w_i \cdot M_\lambda = M_{\lambda-\alpha_i}, \quad \lambda \in \Lambda, i \in \mathbb{I}. \tag{4.19}$$

Morphisms in $\mathcal{C}_{\mathfrak{u}}$ preserve both the action of $\mathfrak{u}$ and the grading by Λ. The category $\mathcal{C}_{\mathfrak{u}^{\mathfrak{b}}}$ is defined analogously. Both categories $\mathcal{C}_{\mathfrak{u}}$ and $\mathcal{C}_{\mathfrak{u}^{\mathfrak{b}}}$ are spherical tensor categories (up to an appropriate choice of Γ), with duals defined in the obvious way. There are functors $\operatorname{Res}^{\mathfrak{u}}_{\mathfrak{u}^{\mathfrak{b}}}$, $\operatorname{Ind}^{\mathfrak{u}}_{\mathfrak{u}^{\mathfrak{b}}}$ and $\operatorname{Pro}^{\mathfrak{u}}_{\mathfrak{u}^{\mathfrak{b}}}$ between the categories $\mathcal{C}_{\mathfrak{u}}$ and $\mathcal{C}_{\mathfrak{u}^{\mathfrak{b}}}$; indeed

$$\operatorname{Ind}^{\mathfrak{u}}_{\mathfrak{u}^{\mathfrak{b}}} T = \mathfrak{u} \otimes_{\mathfrak{u}^{\mathfrak{b}}} T \simeq \mathcal{B}(W) \otimes T,$$

$$\operatorname{Pro}^{\mathfrak{u}}_{\mathfrak{u}^{\mathfrak{b}}} T = \operatorname{Hom}_{\mathfrak{u}^{\mathfrak{b}}}(\mathfrak{u}, T) \simeq \operatorname{Hom}(\mathcal{B}(W), T),$$

so that the grading in $\operatorname{Ind}^{\mathfrak{u}}_{\mathfrak{u}^{\mathfrak{b}}} T$, resp. $\operatorname{Pro}^{\mathfrak{u}}_{\mathfrak{u}^{\mathfrak{b}}} T$, arises from that of $\mathcal{B}(W) \otimes T$, resp. $\operatorname{Hom}(\mathcal{B}(W), T)$.

Given $\lambda \in \Lambda$, we denote by $\Bbbk_\lambda$ the vector space with generator $\mathbf{1}_\lambda$, considered as object in $\mathcal{C}_{\mathfrak{u}^{\mathfrak{b}}}$ by

$$\deg \mathbf{1}_\lambda = \lambda, \qquad g \cdot \mathbf{1}_\lambda = \langle g, \lambda \rangle \mathbf{1}_\lambda, \quad g \in \Gamma, \qquad v \cdot \mathbf{1}_\lambda = 0, \quad v \in V.$$

Note that $\Bbbk_\lambda \simeq \Bbbk_\mu$ in $\mathfrak{u}^{\mathfrak{b}}$-mod whenever $\lambda - \mu \in \Gamma^{\perp}$, but they are not isomorphic as objects in $\mathcal{C}_{\mathfrak{u}^{\mathfrak{b}}}$ unless $\lambda = \mu$. Clearly, $\operatorname{Irr} \mathcal{C}_{\mathfrak{u}^{\mathfrak{b}}} = \{\Bbbk_\lambda : \lambda \in \Lambda\}$.

Consider the modules $\operatorname{Pro}^{\mathfrak{u}}_{\mathfrak{u}^{\mathfrak{b}}}(\Bbbk_\lambda)$ and $\Delta(\lambda) := \operatorname{Ind}^{\mathfrak{u}}_{\mathfrak{u}^{\mathfrak{b}}}(\Bbbk_\lambda)$. We know

$$\Pi(\Delta(\lambda)) = \{\lambda - \alpha : \alpha \in \Pi(\mathcal{B}(V))\}, \tag{4.20}$$

$$\Pi(\operatorname{Pro}^{\mathfrak{u}}_{\mathfrak{u}^{\mathfrak{b}}}(\Bbbk_\lambda)) = \{\lambda + \alpha : \alpha \in \Pi(\mathcal{B}(V))\}. \tag{4.21}$$

Thus, $\Delta(\lambda)$ has a highest weight λ and a lowest weight $\lambda - \varrho$, both of multiplicity 1; and $\operatorname{Pro}^{\mathfrak{u}}_{\mathfrak{u}^{\mathfrak{b}}}(\Bbbk_\lambda)$ has a highest weight $\lambda + \varrho$ and a lowest weight λ, both of multiplicity 1. For convenience, set $\nabla(\lambda) = \operatorname{Pro}^{\mathfrak{u}}_{\mathfrak{u}^{\mathfrak{b}}}(\Bbbk_{\lambda-\varrho}) \simeq \Delta(-\lambda + \varrho)^*$, since $\Bbbk_\lambda^* \simeq \Bbbk_{-\lambda}$.

The statements (a), (b), (c) and (d) in Sect. 4.2.1 above carry over to the present setting. We claim that (4.5), (4.6), (4.7) and (4.8) also hold here.

Indeed, let $S \in \operatorname{Irr} \mathcal{C}_{\mathfrak{u}}$ and $\lambda, \mu \in \Pi(S)$ such that $\mu \preceq \tau \preceq \lambda$ for all $\tau \in \Pi(S)$. If $m \in S_\lambda - 0$, then $v_i \cdot m = 0$ for all $i \in \mathbb{I}$ by (4.19), hence $\Bbbk m \simeq \Bbbk_\lambda$ and we have $\Delta(\lambda) \twoheadrightarrow S$ and $\Pi(S) \subseteq \{\lambda - \alpha : \alpha \in \Pi(\mathcal{B}(V))\}$ by (4.20). Moreover, if $\Delta(\lambda') \twoheadrightarrow S$, then $\lambda \in \Pi(S) \subseteq \{\lambda' - \alpha : \alpha \in \Pi(\mathcal{B}(V))\}$, hence $\lambda' = \lambda$, showing (4.5). The proof of (4.7) is standard: $\Delta(\lambda)$ has a unique maximal submodule, which is the sum of all submodules intersecting trivially $\Delta(\lambda)_\lambda$. Now (4.6) and (4.8) follow by duality, so that $S \hookrightarrow \nabla(\mu)$.

In conclusion we have the following standard result.

Proposition 4.5 *If* $E(\lambda) :=$ *head of* $\Delta(\lambda)$, *then* $\operatorname{Irr} \mathcal{C}_{\mathfrak{u}} = \{E(\lambda) : \lambda \in \Lambda\}$. *If* $\mu \in \Gamma^{\perp}$, *then* $\dim E(\mu) = 1$ *and* $E(\mu) \otimes E(\lambda) \simeq E(\lambda) \otimes E(\mu) \simeq E(\lambda + \mu)$. *There is a bijection* $w_0 : \Lambda \to \Lambda$ *such that* $E(\lambda) :=$ *socle of* $\nabla(w_0(\lambda))$.

The modules $\Delta(\lambda)$, resp. $\nabla(\lambda)$, are called the Weyl modules, resp. the dual Weyl modules. We may then go on and define good and Weyl filtrations, and tilting modules. However, it is likely that tilting modules are projective, thus with 0 quantum dimension, as is the case for $G_1 T$, see Sect. 3.4 in [2], [47].

4.5 *Generalized Quantum Groups*

The next idea is to replace Γ by an infinite abelian group Q, perhaps free of finite rank, and the Nichols algebras $\mathcal{B}(V)$, $\mathcal{B}(W)$ by the distinguished pre-Nichols algebras $\widehat{\mathcal{B}}(V)$, $\widehat{\mathcal{B}}(W)$. Namely, we assume that Q is provided with elements $K_1, \dots, K_\theta$ and characters $\Upsilon_1, \dots, \Upsilon_\theta \in \operatorname{Hom}_{\mathbb{Z}}(Q, \Bbbk^\times)$ such that $\Upsilon_j(K_i) = q_{ij}$, $i, j \in \mathbb{I}$. Then $W \oplus V$ is also a Yetter-Drinfeld module over $\Bbbk Q$ by $v_i \in V_{K_i}^{\Upsilon_i}$, $w_i \in W_{K_i}^{\Upsilon_i^{-1}}$, $i \in \mathbb{I}$. Let $\mathrm{U}(V) = T(W \oplus V) \# \Bbbk Q / \widehat{\mathfrak{I}}$ where $\widehat{\mathfrak{I}}$ is the ideal generated by $\widehat{\mathcal{J}}(V)$, $\widehat{\mathcal{J}}(W)$ and the relations

$$v_i w_j - \chi_j^{-1}(g_i) w_j v_i - \delta_{ij}\big(g_i^2 - 1\big) \quad i, j \in \mathbb{I}. \tag{4.22}$$

This Hopf algebra, for a suitable Q, was introduced in [15]; it is the analogue of the quantized enveloping algebra at a root of one for $(q_{ij})_{i,j \in \mathbb{I}}$ in the version of [26]. It also has a triangular decomposition similar as in the case of $\mathfrak{u}$. Furthermore, there are so-called Lusztig isomorphisms, because they generalize the braid group representations defined by Lusztig, see e.g. [59]. Actually, the definition of the ideal $\widehat{\mathfrak{I}}$ in [15] was designed to have (a) a braided bi-ideal, and (b) the Lusztig automorphisms at the level of $\mathrm{U}(V)$, generalizing results from [42]. More precisely, the situation is as follows.

We assume that Q and Γ are accurately chosen and that there is a group epimorphism $Q \to \Gamma$. Given $i \in \mathbb{I}$, we define the i-th reflection of (V, c). Define $s_i \in \operatorname{Aut} \Lambda$

by $s_i(\alpha_j) = \alpha_j - a_{ij}\alpha_i$, see (4.12). Then $s_i(V, c) = (V, c')$, where C' is the braiding of diagonal type with matrix $(\widetilde{q}_{rs})_{r,s\in\mathbb{I}}$. Here $\widetilde{q}_{rs} = (s_i(\alpha_r)|s_i(\alpha_s))$; we omit the mention to the braidings c, c', etc. Then

- There are algebra isomorphisms $T_i, T_i^- : \mathfrak{u}(V) \to \mathfrak{u}(s_i V)$, such that $T_i T_i^- = T_i^- T_i = \mathrm{id}_{\mathfrak{u}(V)}$ (Theorem 6.12 in [42]).
- There are algebra isomorphisms $T_i, T_i^- : \mathrm{U}(V) \to \mathrm{U}(s_i V)$, such that $T_i T_i^- = T_i^- T_i = \mathrm{id}_{\mathrm{U}(V)}$ (Proposition 3.26 in [15]), compatible with those of $\mathfrak{u}(V)$.

There is a Λ-grading on $T(W \oplus V)\#\Bbbk Q$ given by $\deg\gamma = 0$, $\gamma \in Q$, $\deg v_i = \alpha_i = -\deg w_i$, $i \in \mathbb{I}$; it extends to gradings of $\mathfrak{u}(V)$ and $\mathrm{U}(V)$. Hence we may consider categories $\mathcal{C}_{\mathrm{U}}$ and so on. However, the Hopf algebra $\mathrm{U}(V)$ has a large center Z and is actually finite over it. Thus, it seems that its representation theory should be addressed with the methods of [26, 27].

It remains a third tentative: to repeat the above considerations replacing the distinguished pre-Nichols algebras $\widehat{\mathcal{B}}(V)$, $\widehat{\mathcal{B}}(W)$ by the Lusztig algebras $\mathcal{L}(V)$, $\mathcal{L}(W)$. We hope to address this in future publications.

Acknowledgements The work of N.A., I.A., A.G.I. and C.V. was partially supported by CONICET, FONCyT-ANPCyT and Secyt (UNC). The work of B.T. was partially supported by MTM 2011-27090, Ministerio de Educación y Ciencia (España) and the project FQM 3128 from Junta Andalucía (FEDER).

We thank Olivier Mathieu for a course on tilting modules given at the University of Córdoba in November 2009; and also Alain Bruguières, Flavio Coelho, Juan Cuadra, Michael Müger, Dmitri Nikschych, Claus Ringel and Yorck Sommerhäuser for very useful conversations. We are also grateful the referee for her/his gentle comments.

References

1. Andersen, H.H.: Tensor products of quantized tilting modules. Commun. Math. Phys. **149**, 149–159 (1992)
2. Andersen, H.H.: Quotient categories of modular representations. In: Gyoja, A., et al. (eds.) Representation Theory of Algebraic Groups and Quantum Groups. Progress in Mathematics, vol. 284, pp. 1–16. Birkhäuser, Boston (2010)
3. Andersen, H.H., Paradowski, J.: Fusion categories arising from semisimple Lie algebras. Commun. Math. Phys. **169**, 563–588 (1995)
4. Andersen, H.H., Polo, P., Wen, K.: Representations of quantum algebras. Invent. Math. **104**, 1–59 (1991)
5. Andruskiewitsch, N., Devoto, J.: Extensions of Hopf algebras. Algebra Anal. **7**, 17–52 (1995)
6. Andruskiewitsch, N., Graña, M.: Braided Hopf algebras over non abelian finite group. Bol. Acad. Nac. Cienc. (Córdoba) **63**, 45–78 (1999)
7. Andruskiewitsch, N., Schneider, H.J.: Finite quantum groups and Cartan matrices. Adv. Math. **154**, 1–45 (2000)
8. Andruskiewitsch, N., Schneider, H.J.: Pointed Hopf algebras In: New Directions in Hopf Algebras. MSRI Series, pp. 1–68. Cambridge University Press, Cambridge (2002)
9. Andruskiewitsch, N., Vay, C.: Finite dimensional Hopf algebras over the dual group algebra of the symmetric group in three letters. Commun. Algebra **39**, 4507–4517 (2011)
10. Andruskiewitsch, N., Vay, C.: On a family of Hopf algebras of dimension 72. Bull. Belg. Math. Soc. Simon Stevin **19**, 415–443 (2012)

11. Andruskiewitsch, N., Radford, D.E., Schneider, H.-J.: Complete reducibility theorems for modules over pointed Hopf algebras. J. Algebra **324**, 2932–2970 (2010)
12. Andruskiewitsch, N., Angiono, I., Yamane, H.: On pointed Hopf superalgebras. Contemp. Math. **544**, 123–140 (2011)
13. Angiono, I.: On Nichols algebras with standard braiding. Algebra Number Theory **3**, 35–106 (2009)
14. Angiono, I.: A presentation by generators and relations of Nichols algebras of diagonal type and convex orders on root systems. arXiv:1008.4144
15. Angiono, I.: On Nichols algebras of diagonal type. J. Reine Angew. Math. (2000). doi:10.1515/crelle-2011-0008
16. Angiono, I.: Nichols algebras of unidentified diagonal type. arXiv:1108.5157
17. Angiono, I., García Iglesias, A.: Pointed Hopf algebras with standard braiding are generated in degree one. Contemp. Math. **537**, 57–70 (2011)
18. Bakalov, B., Kirillov, A. Jr.: Lectures on Tensor Categories and Modular Functors. Am. Math. Soc., Providence (2000)
19. Barrett, J.W., Westbury, B.W.: Invariants of piecewise-linear 3-manifolds. Trans. Am. Math. Soc. **348**, 3997–4022 (1996)
20. Barrett, J.W., Westbury, B.W.: Spherical categories. Adv. Math. **143**, 357–375 (1999)
21. Bichon, J.: Cosovereign Hopf algebras. J. Pure Appl. Algebra **157**, 121–133 (2001)
22. Brenner, S., Butler, M.C.R.: Generalizations of the Bernstein-Gelfand-Ponomarev reflection functors. Lect. Notes Math. **832**, 103–169 (1980)
23. Cline, E., Parshall, B., Scott, L.: Finite dimensional algebras and highest weight categories. J. Reine Angew. Math. **391**, 85–99 (1988)
24. Cohen, M., Westreich, S.: Characters and a Verlinde-type formula for symmetric Hopf algebras. J. Algebra **320**, 4300–4316 (2008)
25. Cohen, M., Westreich, S.: Structure constants related to symmetric Hopf algebras. J. Algebra **324**, 3219–3240 (2010)
26. de Concini, C., Procesi, C.: Quantum groups. Lect. Notes Math. **1565**, 31–140 (1993)
27. de Concini, C., Procesi, C., Reshetikhin, N., Rosso, M.: Hopf algebras with trace and representations. Invent. Math. **161**, 1–44 (2005)
28. Donkin, S.: On tilting modules for algebraic groups. Math. Z. **212**, 39–60 (1993)
29. Donkin, S.: The q-Schur Algebra. London Mathematical Society Lecture Note Series, vol. 253. Cambridge University Press, Cambridge (1998)
30. Drinfeld, V.: Quantum groups. In: Proceedings of the International Congress of Mathematicians, Berkeley, CA, vol. 1, pp. 798–820. Academic Press, San Diego (1986)
31. Etingof, P., Ostrik, V.: Finite tensor categories. Mosc. Math. J. **4**(3), 627–654 (2004)
32. Etingof, P., Nikshych, D., Ostrik, V.: On fusion categories. Ann. Math. **162**, 581–642 (2005)
33. Etingof, P., Gelaki, S., Nikshych, D., Ostrik, V.: Tensor categories. http://www-math.mit.edu/~etingof/tenscat1.pdf
34. Freyd, P., Yetter, D.N.: Braided compact closed categories with applications to low dimensional topology. Adv. Math. **77**, 156–182 (1989)
35. Gelfand, S., Kazhdan, D.: Examples of tensor categories. Invent. Math. **109**, 595–617 (1992)
36. Georgiev, G., Mathieu, O.: Fusion rings for modular representations of Chevalley groups. Contemp. Math. **175**, 89–100 (1994)
37. Graña, M., Heckenberger, I., Vendramin, L.: Nichols algebras of group type with many quadratic relations. Adv. Math. **227**, 1956–1989 (2011)
38. Hai, P.H.: Characters of quantum groups of type A_n. Commun. Algebra **30**, 1085–1117 (2002)
39. Harada, M.: On semi-simple abelian category. Osaka J. Math. **7**, 89–95 (1970)
40. Hayashi, T.: Face algebras and unitarity of $\mathrm{SU}(N)_L$-TQFT. Commun. Math. Phys. **203**, 211–247 (1999)
41. Heckenberger, I.: Classification of arithmetic root systems. Adv. Math. **220**, 59–124 (2009)
42. Heckenberger, I.: Lusztig isomorphisms for Drinfel'd doubles of bosonizations of Nichols algebras of diagonal type. J. Algebra **323**, 2130–2180 (2010)

43. Heckenberger, I., Schneider, H.-J.: Right coideal subalgebras of Nichols algebras and the Duflo order on the Weyl groupoid. arXiv:0909.0293
44. Heckenberger, I., Lochmann, A., Vendramin, L.: Braided racks, Hurwitz actions and Nichols algebras with many cubic relations. Transform. Groups **17**, 157–194 (2012)
45. Huang, Y.-Z.: Vertex operator algebras, the Verlinde conjecture and modular tensor categories. Proc. Natl. Acad. Sci. USA **102**, 5352–5356 (2005)
46. Huang, Y.-Z., Lepowsky, J.: Intertwining operator algebras and vertex tensor categories for affine Lie algebras. Duke Math. J. **99**, 113–134 (1999)
47. Jantzen, J.C.: Representations of Algebraic Groups. Mathematical Surveys and Monographs, vol. 107. Am. Math. Soc., Providence (2003)
48. Kassel, C.: Quantum Groups. Springer, Berlin (1995)
49. Kauffman, L.H., Radford, D.E.: A necessary and sufficient condition for a finite-dimensional Drinfel'd double to be a ribbon Hopf algebra. J. Algebra **159**, 98–114 (1993)
50. Kazhdan, D., Lusztig, G.: Tensor structures arising from affine Lie algebras. I. J. Am. Math. Soc. **6**, 905–947 (1993)
51. Kazhdan, D., Lusztig, G.: Tensor structures arising from affine Lie algebras. II. J. Am. Math. Soc. **6**, 949–1011 (1993)
52. Kazhdan, D., Lusztig, G.: Tensor structures arising from affine Lie algebras. III. J. Am. Math. Soc. **6**, 335–381 (1993)
53. Kazhdan, D., Lusztig, G.: Tensor structures arising from affine Lie algebras. IV. J. Am. Math. Soc. **6**, 383–453 (1993)
54. Kharchenko, V.: A quantum analog of the Poincare-Birkhoff-Witt theorem. Algebra Log. **38**, 259–276 (1999)
55. Knop, K.: Tensor envelopes of regular categories. Adv. Math. **214**, 571–617 (2007)
56. Larson, R.G., Towber, J.: Two dual classes of bialgebras related to the concepts of 'quantum group' and 'quantum Lie algebra'. Commun. Algebra **19**, 3295–3345 (1991)
57. Lusztig, G.: Finite dimensional Hopf algebras arising from quantized universal enveloping algebras. J. Am. Math. Soc. **3**, 257–296 (1990)
58. Lusztig, G.: Quantum groups at roots of 1. Geom. Dedic. **35**, 89–114 (1990)
59. Lusztig, G.: Introduction to Quantum Groups. Birkhäuser, Basel (1993)
60. Mac Lane, S.: Categories for the Working Mathematician, 2nd edn. Springer, Berlin (1998)
61. Masuoka, A.: Construction of quantized enveloping algebras by cocycle deformation. Arab. J. Sci. Eng., Sect. C, Theme Issues **33**, 387–406 (2008)
62. Mathieu, O.: Tilting modules and their applications. Adv. Stud. Pure Math. **26**, 145–212 (2000)
63. Montgomery, S.: Hopf Algebras and Their Action on Rings. CBMS Regional Conference Series, vol. 82 (1993)
64. Müger, M.: Tensor categories: a selective guided tour. Rev. Uni6n Mat. Argent. **51**, 95–163 (2010)
65. Oeckl, R.: Generalized lattice gauge theory, spin foams and state sum invariants. J. Geom. Phys. **46**, 308–354 (2003)
66. Ostrik, V.: Module categories over the Drinfeld double of a finite group. Int. Math. Res. Not. **2003**(27), 1507–1520 (2003)
67. Pfeiffer, H.: Finitely semisimple spherical categories and modular categories are self-dual. Adv. Math. **221**, 1608–1652 (2009)
68. Radford, D.: The order of the antipode of a finite dimensional Hopf algebra is finite. Am. J. Math. **98**, 333–355 (1976)
69. Reshetikhin, N.Y., Turaev, V.G.: Ribbon graphs and their invariants derived from quantum groups. Commun. Math. Phys. **127**, 1–26 (1990)
70. Reshetikhin, N.Y., Turaev, V.G.: Invariants of 3-manifolds via link polynomials and quantum groups. Invent. Math. **103**, 547–597 (1991)
71. Ringel, C.M.: The category of modules with good filtrations over a quasi-hereditary algebra has almost split sequences. Math. Z. **208**, 209–223 (1991)

72. Sawin, S.: Quantum groups at roots of unity and modularity. J. Knot Theory Ramif. **15**, 1245–1277 (2006)
73. Sommerhäuser, Y.: On Kaplansky's fifth conjecture. J. Algebra **204**, 202–224 (1998)
74. Turaev, V.G.: Modular categories and 3-manifold invariants. Int. J. Mod. Phys. B **6**, 1807–1824 (1992)
75. Verlinde, E.: Fusion rules and modular transformations in 2D conformal field theory. Nucl. Phys. B **300**, 360–375 (1988)
76. Witten, E.: Quantum field theory and the Jones polynomial. Commun. Math. Phys. **121**, 351–399 (1989)

Pattern-of-Zeros Approach to Fractional Quantum Hall States and a Classification of Symmetric Polynomial of Infinite Variables

Xiao-Gang Wen and Zhenghan Wang

Abstract Some purely chiral fractional quantum Hall states are described by symmetric or anti-symmetric polynomials of infinite variables. In this article, we review a systematic construction and classification of those fractional quantum Hall states and the corresponding polynomials of infinite variables, using the pattern-of-zeros approach. We discuss how to use patterns of zeros to label different fractional quantum Hall states and the corresponding polynomials. We also discuss how to calculate various universal properties (i.e. the quantum topological invariants) from the pattern of zeros.

1 Introduction

To readers who are interested in physics, this is a review article on the pattern-of-zeros approach to fractional quantum Hall (FQH) states. To readers who are interested in mathematics, this is an attempt to classify symmetric polynomials of infinite variables and Z_n vertex algebra. To those interested in mathematical physics, this article tries to provide a way to systematically study pure chiral topological quantum field theories that can be realized by interacting bosons. In the next two subsections, we will review briefly the definition of quantum many-boson systems, and the definition of quantum phase for non-physicists. Then, we will give an introduction of the problems studied in this paper.

X.-G. Wen
Department of Physics, Massachusetts Institute of Technology, Cambridge, MA 02139, USA

X.-G. Wen
Institute for Advanced Study, Tsinghua University, Beijing, 100084, P.R. China

Z. Wang
Microsoft Station Q, University of California, CNSI Bldg. Rm 2237, Santa Barbara, CA 93106, USA

C. Bai et al. (eds.), *Conformal Field Theories and Tensor Categories*,
Mathematical Lectures from Peking University, DOI 10.1007/978-3-642-39383-9_2,

1.1 What Is a Quantum Many-Boson System

The fermionic FQH states [1, 2] are described by anti-symmetric wave functions, while the bosonic FQH states are described by symmetric wave functions. Since there is an one-to-one correspondence between the anti-symmetric wave functions and the symmetric wave functions, in this article, we will only discuss bosonic FQH states and their symmetric wave functions.

Bosonic FQH systems are quantum many-boson systems. Let us first define mathematically what is a quantum many-boson system, using an N-boson system in two spatial dimensions as an example. A many-body state of N bosons is a symmetric complex function of N variables

$$\begin{aligned}&\Psi(\boldsymbol{r}_1,\dots,\boldsymbol{r}_i,\dots,\boldsymbol{r}_j,\dots,\boldsymbol{r}_N)\\&\quad=\Psi(\boldsymbol{r}_1,\dots,\boldsymbol{r}_j,\dots,\boldsymbol{r}_i,\dots,\boldsymbol{r}_N)\end{aligned} \tag{1}$$

where the ith variable $\boldsymbol{r}_i=(x_i,y_i)$ describes the coordinates of the ith boson. All such symmetric functions form a Hilbert space where the normal is defined as

$$\langle\Psi|\Psi\rangle=\int\prod_i \mathrm{d}x_i\mathrm{d}y_i\Psi^*\Psi \tag{2}$$

A quantum system of N bosons is described by a Hamiltonian, which is a Hermitian operator in the above Hilbert space. It may have a form

$$H(g_1,g_2)=\sum_{i=1}^{N}-\frac{1}{2}(\partial_{x_i}^2+\partial_{y_i}^2)+\sum_{i<j}V_{g_1,g_2}(\boldsymbol{r}_i-\boldsymbol{r}_j) \tag{3}$$

Here $V_{g_1,g_2}(\boldsymbol{r}_i-\boldsymbol{r}_j)$ is the interaction potential between two bosons. We require the interaction potential to be short ranged:

$$V_{g_1,g_2}(x,y)=0,\quad \text{if}\sqrt{x^2+y^2}>\xi, \tag{4}$$

where ξ describes the interaction range. Hamiltonians with short-ranged interactions are called local Hamiltonians.

The ground state of the N boson system is an eigenvector of H:

$$H(g_1,g_2)\Psi_{g_1,g_2}(\boldsymbol{r}_1,\dots,\boldsymbol{r}_N)=E_{\text{grnd}}(g_1,g_2)\Psi_{g_1,g_2}(\boldsymbol{r}_1,\dots,\boldsymbol{r}_N) \tag{5}$$

with the minimal eigenvalue $E_{\text{grnd}}(g_1,g_2)$. The eigenvalues of the Hamiltonian are called energies.

Here we assume that the interaction potential may depend on some parameters g_1,g_2. As we change g_1,g_2, the ground states Ψ_{g_1,g_2} for different g_1,g_2's can some times have similar properties. We say that those states belong to the same phase. Some other times, they may have very different properties. Then we regard those states to belong to the different phases.

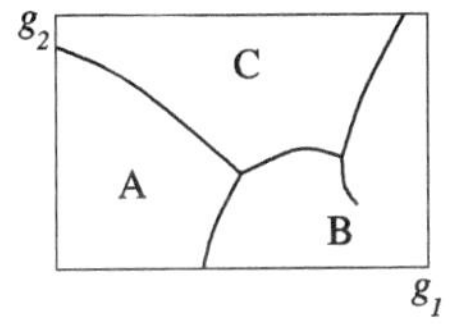

Fig. 1 *The curves* mark the position of singularities in functions $E_{\mathrm{grnd}}(g_1, g_2)/N$ and $\langle O \rangle(g_1, g_2)$. They also represent phase transitions. *The regions*, *A*, *B*, and *C*, separated by phase transitions correspond to different phases

1.2 What Are Quantum Phases

More precisely, quantum phases are defined through quantum phase transitions. So we first need to define *what quantum phase transitions are.*

As we change the parameters g_1, g_2 in the Hamiltonian $H(g_1, g_2)$, if the average of ground state energy per particle $E_{\mathrm{grnd}}(g_1, g_2)/N$ has a singularity in $N \to \infty$ limit, then the system has a phase transition. More generally, if the average of any local operator O on the ground state

$$\langle O \rangle(g_1, g_2) = \int \prod_i \mathrm{d}x_i \mathrm{d}y_i \Psi^*_{g_1,g_2} O \Psi_{g_1,g_2} \tag{6}$$

has a singularity in $N \to \infty$ limit as we change g_1, g_2, then the system has a phase transition (see Fig. 1).

Using the quantum phase transition, we can define an equivalence relation between quantum ground states Ψ_{g_1,g_2} in $N \to \infty$ limit: Two quantum ground states Ψ_{g_1,g_2} and $\Psi_{g'_1,g'_2}$ are equivalent if we can find a path that connect (g_1, g_2) and (g'_1, g'_2) such that we can change Ψ_{g_1,g_2} into $\Psi_{g'_1,g'_2}$ without encountering a phase transition. The quantum phases are nothing but the equivalent classes of such an equivalence relation [3]. In short, the quantum phases are regions of (g_1, g_2) space which are separated by phase transitions (see Fig. 1).

1.3 How to Classify Quantum Phases of Matter

One of the most important questions in condensed matter physics is how to classify the many different quantum phases of matter. One attempt is the theory of symmetry breaking [4–6], which tells us that we should classify various phases based on the symmetries of the ground state wave function. Yet with the discovery of the FQH states [1, 2] came also the understanding that there are many distinct and fascinating quantum phases of matter, called topologically ordered phases [7, 8], whose characterization has nothing at all to do with symmetry. How should we systematically classify the different possible topological phases that may occur in a FQH system? In this paper, we will try to address this issue.

We know that the FQH states contain topology-dependent degenerate ground states, which are topologically stable (i.e. robust against any *local* perturbations of the Hamiltonians). This allows us to introduce the concept of topological order in FQH states [9, 10]. Such topology-dependent degenerate ground states suggest that the low energy theories describing the FQH states are topological quantum field theories [11–13], which take a form of pure Chern-Simons theory in $2 + 1$ dimensions [14–19]. So one possibility is that we may try to classify the different FQH phases by classifying all of the different possible pure Chern-Simons theories. Although such a line of thinking leads to a classification of Abelian FQH states in terms of integer K-matrices [15–20], it is not a satisfactory approach for non-Abelian FQH states [21, 22] because we do not have a good way of knowing which pure Chern-Simons theories can possibly correspond to a physical system made of bosons and which cannot.

Another way to classify FQH states is through the connection between FQH wave functions and conformal field theory (CFT). It was discovered around 1990 that correlation functions in certain two-dimensional conformal field theories may serve as good model wave functions for FQH states [21, 23, 24]. Thus perhaps we may classify FQH states by classifying all of the different CFTs. However, the relation between CFTs and FQH states is not one-to-one. If a CFT produces a FQH wave function, then any other CFTs that contain the first CFT can also produce the FQH wave function [24].

Following the ideas of CFT and in an attempt to obtain a systematic classification of FQH states without using conformal invariance, it was shown recently that a wide class of FQH states and their topological excitations can be classified by their *patterns of zeros*, which describe the way ideal FQH wave functions go to zero when various clusters of particles are brought together [25–28]. (We would like to point out that the "1D charge-density-wave" characterization of FQH states [29–34] is closely related to the pattern-of-zeros approach.) This analysis led to the discovery of some new non-Abelian FQH states whose corresponding CFT has not yet been identified. It also helped to elucidate the role of CFT in constructing FQH wave functions: *The CFT encodes the way the wave function goes to zero as various clusters of bosons are brought together.* The order of these zeros must satisfy certain conditions and the solutions to these conditions correspond to particular CFTs. Thus in classifying and characterizing FQH states, one can bypass the CFT altogether and proceed directly to classifying the different allowed pattern of zeros and subsequently obtaining the topological properties of the quasiparticles from the pattern of zeros [26–28]. This construction can then even be thought of as a classification of the allowed CFTs that can be used to construct FQH states [35]. Furthermore, these considerations give a natural notion of which pattern of zeros solutions are simpler than other ones. In this sense, then, one can see that the Moore-Read Pfaffian quantum Hall state [21] is the "simplest" non-Abelian generalization of the Laughlin state.

We would like to point that in the pattern-of-zeros classification of FQH states, we do not try to study the phase transition and equivalence classes. Instead, we just try to classify some special complex functions of infinite variables. We hope those

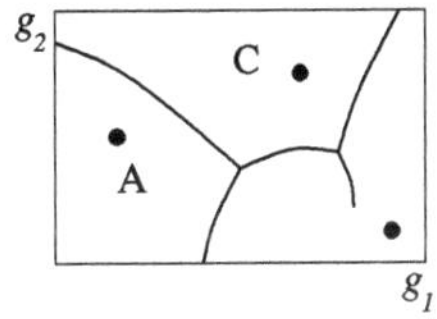

Fig. 2 *The black dots* represent the ideal wave functions that can represent each quantum phase

special complex functions can represent each equivalence class (i.e. represent each quantum phase) (see Fig. 2).

2 Examples of Fractional Quantum Hall States

Before trying to classify a type of quantum phases—FQH phases, let us study some examples of ideal FQH wave functions to gain some intuitions.

2.1 *The Hamiltonian for FQH Systems*

A FQH state of N-bosons is described by the following Hamiltonian:

$$H(g_1, g_2) = \sum_{i=1}^{N} \left(\mathrm{i}\partial_{z_i} - \mathrm{i}\frac{1}{4} z_i^* \right) \left(\mathrm{i}\partial_{z_i^*} + \mathrm{i}\frac{1}{4} z_i \right) + \sum_{i<j} V_{g_1,g_2}(z_i - z_j) \tag{7}$$

where the two dimensional plane is parametrized by $z = x + \mathrm{i}y$. When $V_{g_1,g_2} = 0$, there are many wave functions

$$\begin{aligned} \Psi(z_1, \ldots, z_N) &= P(z_1, \ldots, z_N) e^{-(1/4)\sum_{i=1}^{N} z_i z_i^*}, \\ P &= \text{a symmetric polynomial} \end{aligned} \tag{8}$$

that all have the minimal zero eigenvalue (or energy) for any P:

$$\left[\sum_{i=1}^{N} \left(\mathrm{i}\partial_{z_i} - \mathrm{i}\frac{1}{4} z_i^* \right) \left(i\partial_{z_i^*} + \mathrm{i}\frac{1}{4} z_i \right) \right] P(z_1, \ldots, z_N) e^{-(1/4)\sum_{i=1}^{N} z_i z_i^*} = 0, \tag{9}$$

since

$$e^{(1/4)zz^*} \left(\mathrm{i}\partial_z - \mathrm{i}\frac{1}{4} z^* \right) \left(\mathrm{i}\partial_{z^*} + \mathrm{i}\frac{1}{4} z \right) e^{-(1/4)zz^*} = \left(\mathrm{i}\partial_z - \mathrm{i}\frac{1}{2} z^* \right) \mathrm{i}\partial_{z^*} \tag{10}$$

For small non-zero V_{g_1,g_2}, there is only one minimal energy wave function described by a particular polynomial P whose form is determined by V_{g_1,g_2}. In general, it is very hard to calculate this unique ground state wave function. In the following, we will show that for some special interaction potential V_{g_1,g_2}, the ground state wave function can be obtained exactly.

2.2 Three Ideal FQH States: The Exact Zero-Energy Ground States

For interaction

$$V_{1/2}(z_1,z_2)=\delta(z_1-z_2), \tag{11}$$

the wave function $P_{1/2}(z_1,\dots,z_N)e^{-(1/4)\sum_{i=1}^N z_i z_i^*}$ with

$$P_{1/2}=\prod_{i<j}(z_i-z_j)^2 \tag{12}$$

is the only zero energy state with minimal total power of z_i's. This is because

$$\int\prod_i \mathrm{d}^2 z_i e^{-(1/4)\sum_i |z_i|^2} P_{1/2}^*\Big[\sum_{i<j} V_{1/2}(z_i,z_j)\Big]P_{1/2}e^{-(1/4)\sum_i |z_i|^2}=0. \tag{13}$$

Such a state is called $\nu=1/2$ Laughlin state.

For interaction

$$V_{1/4}(z_1,z_2)=v_0\delta(z_1-z_2)+v_2\partial_{z_1^*}^2\delta(z_1-z_2)\partial_{z_1}^2, \tag{14}$$

the wave function $P_{1/4}(z_1,\dots,z_N)e^{-(1/4)\sum_{i=1}^N z_i z_i^*}$ with

$$P_{1/4}=\prod_{i<j}(z_i-z_j)^4 \tag{15}$$

is the only zero energy state with minimal total power of z_i's, since

$$\int\prod_i \mathrm{d}^2 z_i e^{-(1/4)\sum_i |z_i|^2} P_{1/4}^*\Big[\sum_{i<j} V_{1/4}(z_i,z_j)\Big]P_{1/4}e^{-(1/4)\sum_i |z_i|^2}=0. \tag{16}$$

Such a state is called $\nu=1/4$ Laughlin state.

Now let us consider interaction [36, 37]

$$V_{\text{Pf}}(z_1,z_2,z_3)=\mathcal{S}\big[v_0\delta(z_1-z_2)\delta(z_2-z_3)-v_1\delta(z_1-z_2)\partial_{z_3^*}\delta(z_2-z_3)\partial_{z_3}\big] \tag{17}$$

where $\mathcal{S}$ symmetrizes among z_1,z_2,z_3 to make $V_{\text{Pf}}(z_1,z_2,z_3)$ a symmetric function. Then the wave function $P_{\text{Pf}}(z_1,\dots,z_N)e^{-(1/4)\sum_{i=1}^N z_i z_i^*}$ with

$$\begin{aligned} P_{\text{Pf}}&=\mathcal{A}\Big(\frac{1}{z_1-z_2}\frac{1}{z_3-z_4}\cdots\frac{1}{z_{N-1}-z_N}\Big)\prod_{i<j}(z_i-z_j)\\ &=\text{Pf}\Big(\frac{1}{z_i-z_j}\Big)\prod_{i<j}(z_i-z_j) \end{aligned} \tag{18}$$

is the only zero energy state with minimal total power of z_i's, where $\mathcal{A}$ anti-symmetrizes among $z_1, \ldots, z_N$. This is because

$$\int \prod_i \mathrm{d}^2 z_i e^{-(1/4)\sum_i |z_i|^2} P_{\mathrm{Pf}}^* \Big[\sum_{i<j<k} V_{\mathrm{Pf}}(z_i, z_j, z_k) \Big] P_{\mathrm{Pf}} e^{-(1/4)\sum_i |z_i|^2} = 0. \quad (19)$$

Such a state is called the Pfaffian state [21].

3 The Universal Properties of FQH Phases

The three many-body wave functions $P_{1/2}e^{-(1/4)\sum_i |z_i|^2}$, $P_{1/4}e^{-(1/4)\sum_i |z_i|^2}$, and $P_{\mathrm{Pf}}e^{-(1/4)\sum_i |z_i|^2}$ have some amazing exact properties in $N \to \infty$ limit. We believe that those properties do not depend on any local deformations of the wave functions.[1] In other words, those properties are shared by all the wave functions in the same phase. We call such kind of properties universal properties.

The universal properties can be viewed as quantum topological invariants in mathematics, since they do not change under any perturbations of the local Hamiltonian. Thus, from mathematical point of view, the symmetric polynomials of infinite variables, such as $P_{1/2}$, $P_{1/2}$, and P_{Pf}, can have many quantum topological invariants (i.e. the universal properties) once we define their norm to be

$$\langle P|P\rangle = \int \prod_{i=1}^{N} \mathrm{d}^2 z_i \big| P(z_1, \ldots, z_N) \big|^2 \mathrm{e}^{-(1/2)\sum |z_i|^2}. \quad (20)$$

Since the three wave functions have different universal properties, this implies that the three wave functions belong to three different quantum phases. In this section, we will discuss some of the universal properties, by first listing them in bold-face. Then we will give an understanding of them from physics point of view. Those conjectured universal properties are exact, but not rigorously proven to be true.

3.1 The Filling Fractions of FQH Phases

The density profile of a FQH wave function is given by

$$\rho(z) = \frac{\int d^2 z_2 \cdots d^2 z_N |P(z, z_2, \ldots, z_N)|^2 e^{-(1/2)\sum |z_i|^2}}{\int d^2 z_1 d^2 z_2 \cdots d^2 z_N |P(z_1, z_2, \ldots, z_N)|^2 e^{-(1/2)\sum |z_i|^2}} \quad (21)$$

[1] A local deformation of a many-body wave function Ψ is generated as $\Psi \to \Psi' = \mathrm{e}^{\mathrm{i}\delta H}\Psi$ where δH is a hermitian operator that can be viewed as an local Hamiltonian.

Fig. 3 The shape of the density profile $\rho(z)$

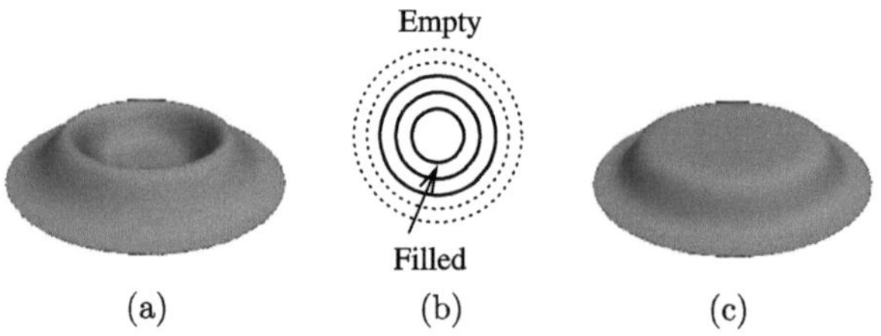

Fig. 4 **a** The density profile of the lth orbital. **b** The filling of the orbitals gives rise to a disk-like density profile in **c**

We believe that

$$\nu \equiv 2\pi\rho(0) \tag{22}$$

is a rational number in $N \to \infty$ *limit.* ν is called the filling fraction of the corresponding FQH state. We find that

$$\begin{aligned} P_1 &= \prod(z_i - z_j) \to \nu = 1, \qquad P_{1/2} = \prod(z_i - z_j)^2 \to \nu = 1/2, \\ P_{1/4} &= \prod(z_i - z_j)^4 \to \nu = 1/4, \\ P_{\mathrm{Pf}} &= \mathrm{Pf}\Big(\frac{1}{z_i - z_j}\Big)\prod(z_i - z_j) \to \nu = 1. \end{aligned} \tag{23}$$

Note that P_1 is anti-symmetric and describe a many-fermion state, while $P_{1/2}$, $P_{1/4}$, and P_{Pf} are symmetric and describe many-boson states.

We also believe that the density profile $\rho(z)$ has disk shape (see Fig. 3) in large N limit: $\rho(z)$ is almost a constant $\nu/2\pi$ for $|z| < \sqrt{2N/\nu}$ and quickly drop to almost zero for $|z| > \sqrt{2N/\nu}$.

3.1.1 Why $\nu = 1$ for State $\Psi_1 = \prod_{i<j}(z_i - z_j)e^{-\sum|z_i|^2/4}$

We note that the one-particle eigenstates (the orbitals) for one-particle Hamiltonian $H_0 = -\sum(\partial_z - \frac{1}{4}z^*)(\partial_{z^*} + (1/4)z)$ can be labeled by the angular momentum l, which is given by $z^l e^{-(1/4)|z|^2}$. The one-particle eigenstate has a ring-like shape with maximum at $|z| = r_l = \sqrt{2l}$ (see Fig. 4a). The $\nu = 1$ many-fermion state is obtained by filling the orbitals (see Fig. 4b):

$$\Psi = \prod_{i<j}(z_i - z_j)e^{-(1/4)\sum|z_i|^2} = \mathcal{A}\big[(z_1)^0(z_2)^1\cdots\big]e^{-(1/4)\sum|z_i|^2} \tag{24}$$

We see that there are l fermions within radius r_l. So there is one fermion per $\pi r_l^2/l = 2\pi$ area, and thus $\nu = 1$ (see Fig. 4c).

3.1.2 Why $\nu = 1/m$ for the Laughlin State $\Psi_{1/m} = \prod_{i<j}(z_i - z_j)^m e^{-\sum |z_i|^2/4}$

Let us consider the joint probability distribution of boson positions, which is given by the absolute-value-square of the ground state wave function:

$$p(z_1 \cdots z_N) \propto \left|\Psi_{1/m}(z_1 \cdots z_N)\right|^2$$
$$= e^{-2m \sum_{i<j} \ln|z_i - z_j| - (m/2) \sum_i |z_i|^2} = e^{-\beta V(z_1 \cdots z_N)} \tag{25}$$

Choosing $T = \frac{1}{\beta} = \frac{m}{2}$, we can view $e^{-\beta V(z_1 \cdots z_N)}$ as the probability distribution for N particles with potential energy $V(z_1 \cdots z_N)$ at temperature $T = \frac{m}{2}$. The potential has a form

$$V = -m^2 \sum_{i<j} \ln|z_i - z_j| + \frac{m}{4} \sum_i |z_i|^2 \tag{26}$$

which is the potential for a two-dimensional plasma of 'charge' m particles [2]. The two-body term $-m^2 \ln|z - z'|$ represents the interaction between two particles and the one-body term $\frac{m}{4}|z|^2$ represents the interaction of a particle with the background "charge".

For a uniform background "charge" distribution with charge density ρ_ϕ, a charge m particle at z feel a force, $F = (\pi |z|^2 \rho_\phi)(m)/|z|$. The corresponding background potential energy is $-\rho_\phi m \frac{\pi}{2}|z|^2$. We see that to produce the one-body potential energy $\frac{m}{4}|z|^2$ we need to set $\rho_\phi = -1/2\pi$. Since the plasma must be "charge" neutral: $m\rho + \rho_\phi = 0$, we find that $\rho = \frac{1}{m}\frac{1}{2\pi}$. So $\nu = 1/m$.

3.2 *Quasiparticle and Fractional Charge in $\nu = 1/m$ Laughlin States*

If we remove a boson at position ξ from the Laughlin wave function $\prod_{i<j}(z_i - z_j)^m e^{-\sum |z_i|^2/4}$, we create a hole-like excitation described by the wave function $\Psi_\xi^{\text{hole}}(z_1, \ldots, z_N)$:

$$\Psi_\xi^{\text{hole}}(z_1, \ldots, z_N) \propto \prod_i (\xi - z_i)^m \prod_{i<j} (z_i - z_j)^m e^{-\sum |z_i|^2/4} \tag{27}$$

Despite the hole-like excitation has a charge $= 1$, the minimal value for non-zero integers, it is not the minimally charged excitation. The minimally charged excitation corresponds to a quasi-hole excitation, which is described by the wave function

$$\Psi_\xi^{\text{quasi-hole}}(z_1, \ldots, z_N) \propto \prod_i (\xi - z_i) \prod_{i<j} (z_i - z_j)^m e^{-\sum |z_i|^2/4} \tag{28}$$

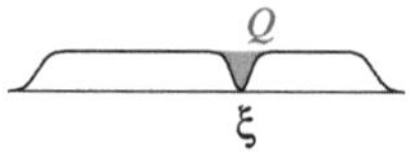

Fig. 5 The density profile of a many-boson wave function with a quasi-hole excitation at ξ

The density profile for the quasi-hole wave function $\Psi_\xi^{\text{quasi-hole}}(z_1, \ldots, z_N)$ is given by

$$\rho_\xi(z) = \frac{\int \prod_{i=2}^N \mathrm{d}^2 z_i |\Psi_\xi^{\text{quasi-hole}}(z, z_2, \ldots, z_N)|^2}{\int \prod_{i=1}^N \mathrm{d}^2 z_i |\Psi_\xi^{\text{quasi-hole}}(z_1, z_2, \ldots, z_N)|^2} \tag{29}$$

$\rho_\xi(z)$ has a shape as in Fig. 5. The quasi-particle charge is defined as

$$Q = \int_{D_\xi} \mathrm{d}^2 z \left(\frac{\nu}{2\pi} - \rho_\xi(z) \right) \tag{30}$$

in the $N \to \infty$ limit, where D_ξ is a big disk covering ξ. (Note that, away from the quasi-hole, $\rho_\xi(z) = \frac{\nu}{2\pi}$.) We believe that *the quasi-hole charge is a rational number* $Q = 1/m$ [2].

One way to understand the above result is to note that m quasi-holes correspond to a missing boson: $[\prod_i(\xi - z_i)]^m = \prod_i(\xi - z_i)^m$. So a quasi-hole excitation has a fractional charge $1/m$ although the FQH state is formed by particles of charge 1!

We can also calculate the quasi-hole charge directly. Note that, for the Laughlin state $\Psi_\xi^{\text{quasi-hole}}(z_1, \ldots, z_N)$ with a quasi-hole at ξ, the corresponding joint probability distribution of boson positions is given by $p(\{z_i\}) \propto |\Psi_\xi^{\text{quasi-hole}}(\{z_i\})| = \mathrm{e}^{-\beta V}$ with

$$V = -m^2 \sum_{i<j} \ln|z_i - z_j| - m \sum_i \ln|z_i - \xi| + \frac{m}{4} \sum_i |z_i|^2 \tag{31}$$

Now, the one-body potential term $-m \ln|z - \xi| + \frac{m}{4}|z|^2$ is produced by background charge density: $\rho_\phi = -\frac{1}{2\pi} + \delta(\xi)$. The "charge" neutral condition $m\rho_\xi(z) + \rho_\phi(z) \approx 0$ allows us to show that $\rho_\xi(z)$ has a shape as in Fig. 5 and satisfies Eq. (30) with $Q = 1/m$.

3.3 The Concept of Quasiparticle Type

We would like to point out that the wave function $\Psi_\xi^{\text{quasi-hole}}(z_1, \ldots, z_N) \propto \prod_i(\xi - z_i) \prod_{i<j}(z_i - z_j)^m \mathrm{e}^{-\sum |z_i|^2/4}$ just describes a particular kind of quasiparticle excitation. More general quasiparticle excitations can be constructed as

$$\Psi_\xi^{\text{quasi-hole-}k}(z_1, \ldots, z_N) \propto \prod_i (\xi - z_i)^k \prod_{i<j} (z_i - z_j)^m \mathrm{e}^{-\sum |z_i|^2/4} \tag{32}$$

which can be viewed as a bound state of k charge-$1/m$ quasi-holes. So it appears that different types of quasiparticles are labeled by integer k.

Here we would like to introduce a concept of quasiparticle type: *two quasiparticles belong to the same type if they only differ by a number of bosons that form the FQH state.* Since the quasiparticle labeled by $k = m$ correspond to a boson, so the different types of quasiparticles in the $\nu = 1/m$ Laughlin state are labeled by k mod m. *There are m types of quasiparticles in the $\nu = 1/m$ Laughlin state* (including the trivial type labeled by $k = 0$).

There is an amazing relation between the number of quasiparticle type and the ground state degeneracy of the FQH state on torus: *the number of quasiparticle type always equal to the ground state degeneracy on torus, in the $N \to \infty$ limit.*

3.4 Fractional Statistics in Laughlin States

We note that the normalized state with a quasi-hole at ξ is described by an N-boson wave function parameterized by ξ:

$$\Psi_{\xi}^{\text{quasi-hole}} = \left[N\left(\xi, \xi^*\right)\right]^{-1/2} \prod_i (\xi - z_i) \prod_{i<j} (z_i - z_j)^2 \mathrm{e}^{-\sum |z_i|^2/4} \tag{33}$$

where $N(\xi, \xi^*)$ is the normalization factor. The normalized two quasi-hole wave function is given by

$$\begin{aligned}\Psi_{\xi,\xi'}^{\text{quasi-hole}} &= \left[N\left(\xi, \xi^*, \xi', \xi'^*\right)\right]^{-1/2} \\ &\times \prod_i (\xi - z_i) \prod_i (\xi' - z_i) \prod_{i<j} (z_i - z_j)^2 \mathrm{e}^{-\sum |z_i|^2/4}\end{aligned} \tag{34}$$

We conjecture that the above two normalization factors are given by

$$N\left(\xi, \xi^*\right) = \mathrm{e}^{(1/(2m))|\xi|^2} \times \text{Const.} \tag{35}$$

and

$$N\left(\xi, \xi^*, \xi', \xi'^*\right) = \mathrm{e}^{(1/(2m))(|\xi|^2+|\xi'|^2)+(1/m)\ln|\xi-\xi'|^2} \times \text{Const.} \tag{36}$$

in the $N \to \infty$ limit, where ξ and ξ' are hold fixed in the limit.

The quasi-holes in the Laughlin states also have fractional statistics [38–41]. We can calculate the fractional statistics by calculating the Berry phase [42] of moving the quasi-holes. It turns out that the Berry phase of moving the quasi-holes can be calculated from the above normalization factors. Let us first calculate the Berry phase for one quasi-hole and the normalization factor $N(\xi, \xi^*)$. The Berry's phase $\Delta\varphi$ induced by moving ξ is defined as $\mathrm{e}^{\mathrm{i}\Delta\varphi} = \langle \Psi_{\xi}^{\text{quasi-hole}} | \Psi_{\xi+d\xi}^{\text{quasi-hole}} \rangle$. It is given by

$$\Delta\varphi = a_\xi d\xi + a_{\xi^*} d\xi^*, \quad a_\xi = -\mathrm{i}\langle \Psi_\xi | \frac{\partial}{\partial \xi} | \Psi_\xi \rangle, \qquad a_{\xi^*} = -\mathrm{i}\langle \Psi_\xi | \frac{\partial}{\partial \xi^*} | \Psi_\xi \rangle, \tag{37}$$

where a_ξ and a_{ξ^*} are Berry connections. Since the unnormalized state $\prod_i(\xi - z_i)\prod_{i<j}(z_i - z_j)^2 \mathrm{e}^{-\sum |z_i|^2/4}$ has a special property that it only depends only on ξ (holomorphic), the Berry connection (a_ξ, a_{ξ^*}) can be calculated from the normalization $N(\xi,\xi^*)$ of the holomorphic state:

$$a_\xi = -\frac{\mathrm{i}}{2}\frac{\partial}{\partial \xi}\ln\left[N(\xi,\xi^*)\right], \qquad a_{\xi^*} = \frac{\mathrm{i}}{2}\frac{\partial}{\partial \xi^*}\ln\left[N(\xi,\xi^*)\right]. \tag{38}$$

Now let us calculate $N(\xi,\xi^*)$. Let us guess that $N(\xi,\xi^*)$ is given by Eq. (35). To show the guess to be right, we need to show that the norm of $|\Psi_\xi^{\text{quasi-hole}}\rangle$ does not depend on ξ. We note that $|\Psi_\xi^{\text{quasi-hole}}|^2 = \mathrm{e}^{-\beta V_\xi}$ with

$$\begin{aligned} V_\xi(z_1,\ldots,z_N) = &-m^2\sum_{i<j}\ln|z_i - z_j| - m\sum_i \ln|z_i - \xi| \\ &+ \frac{1}{4}|\xi|^2 + \frac{m}{4}\sum_i |z_i|^2. \end{aligned} \tag{39}$$

Here V_ξ can be viewed as the total energy of a plasma of N 'charge'-m particles at z_i and one 'charge'-1 particle hold fixed at ξ. Both particles interact with the same background charge. Note that the norm $\langle\Psi_\xi^{\text{quasi-hole}}|\Psi_\xi^{\text{quasi-hole}}\rangle$ is given by

$$\left\langle\Psi_\xi^{\text{quasi-hole}}\middle|\Psi_\xi^{\text{quasi-hole}}\right\rangle = \int \prod \mathrm{d}^2 z_i \mathrm{e}^{-\beta V_\xi} \tag{40}$$

Due to the screening of the plasma, we argue that $\int \prod \mathrm{d}^2 z_i \mathrm{e}^{-\beta V_\xi}$ does not depend on ξ in $N \to \infty$ limit, which implies that $\langle\Psi_\xi^{\text{quasi-hole}}|\Psi_\xi^{\text{quasi-hole}}\rangle$ does not depend on ξ. Thus $N(\xi,\xi^*)$ is indeed given by Eq. (35).

This allows us to find

$$a_\xi = -\mathrm{i}\frac{1}{4m}\xi^*, \qquad a_{\xi^*} = \mathrm{i}\frac{1}{4m}\xi \tag{41}$$

Using such a Berry connection, let us calculate the Berry's phase for moving ξ around a circle C of radius r center at $z = 0$:

$$\begin{aligned} \Delta\varphi &= \oint_C \left(a_\xi d\xi + a_{\xi^*} d\xi^*\right) \\ &= 2\pi\frac{r^2}{4m} \times 2 = 2\pi\frac{\text{Area enclosed by } C}{2\pi m} \\ &= 2\pi \times \text{number of enclosed bosons by } C. \end{aligned} \tag{42}$$

We see that the Berry connection describes a uniform 'magnetic' field. The above result can also be understood directly from the wave function $\prod_i(\xi - z_i)\prod_{i<j}(z_i - z_j)^2 \mathrm{e}^{-\sum |z_i|^2/4}$.

Similarly, we can calculate the Berry connection for two quasi-holes. Let us guess that $N(\xi, \xi^*, \xi', \xi'^*)$ is given by Eq. (36). For such a normalization factor, we find that $|\Psi_{\xi,\xi'}^{\text{quasi-hole}}|^2 = \mathrm{e}^{-\beta V_{\xi,\xi'}}$ with

$$\begin{aligned} V_{\xi,\xi'}(z_1, \ldots, z_N) = &-m \sum_i \left[\ln|z_i - \xi| + \ln|z_i - \xi'|\right] \\ &+ \frac{1}{4}\left[|\xi|^2 + |\xi'|^2\right] - \ln|\xi - \xi'| \\ &- m^2 \sum_{i<j} \ln|z_i - z_j| + \frac{m}{4} \sum_i |z_i|^2 \end{aligned} \tag{43}$$

Such a $V_{\xi,\xi'}$ can be viewed as the total energy of a plasma of N 'charge'-m particles at z_i and two 'charge'-1 particles at ξ and ξ'. Due to the screening, $\int \prod \mathrm{d}^2 z_i \mathrm{e}^{-\beta V_{\xi,\xi'}}$ does not depend on ξ and ξ' in $N \to \infty$ limit, which implies that $\langle \Psi_{\xi,\xi'}^{\text{quasi-hole}} | \Psi_{\xi,\xi'}^{\text{quasi-hole}} \rangle$ does not depend on ξ and ξ'. So our guess is correct. Using the normalization factor (36), we find the Berry connection to be

$$a_\xi = -\mathrm{i}\frac{1}{4m}\xi^* + \frac{\mathrm{i}}{2m}\frac{1}{\xi - \xi'}, \qquad a_{\xi^*} = \mathrm{i}\frac{1}{4m}\xi - \frac{\mathrm{i}}{2m}\frac{1}{\xi^* - \xi'^*} \tag{44}$$

Using such a Berry connection, we can calculate the fractional statistics of the quasi-holes in the $\nu = 1/m$ Laughlin state. Moving a quasi-hole around another, we find the Berry phase to be $\Delta\varphi = \frac{\text{enclosed area}}{m} - \frac{2\pi}{m}$ (see Eq. (42) for comparison). If we only look at the sub-leading term $-2\pi/m$, we find that exchanging two quasi-holes give rise to phase $\theta = -\pi/m$, since exchanging two quasi-holes correspond to moving a quasi-hole half way around another and we get the half of $-2\pi/m$. We find that *quasi-holes in the* $\nu = 1/m$ *Laughlin state have a fractional statistics described by the phase factor* $\mathrm{e}^{-\mathrm{i}\pi/m}$ [40, 41].

The term $\frac{\text{enclosed area}}{m}$ implies that the quasi-holes sees a uniform magnetic field. So the quasi-holes in the $\nu = 1/m$ Laughlin state are anyons in magnetic field.

3.5 *Quasi-holes in the* $\nu = 1$ *Pfaffian State*

3.5.1 Charge-1 and Charge-1/2 Quasi-holes

Ground state wave function for the $\nu = 1$ Pfaffian state is given by

$$\Psi_{\text{Pf}} = \mathcal{A}\left(\frac{1}{z_1 - z_2}\frac{1}{z_3 - z_4} \cdots \frac{1}{z_{N-1} - z_N}\right)\Psi_1 = \text{Pf}\left(\frac{1}{z_i - z_j}\right)\Psi_1 \tag{45}$$

where Ψ_1 is given by $\prod_{i<j}(z_i - z_j)\mathrm{e}^{-(1/4)\sum_i |z_i|^2}$. A simple quasi-hole state is given by

$$\begin{aligned}\Psi_\xi^{\text{charge-1}} &= \prod(\xi - z_i)\Psi_{\text{Pf}}\\ &= \mathcal{A}\left(\frac{(\xi - z_1)(\xi - z_2)}{z_1 - z_2}\frac{(\xi - z_3)(\xi - z_4)}{z_3 - z_4}\cdots\right)\Psi_1\\ &= \text{Pf}\left(\frac{(\xi - z_i)(\xi - z_j)}{z_i - z_j}\right)\Psi_1\end{aligned} \tag{46}$$

which is created by multiplying the factor $\prod(\xi - z_i)$ to the ground state wave function. *Such a quasi-hole has a charge 1.* The above quasi-hole can be splitted into two fractionalized quasi-holes. A state with two fractionalized quasi-holes at ξ and ξ' is given by

$$\begin{aligned}\Psi_{\xi,\xi'}^{\text{charge-1/2}} &= \mathcal{A}\left(\frac{(\xi - z_1)(\xi' - z_2) + (1 \leftrightarrow 2)}{z_1 - z_2}\frac{(\xi - z_3)(\xi' - z_4) + (3 \leftrightarrow 4)}{z_3 - z_4}\cdots\right)\Psi_1\\ &= \text{Pf}\left(\frac{(\xi - z_i)(\xi' - z_j) + (\xi - z_j)(\xi' - z_i)}{z_i - z_j}\right)\Psi_1\end{aligned} \tag{47}$$

Such a fractionalized quasi-hole has a charge 1/2. We note that combining two charge-1/2 quasi-holes gives us one charge-1 quasi-hole: $\Psi_{\xi,\xi}^{\text{charge-1/2}} \propto \Psi_\xi^{\text{charge-1}}$.

3.5.2 How Many States with Four Charge-1/2 Quasi-holes?

One of the state with four charge-1/2 quasi-holes at ξ_1, ξ_2, ξ_3, and ξ_4 is given by

$$\begin{aligned}P_{(12)(34)} &= \text{Pf}\left(\frac{(\xi_1 - z_i)(\xi_2 - z_i)(\xi_3 - z_j)(\xi_4 - z_j) + (i \leftrightarrow j)}{z_i - z_j}\right)\Psi_1\\ &= \text{Pf}\left(\frac{[12, 34]_{z_i z_j}}{z_i - z_j}\right)\Psi_1\end{aligned} \tag{48}$$

The other two are $P_{(13)(14)}$, $P_{(14)(23)}$. But only two of them are linearly independent [43]. Using the relation

$$[12, 34]_{z_i z_j} - [13, 24]_{z_i z_j} = (z_i - z_j)^2(\xi_1 - \xi_4)(\xi_2 - \xi_3) = z_{ij}^2\xi_{14}\xi_{23} \tag{49}$$

we find (with $z_{12} = z_1 - z_2$, $\xi_{12} = \xi_1 - \xi_2$, etc.)

$$\begin{aligned}P_{(13)(24)} &= \mathcal{A}\left(\frac{[12, 34]_{z_1 z_2} - z_{12}^2\xi_{14}\xi_{23}}{z_{12}}\frac{[12, 34]_{z_3 z_4} - z_{34}^2\xi_{14}\xi_{23}}{z_{34}}\cdots\right)\Psi_1\\ &= P_{(12)(34)} - N_{pair}\mathcal{A}\left(z_{12}\xi_{14}\xi_{23}\frac{[12, 34]_{z_3 z_4}}{z_{34}}\cdots\right)\Psi_1\end{aligned} \tag{50}$$

So

$$P_{(12)(34)} - P_{(13)(24)} = N_{pair}\xi_{14}\xi_{23}\mathcal{A}\left(z_{12}\frac{[12, 34]_{z_3 z_4}}{z_{34}}\cdots\right)\Psi_1 \tag{51}$$

Similarly

$$P_{(12)(34)} - P_{(14)(23)} = N_{pair}\xi_{13}\xi_{24}\mathcal{A}\left(z_{12}\frac{[12,34]_{z_3z_4}}{z_{34}}\cdots \right)\Psi_1 \tag{52}$$

Thus

$$\frac{P_{(12)(34)} - P_{(13)(24)}}{\xi_{14}\xi_{23}} = \frac{P_{(12)(34)} - P_{(14)(23)}}{\xi_{13}\xi_{24}} \tag{53}$$

We find that *there are two states for four charge-1/2 quasi-holes, even if we fixed their positions. The two states are topologically degenerate (have the same energy in $N \to \infty$ limit)* [43]. The appearance of the topological degeneracy even with fixed quasi-hole positions is a defining property of the non-Abelian statistics. In the presence of the topological degeneracy, as we exchange quasi-holes, we will generate non-Abelian Berry phases which also describe non-Abelian statistics.

More generally we find that there are $D_n = \frac{1}{2}(\sqrt{2})^n$ topologically degenerate states for n charge-1/2 quasi-holes, even if we fixed their positions [43]. We see that there are $\sqrt{2}$ states per charge-1/2 quasi-hole! The $\sqrt{2}$ is called the quantum dimension for the charge-1/2 quasi-hole. We see that the charge-1/2 quasi-hole has a non-Abelian statistics, since for Abelian anyons, the quantum dimension is always 1.

3.6 Edge Excitations and Conformal Field Theory

Under the $z \to \mathrm{e}^{\mathrm{i}\theta} z$ transformation, the N-particle $\nu = 1/2$ Laughlin wave function $\Psi_{1/2} = P_{1/2}(z_1, \ldots, z_N)e^{-\sum |z_i|^2/4} = \prod_{1\le i<j\le N}(z_i - z_j)^2 e^{-\sum |z_i|^2/4}$ transforms as $\Psi_{1/2} \to \mathrm{e}^{\mathrm{i}S_N\theta}\Psi_{1/2}$, with $S_N = N(N-1)$. We call S_N the angular momentum of the Laughlin wave function (which is also the total power of z_i's of the polynomial $P_{1/2}(z_1, \ldots, z_N)$). For interaction $V_{1/2} = \sum \delta(z_i - z_j)$, the $\nu = 1/2$ Laughlin wave function is the only zero energy state with angular momentum $N(N-1)$ since $\Psi_{1/2}(z_1, \ldots, z_N)$ vanishes as $z_i \to z_j$. There are no zero energy states with angular momentum less than S_N. In fact, we believe that, *for wave functions Ψ with angular momentum less then S_N,*

$$\langle V_{1/2}\rangle = \frac{\int \prod \mathrm{d}^2 z_i V_{1/2}|\Psi(z_1, \ldots, z_N)|^2}{\int \prod \mathrm{d}^2 z_i |\Psi(z_1, \ldots, z_N)|^2} \ge \Delta \tag{54}$$

for a positive Δ and any N. The maximal Δ is called the energy gap for the interaction $V_{1/2}$.

On the other hand, there are many zero energy states ($\langle V_{1/2}\rangle = 0$) with angular momentum bigger than S_N. We call those zero energy states edge states, and denote them as Ψ_{edge}. We can introduce a sequence of integers D_L^{edge} to denote the number of zero energy states with angular momentum $S_N + L$. We will call D_L^{edge} the edge spectrum.

To obtain the edge spectrum for the $\nu = 1/2$ Laughlin state with interaction $V_{1/2}$, we note that the zero-energy edge states can be obtained by multiplying the Laughlin wave function by a symmetric polynomial which does not reduce the order of zeros:

$$\Psi_{\text{edge}} = P_{\text{sym}}\big(\{z_i\}\big)\Psi_{1/2}. \tag{55}$$

Since the number of the symmetric polynomials with the total power of z_i's equal to L is given by the partition number p_L, we find $D_L^{\text{edge}} = p_L$. Such an argument applies to any Laughlin states. So we believe that *for $\nu = 1/m$ Laughlin the edge spectrum is given by the partition numbers: $D_L^{\text{edge}} = p_L$* [44]:

L	0	1	2	3	4	5	6
D_L^{edge}	1	1	2	3	5	7	11
P_{sym}	1	$\sum z_i$	$(\sum z_i)^2$	$\cdots$	$\cdots$	$\cdots$	$\cdots$
			$\sum z_i^2$	$\cdots$	$\cdots$	$\cdots$	$\cdots$

(56)

In large L limit, $D_L^{\text{edge}} \approx \frac{1}{4\sqrt{3}L} e^{\pi\sqrt{2L/3}} \approx e^{\pi\sqrt{2L/3}}$.

For the $\nu = 1$ Pfaffian state with the ideal Hamiltonian $\mathcal{S}[v_0\delta(z_1 - z_2)\delta(z_2 - z_3) - v_1\delta(z_1 - z_2)\partial_{z_3^*}\delta(z_2 - z_3)\partial_{z_3}]$, $\Psi_{\text{Pf}} = \mathcal{A}(\frac{1}{z_1-z_2}\frac{1}{z_3-z_4}\cdots)\prod_{i<j}(z_i - z_j)$, is the zero-energy state with the minimal total angular momentum S_N. Other zero-energy states with higher angular momenta are given by

$$\Psi_{\text{edge}} = \mathcal{A}\left(P_{\text{any}}\big(\{z_i\}\big)\frac{1}{z_1 - z_2}\frac{1}{z_3 - z_4}\cdots\right)\Psi_1, \tag{57}$$

where P_{any} is any polynomial. Now the counting is much more difficult, since linearly independent P_{any}'s may generate linearly dependent wave functions. We find, for large even total boson number N, the edge spectrum is given by [45]

L	0	1	2	3	4	5	6
D_L^{edge}	1	1	3	5	10	16	28

(58)

We believe that, *for the $\nu = 1$ Pfaffian state, the edge spectrum in large L limit is given by $D_L^{\text{edge}} \approx e^{\pi\sqrt{2L/3}\sqrt{c}}$ with $c = 3/2$, if $N \to \infty$ and $L \ll N$.*

It turns out that the edge spectrum for $\nu = 1/m$ Laughlin state can be produced by a central charge $c = 1$ CFT and the edge spectrum for $\nu = 1$ Pfaffian state can be produced by a central charge $c = 3/2$ CFT [44, 45]. This allows us to connect the edge excitations of a FQH state to a CFT.

Using the quasi-hole wave function $\Psi_\xi^{\text{quasi-hole}}(z_1, \ldots, z_N)$ that describes a quasi-hole at ξ, we can even calculate the correlation function of the quasi-hole operator. We know that the circular quantum Hall droplet has a radius $R = \sqrt{2N/\nu}$. The quasi-hole correlation function on the edge of the droplet is given by

$$G^{\text{quasi-hole}}(\theta' - \theta) \propto \int \prod d^2 z_i \big[\Psi_{\xi'}^{\text{quasi-hole}}(z_1, \ldots, z_N)\big]^* \times \Psi_{\xi}^{\text{quasi-hole}}(z_1, \ldots, z_N)\Big|_{\xi = R e^{i\theta}; \xi' = R e^{i\theta'}}. \tag{59}$$

We find that $G^{\text{quasi-hole}}(\theta - \theta')$ *has a form*

$$G^{\text{quasi-hole}}(\theta) \propto e^{iQ\nu^{-1}N\theta} \left(\frac{1}{1 - e^{-i\theta}}\right)^{2h} \tag{60}$$

where Q is the quasi-hole charge and h is a rational number. We will call h the scaling dimension of the quasi-hole. *For the $\nu = 1/m$ Laughlin state, we find that $h = \frac{1}{2m}$ for the charge $Q = 1/m$ quasi-hole. For the $\nu = 1$ Pfaffian state, we find that $h = \frac{1}{2}$ for the charge-1 quasi-hole, and $h = \frac{3}{16}$ for the charge-1/2 quasi-hole, all in $N \to \infty$ limit* [45, 46].

4 Pattern-of-Zeros Approach to FQH States and Symmetric Polynomials

Using $P_{1/2}$, $P_{1/4}$, and P_{Pf} as examples, we have seen that symmetric polynomials with infinite variables can have some amazing universal properties, once we defined the norm of the infinite-variable polynomials to be

$$\langle P|P\rangle = \int \prod d^2 z_i |P|^2 e^{-(1/2)\sum |z_i|^2}. \tag{61}$$

This suggests that it may be possible to come up with a definition of "infinite-variable symmetric polynomials". Such properly defined infinite-variable symmetric polynomials should have those amazing universal properties. The proper definition also allow us to classify infinite-variable symmetric polynomials, which will lead to a classification of FQH phases.

In this section, we will first discuss an attempt to define infinite-variable symmetric polynomials through pattern of zeros. Then, we will try to provide a classification of patterns of zeros. After that, we will use the patterns of zeros to calculate the universal properties of the corresponding infinite-variable symmetric polynomials.

4.1 What Is Infinite-Variable Symmetric Polynomial

The main difficulty to define symmetric polynomial with infinite variables is that the number of the variables is not fixed. To overcome this difficulty, we will characterize the symmetric polynomials through their "local properties" that do not depend on the number of the variables. One such "local property" is pattern of zeros.

4.1.1 What Is Pattern of Zeros?

We have seen that the different short-range interactions $V(z_i - z_j)$ in Hamiltonian

$$H = \sum_{i=1}^{N} -\left(\partial_{z_i} - \frac{B}{4} z_i{}^*\right)\left(\partial_{z_i{}^*} + \frac{B}{4} z_i\right) + \sum_{i<j} V(z_i - z_j) \tag{62}$$

leads to different FQH states $P(z_1, \ldots, z_N)\mathrm{e}^{-(1/4)\sum_{i=1}^{N} |z_i|^2}$, which in turn leads to different symmetric polynomials $P(z_1, \ldots, z_N)$.

One of the resulting polynomial $P_{1/2} = \prod_{i<j}(z_i - z_j)^2$ has a property that as $z_1 \approx z_2$, it has a second-order zero $P_{1/2} \propto (z_1 - z_2)^2$. Another resulting polynomial $P_{1/4} = \prod_{i<j}(z_i - z_j)^4$ has a property that as $z_1 \approx z_2$, it has a fourth-order zero $P_{1/4} \propto (z_1 - z_2)^4$. The third resulting polynomial

$$P_{\mathrm{Pf}} = \mathcal{A}\left(\frac{1}{z_1 - z_2}\frac{1}{z_3 - z_4} \cdots \frac{1}{z_{N-1} - z_N}\right)\prod_{i<j}(z_i - z_j) \tag{63}$$

has a property that as $z_1 \approx z_2$, P_{Pf} has no zero, while as $z_1 \approx z_2 \approx z_3$, P_{Pf} has a second-order zero. We see that different polynomials can be characterized by different patterns of zeros.

The above examples suggest the following general definition of pattern of zeros for a symmetric polynomial $P(\{z_i\})$. Let $z_i = \lambda\eta_i + z^{(a)}$, $i = 1, 2, \ldots, a$. In the small λ limit, we have

$$P(\{z_i\}) = \lambda^{S_a} P(\eta_1, \ldots, \eta_a; z^{(a)}, z_{a+1}, z_{a+2}, \ldots) + O(\lambda^{S_a+1}) \tag{64}$$

The sequence of integers $\{S_a\}$ characterizes the symmetric polynomial $P(\{z_i\})$ and is called the *pattern of zeros* of P. We note that S_N happen to be the total power of z_i (or the total angular momentum) of P if the polynomial has N variables.

4.1.2 The Unique Fusion Condition

If the above induced $P(\{\eta_i\}; z^{(a)}, z_{a+1}, z_{a+2}, \ldots)$, does not depend on the "shape" $\{\eta_i\}$

$$P(\{\eta_i\}; z^{(a)}, z_{a+1}, z_{a+2}, \ldots) \propto P(z^{(a)}, z_{a+1}, z_{a+2}, \ldots), \tag{65}$$

we then say that the symmetric polynomial $P(\{z_i\})$ satisfy the *unique fusion condition*.

4.1.3 Different Encodings of Pattern of Zeros S_a

There are many different ways to encode the sequence of integers S_a. For example, we may use

$$l_a = S_a - S_{a-1}, \quad a = 1, 2, 3, \dots \tag{66}$$

to encode S_a, $a = 1, 2, 3, \dots$:

$$S_a = \sum_{i=1}^{a} l_i . \tag{67}$$

Here we have assumed that $S_0 = 0$. It turns out that $l_i \geq 0$ and $l_i \leq l_{i+1}$.

We may also use n_l, $l = 0, 1, 2, \dots$ to encode S_a. Here n_l is the number of times that the value l appears in the sequence l_i:

$$n_l = \sum_{i=1}^{\infty} \delta_{l,l_i} . \tag{68}$$

Let us list the pattern of zeros for some simple polynomials. For the $\nu = 1$ integer quantum Hall state $P_1 = \prod_{i<j}(z_i - z_j)$, the pattern of zeros is given by

$$\begin{aligned} &S_1, S_2, \dots : 0, 1, 3, 6, 10, 15, \dots \\ &l_1, l_2, \dots : 0, 1, 2, 3, 4, 5, \dots \\ &n_0 n_1 n_2 \cdots : 11111111 \cdots \end{aligned} \tag{69}$$

We see that we can view l in n_l as the label for the orbital $z^l e^{-(1/4)|z|^2}$, and n_l as the occupation number on the lth orbital (see Sect. 3.1.1 and Fig. 4b).

The pattern of zeros of $\nu = 1/2$ Laughlin state $P_{1/2}$ is described by

$$\begin{aligned} &S_1, S_2, \dots : 0, 2, 6, 12, 20, 30, \dots \\ &l_1, l_2, \dots : 0, 2, 4, 6, 8, 10, \dots \\ &n_0 n_1 n_2 \cdots : 1010101010101010 \cdots \end{aligned} \tag{70}$$

We see that n_l has a periodic structure. Each unit cell (each cluster) has 1 particle and 2 orbitals.

The pattern of zeros of $\nu = 1/4$ Laughlin state $P_{1/4}$ is described by

$$\begin{aligned} &S_1, S_2, \dots : 0, 4, 12, 24, 40, 60, 84, \dots \\ &l_1, l_2, \dots : 0, 4, 8, 12, 16, 20, \dots \\ &n_0 n_1 n_2 \cdots : 10001000100010001 \cdots \end{aligned} \tag{71}$$

Again, n_l has a periodic structure. Each unit cell (each cluster) has 1 particle and 4 orbitals.

For the $\nu = 1$ Pfaffian state $P_{\text{Pf}} = \mathcal{A}(\frac{1}{z_1 - z_2}\frac{1}{z_3 - z_4}\cdots)\prod_{i<j}(z_i - z_j)$, the pattern of zeros is given by

$$\begin{aligned} &S_1, S_2, \ldots : 0, 0, 2, 4, 8, 12, 18, 24, \ldots \\ &l_1, l_2, \ldots : 0, 0, 2, 2, 4, 4, 6, 6, \ldots \\ &n_0 n_1 n_2 \cdots : 20202020202020202 \cdots \end{aligned} \tag{72}$$

Now a cluster (unit cell) has 2 particles and 2 orbitals.

4.1.4 The Cluster Condition

Motivated by the above examples, here we would like to introduce a cluster condition for symmetric polynomials: *an symmetric polynomial satisfies a cluster condition if n_l is periodic.* Let each unit cell contains n particles and m orbitals. In this case, S_a has a form

$$S_{a+kn} = S_a + kS_n + \frac{k(k-1)nm}{2} + kma \tag{73}$$

Since $S_1 = 0$, we see that we can use a finite sequence $(\frac{m}{n}; S_2, \ldots, S_n)$ to describe the pattern of zeros for symmetric polynomial satisfying the cluster condition.

We note that the filling fraction ν is given by the average number of particles per orbital. Thus $\nu = n/m$. We also call the cluster condition with n particles per unit cell an n-cluster condition.

4.1.5 A Definition of Infinite-Variable Symmetric Polynomial

Now, we are ready to define *the infinite-variable symmetric polynomial as a symmetric polynomial of infinite variables that satisfy the unique fusion condition and the cluster condition.* The cluster condition makes the $N \to \infty$ limit possible. [Or more precisely, the infinite-variable symmetric polynomial is a sequence of symmetric polynomials of N variables (with $N \to \infty$), and those N-variable symmetric polynomials each has the minimal total power of the variables that satisfy the unique fusion condition and the cluster condition. We will loosely refer such a sequence of N-variable symmetric polynomials as an infinite-variable symmetric polynomial.]

From the above discussions, we see that an infinite-variable symmetric polynomial can be described by a finite amount of data $(\frac{m}{n}; S_2, \ldots, S_n)$. The $\nu = 1/2$ Laughlin state, $P_{1/2}$, satisfies the unique fusion condition and cluster condition. So $P_{1/2}$ is an infinite-variable symmetric polynomial described by a pattern of zero: $(\frac{m}{n}; S_2, \ldots, S_n) = (\frac{2}{1};)$. Once we define the norm of those infinite-variable symmetric polynomials as Eq. (61), infinite-variable symmetric polynomials may have some very interesting universal properties discussed in Sect. 3. We like to mention that the infinite-variable symmetric polynomials (also referred as symmetric functions) are studied in mathematics in various contexts, such as representation theory, combinatorics and algebraic topology [47, 48]. It is not clear if there is a relation between our pattern-of-zeros point of view and those previous studies. But we like to point out in our pattern-of-zeros approach, we only interested in symmetric polynomials of $N \to \infty$ variables, and with the total power of the variables of order $O(N^2)$. We are not interested in the infinite-variable symmetric polynomials with all possible total power of the variables.

4.2 *A Classification of Infinite-Variable Symmetric Polynomials*

We have seen that each infinite-variable symmetric polynomial $P(\{z_i\})$ has a sequence of integers $\{S_a\}$—a pattern of zeros. But each sequence of integers $\{S_a\}$ may not correspond to an infinite-variable symmetric polynomial $P(\{z_i\})$. In this subsection, we will try to find all the conditions that a sequence $\{S_a\}$ must satisfy, such that $\{S_a\}$ describes a infinite-variable symmetric polynomial. This may lead to a classification of infinite-variable symmetric polynomials (or FQH states) through pattern of zeros.

4.2.1 Derived Polynomials

To find the conditions on $\{S_a\}$, it is very helpful to introduce the derived polynomials. Let $z_1, \ldots, z_a \to z^{(a)}$ in an infinite-variable symmetric polynomial $P(\{z_i\})$ and use the unique fusion condition:

$$P\big(\{z_i\}\big) \to \lambda^{S_a} P_{\text{derived}}\big(z^{(a)}, z_{a+1}, z_{a+2}, \ldots\big) + O\big(\lambda^{S_a+1}\big), \tag{74}$$

we obtain a derived polynomial $P_{\text{derived}}(z^{(a)}, z_{a+1}, z_{a+2}, \ldots)$ from the original polynomial P. Repeating the process on other variables, we get a more general derived polynomial $P_{\text{derived}}(z^{(a)}, z^{(b)}, z^{(c)}, \ldots)$, where $z^{(a)}$, $z^{(b)}$, etc. are fusions of a variables, b variables, etc.

The zeros in derived polynomials are described by $D_{a,b}$:

$$P_{\text{derived}}\big(z^{(a)}, z^{(b)}, z^{(c)}, \ldots\big) \sim \big(z^{(a)} - z^{(b)}\big)^{D_{a,b}} P'_{\text{derived}}\big(z^{(a+b)} \ldots\big) + \cdots \tag{75}$$

where $z^{(a+b)} = (z^{(a)} + z^{(b)})/2$. $D_{a,b} = D_{b,a}$ also characterize the pattern of zeros. In effect, $D_{a,b}$ and S_a encode the same information:

$$D_{a,b} = S_{a+b} - S_a - S_b, \quad S_a = \sum_{b=1}^{a-1} D_{b,1}. \tag{76}$$

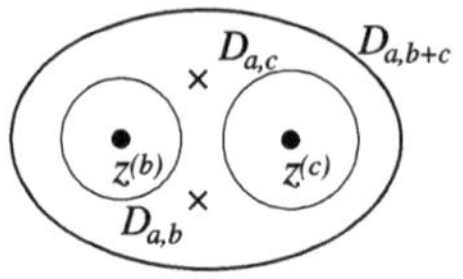

Fig. 6 $W_{a,bc}$ obtained by moving $z^{(a)}$ along *a large loop* around $z^{(b)}$ and $z^{(c)}$ counts the total numbers of zeros of $f(z^{(a)})$ in *the loop*. *The crosses* mark the off-particle zeros of $f(z^{(a)})$ not at $z^{(b)}$ and $z^{(c)}$

4.2.2 The Concave Conditions on Pattern of Zeros

Since $D_{a,b} \geq 0$, we obtain the first concave condition:

$$\Delta_2(a,b) \equiv S_{a+b} - S_a - S_b \geq 0. \tag{77}$$

Such a condition comes from the fusion of two clusters. We also have a second concave condition:

$$\Delta_3(a,b,c) \equiv S_{a+b+c} - S_{a+b} - S_{b+c} - S_{a+c} + S_a + S_b + S_c \geq 0 \tag{78}$$

from the fusion of three clusters.

To derive the second concave condition, let us fix all variables $z^{(b)}, z^{(c)}, \ldots$ except $z^{(a)}$ in the derived polynomial $P_{\text{derived}}(z^{(a)}, z^{(b)}, z^{(c)}, \ldots)$. Then the derived polynomial $P_{\text{derived}}(z^{(a)}, z^{(b)}, z^{(c)}, \ldots)$ can be viewed as a complex function $f(z^{(a)})$, which has isolated on-particle zeros at $z^{(b)}$, $z^{(c)}$, ..., and possibly some other off-particle zeros.

Let us move $z^{(a)}$ around both points $z^{(b)}$ and $z^{(c)}$. The phase of the complex function $f(z^{(a)})$ will change by $2\pi W_{a,bc}$ where $W_{a,bc}$ is an integer (see Fig. 6). Since $f(z^{(a)})$ has an order D_{ab} zero at $z^{(b)}$ and an order D_{ac} zero at $z^{(c)}$, the integer $W_{a,bc}$ satisfy

$$W_{a,bc} \geq D_{ab} + D_{ac}$$

because $f(z^{(a)})$ may also have off-particle zeros. Now let $z^{(b)} \to z^{(c)}$ to fuse into $z^{(b+c)}$. In this limit $W_{a,bc}$ becomes the order of zeros between $z^{(a)}$ and $z^{(b+c)}$: $W_{a,bc} = D_{a,b+c}$. Thus we obtain the following condition on D_{ab}: $D_{a,b+c} \geq D_{ab} + D_{ac}$, which gives us the second concave condition (78).

We like to point out that the n-cluster condition has a very simple meaning in the derived polynomial: $f(z^{(a)})$ has no off-particle zeros if $a = 0 \bmod n$. So $D_{a+b,n} = D_{a,n} + D_{b,n}$ which leads to the cluster condition (73).

4.2.3 Some Additional Conditions

The two concave conditions are the main conditions on $\{S_a\}$. We also have another condition

$$\Delta_2(a,a) = \text{even} \tag{79}$$

since the polynomial is a symmetric polynomial. It turns out that we need yet another a condition

$$\Delta_3(a,b,c) = \text{even}. \tag{80}$$

It is hard to prove this mysterious condition using elementary methods. Using the connection between the symmetry polynomial and CFT (or vertex algebra), we find that the condition $\Delta_3(a,b,c) = \text{even}$ is directly related to the requirement that the fermionic operators have half-integer scaling dimensions and bosonic operators have integer scaling dimensions [35].

We conjecture that *the patterns of zeros* $(\frac{m}{n}; S_2, \ldots, S_n)$ *that satisfy the above conditions describe infinite-variable symmetric polynomials* [25]. Those $(\frac{m}{n}; S_2, \ldots, S_n)$ "classify" infinite-variable symmetric polynomials and FQH states with filling fraction $\nu = n/m$.

4.2.4 Primitive Solutions for Pattern of Zeros

Let us list some patterns of zeros, $(\frac{m}{n}; S_2, \ldots, S_n)$, that satisfy the above conditions. We note that the conditions are semi-linear in $(\frac{m}{n}; S_2, \ldots, S_n)$. So, if $(\frac{m}{n}; S_2, \ldots, S_n)$ and $(\frac{m'}{n'}; S'_2, \ldots, S'_n)$ are solutions, then $(\frac{m''}{n''}; S''_2, \ldots, S''_n) = (\frac{m}{n}; S_2, \ldots, S_n) + (\frac{m'}{n'}; S'_2, \ldots, S'_n)$ is also a solution. Such a result has the following meaning: Let $P(\{z_i\})$, $P'(\{z_i\})$, and $P''(\{z_i\})$ are three symmetric polynomials described by pattern of zeros $(\frac{m}{n}; S_2, \ldots, S_n)$, $(\frac{m'}{n'}; S'_2, \ldots, S'_n)$, and $(\frac{m''}{n''}; S''_2, \ldots, S''_n)$ respectively, we then have $P''(\{z_i\}) = P(\{z_i\})P'(\{z_i\})$. Such a property allow us to introduce the notion of primitive pattern of zeros as the patterns of zeros that cannot to written as the sum of two other patterns of zeros. In this section, we will only list the primitive patterns of zeros.

1-cluster state: $\nu = 1/k$ Laughlin state

$$\begin{gathered} P_{1/k}: \left(\frac{m}{n};\right) = \left(\frac{k}{1};\right), \\ (n_0, \ldots, n_{k-1}) = (1, 0, \ldots, 0). \end{gathered} \tag{81}$$

2-cluster state: Pfaffian state (Z_2 parafermion state)

$$\begin{gathered} P_{2/2;Z_2}: \left(\frac{m}{n}; S_2\right) = \left(\frac{2}{2}; 0\right), \\ (n_0, \ldots, n_{m-1}) = (2, 0) \end{gathered} \tag{82}$$

3-cluster state: Z_3 parafermion state

$$\begin{gathered} P_{3/2;Z_3}: \left(\frac{m}{n}; S_2, S_3\right) = \left(\frac{2}{3}; 0, 0\right), \\ (n_0, \ldots, n_{m-1}) = (3, 0) \end{gathered} \tag{83}$$

4-cluster state: Z_4 parafermion state

$$P_{4/2;Z_4}: \left(\frac{m}{n}; S_2, \ldots, S_4\right) = \left(\frac{2}{4}; 0, 0, 0\right),$$
$$(n_0, \ldots, n_{m-1}) = (4, 0), \tag{84}$$

5-cluster states (we have two of them): Z_5 (generalized) parafermion states

$$P_{5/2;Z_5}: \left(\frac{m}{n}; S_2, \ldots, S_5\right) = \left(\frac{2}{5}; 0, 0, 0, 0\right),$$
$$(n_0, \ldots, n_{m-1}) = (5, 0) \tag{85}$$

$$P_{5/8;Z_5^{(2)}}: \left(\frac{m}{n}; S_2, \ldots, S_5\right) = \left(\frac{8}{5}; 0, 2, 6, 10\right),$$
$$(n_0, \ldots, n_{m-1}) = (2, 0, 1, 0, 2, 0, 0, 0) \tag{86}$$

6-cluster state:

$$P_{6/2;Z_6}: \left(\frac{m}{n}; S_2, \ldots, S_6\right) = \left(\frac{2}{6}; 0, 0, 0, 0, 0\right),$$
$$(n_0, \ldots, n_{m-1}) = (6, 0) \tag{87}$$

7-cluster states (we have four of them):

$$P_{7/2;Z_7}: \left(\frac{m}{n}; S_2, \ldots, S_7\right) = \left(\frac{2}{7}; 0, 0, 0, 0, 0, 0\right),$$
$$(n_0, \ldots, n_{m-1}) = (7, 0) \tag{88}$$

$$P_{7/8;Z_7^{(2)}}: \left(\frac{m}{n}; S_2, \ldots, S_7\right) = \left(\frac{8}{7}; 0, 0, 2, 6, 10, 14\right),$$
$$(n_0, \ldots, n_{m-1}) = (3, 0, 1, 0, 3, 0, 0, 0) \tag{89}$$

$$P_{7/18;Z_7^{(3)}}: \left(\frac{m}{n}; S_2, \ldots, S_7\right) = \left(\frac{18}{7}; 0, 4, 10, 18, 30, 42\right),$$
$$(n_0, \ldots, n_{m-1}) = (2, 0, 0, 0, 0, 1, 0, 0, 0, 2, 0, 0, 0, 0, 0) \tag{90}$$

$$P_{7/14;C_7}: \left(\frac{m}{n}; S_2, \ldots, S_7\right) = \left(\frac{14}{7}; 0, 2, 6, 12, 20, 28\right),$$
$$(n_0, \ldots, n_{m-1}) = (2, 0, 1, 0, 1, 0, 1, 0, 2, 0, 0, 0, 0, 0) \tag{91}$$

4.2.5 How Good Is the Pattern-of-Zeros Classification?

How good is the pattern-of-zeros classification? Not so good, and not so bad.

Clearly, every symmetric polynomial P corresponds to a unique pattern of zeros $\{S_a\}$. But only some patterns of zeros correspond to a unique symmetric polynomial. So the pattern-of-zeros classification is not so good. It appears that all the primitive pattern of zeros correspond to a unique a unique symmetric polynomial. Therefore, the pattern-of-zeros classification is not so bad.

We also know that some composite patterns of zeros correspond a unique symmetric polynomial, while other composite patterns of zeros do not correspond a unique symmetric polynomial. Let P_{n_i} be a symmetric polynomial described by a primitive pattern of zeros with an n_i-cluster. It appear that $P = \prod_i P_{n_i}$ will have a pattern of zeros that corresponds a unique symmetric polynomial if n_i's has no common factor.

So only for certain patterns of zeros, the data $\{\frac{m}{n}; S_2, \dots, S_n\}$ contain all the information to fix the symmetric polynomials. In general, we need more information than $\{\frac{m}{n}; S_2, \dots, S_n\}$ to fully characterize symmetry polynomials of infinite variables.

4.3 *Topological Properties from Pattern of Zeros*

For those patterns of zeros that uniquely characterize the symmetry polynomials of infinite variables (or FQH wave functions), we should be able to calculate the universal properties of the FQH states from the data $(\frac{m}{n}; S_2, \dots, S_n)$. Those universal properties include:

- The filling fraction ν.
- Topological degeneracy on torus and other Riemann surfaces
- Number of quasiparticle types
- Quasiparticle charges
- Quasiparticle scaling dimensions
- Quasiparticle fusion algebra
- Quasiparticle statistics (Abelian and non-Abelian)
- The counting of edge excitations (central charge c and spectrum)

At moment, we can calculate many of the above universal properties from the pattern-of-zeros data $(\frac{m}{n}; S_2, \dots, S_n)$. For example, the filling fraction ν is given by $\nu = n/m$. But we still do not know how to calculate scaling dimensions and statistics for some of the quasiparticles.

In this subsection, we develop a pattern-of-zeros description of the quasiparticle excitations in FQH states. This will allow us to calculate many universal properties from the pattern of zeros.

4.3.1 Pattern of Zeros of Quasiparticle Excitations

A quasiparticle is a defect in the ground state wave function $P(\{z_i\})$. It is a place where we have more power of zeros. For example, the ground state wave function

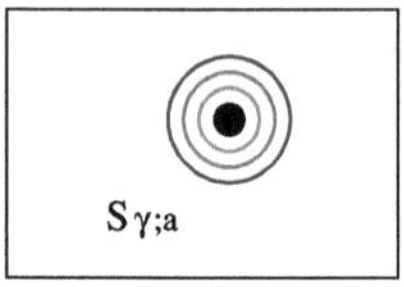

Fig. 7 The graphic picture of the pattern of zeros for a quasiparticle

of $\nu = 1/2$ Laughlin state is given by $\prod_{i<j}(z_i - z_j)^2$. The state with a quasiparticle at ξ is given by $\prod_i (z_i - \xi) \prod_{i<j}(z_i - z_j)^2$ (see Sect. 3.2). As we bring several z_i's to ξ, $\prod_i (z_i - \xi) \prod_{i<j}(z_i - z_j)^2$ vanishes according to a pattern of zeros. In general, each quasiparticle labeled by γ in a FQH state can be quantitatively characterized by distinct pattern of zeros (see Fig. 7).

Let $P_\gamma(\xi; \{z_i\})$ be the wave function with a quasiparticle γ at $z = \xi$. To describe the structure of the zeros as we bring bosons to the quasiparticle, we set $z_i = \lambda \eta_i + \xi$, $i = 1, 2, \ldots, a$ and let $\lambda \to 0$:

$$P_\gamma\left(\xi; \{z_i\}\right) = \lambda^{S_{\gamma;a}} \tilde{P}_\gamma\left(z^{(a)} = \xi, z_{a+1}, z_{a+2}, \ldots\right) + O\left(\lambda^{S_a+1}\right) \tag{92}$$

$S_{\gamma;a}$ is the order of zeros of $P_\gamma(\xi; z_i)$ when we bring a bosons to ξ. *The sequence of integers* $\{S_{\gamma;a}\}$ *is the quasiparticle pattern of zeros that characterizes the quasiparticle* γ. We note that the ground-state pattern of zeros $\{S_a\}$ correspond to the trivial quasiparticle $\gamma = 0$: $\{S_{0;a}\} = \{S_a\}$.

To find the allowed quasiparticles, we simply need to find (i) the conditions that $S_{\gamma;a}$ must satisfy and (ii) all the $S_{\gamma;a}$ that satisfy those conditions.

4.3.2 Conditions on Quasiparticle Pattern of Zeros $S_{\gamma;a}$

The quasiparticle pattern of zeros also satisfy two concave conditions

$$S_{\gamma;a+b} - S_{\gamma;a} - S_b \geq 0, \tag{93}$$

$$S_{\gamma;a+b+c} - S_{\gamma;a+b} - S_{\gamma;a+c} - S_{b+c} + S_{\gamma;a} + S_b + S_c \geq 0 \tag{94}$$

and a cluster condition

$$S_{\gamma;a+kn} = S_{\gamma;a} + k(S_{\gamma;n} + ma) + mn\frac{k(k-1)}{2} \tag{95}$$

The cluster condition implies that a finite sequence $(S_{\gamma;1}, \ldots, S_{\gamma;n})$ determines the infinity sequence $\{S_{\gamma;a}\}$.

We can also use the sequence $l_{\gamma;a} = S_{\gamma,a} - S_{\gamma,a-1}$ or $n_{\gamma;l} = \sum_{i=1} \delta_{l,l_{\gamma;i}}$ to describe the quasiparticle sequence $S_{\gamma;a}$. The $n_{\gamma;l}$ description is simpler and reveals physical picture more clearly than $S_{\gamma;a}$.

4.3.3 The Solutions for the Quasiparticle Patterns of Zeros

We can find all $(S_{\gamma;1}, \ldots, S_{\gamma;n})$ that satisfy the above concave and cluster conditions through numerical calculations. This allow us to obtain all the quasiparticles.

For the $\nu = 1$ Pfaffian state ($n = 2$ and $m = 2$) described by

$$\begin{aligned} &S_1, S_2, \ldots : 0, 0, 2, 4, 8, 12, 18, 24, \ldots \\ &n_0 n_1 n_2 \cdots : 20202020202020202 \cdots , \end{aligned} \tag{96}$$

we find that the quasiparticle patterns of zeros are given by (expressed in terms of $n_{\gamma,l}$)

$$\begin{aligned} &n_{\gamma;0} n_{\gamma;1} n_{\gamma;2} \cdots : 20202020202020202 \cdots \quad Q_\gamma = 0 \\ &n_{\gamma;0} n_{\gamma;1} n_{\gamma;2} \cdots : 02020202020202020 \cdots \quad Q_\gamma = 1 \\ &n_{\gamma;0} n_{\gamma;1} n_{\gamma;2} \cdots : 11111111111111111 \cdots \quad Q_\gamma = 1/2 \end{aligned} \tag{97}$$

The above three pattern of zeros are not all the solutions of the quasiparticle conditions. However, all other quasiparticle solutions can be obtained from the above three by removing some bosons. Those quasiparticle solutions are equivalent to one of the above three solutions. For example $n_{\gamma;0} n_{\gamma;1} \cdots = 102020202 \cdots$, $n_{\gamma;0} n_{\gamma;1} \cdots = 002020202 \cdots$, etc. are also quasiparticle solutions which are equivalent to $n_{\gamma;0} n_{\gamma;1} \cdots = 202020202 \cdots$. Therefore, we find that the $\nu = 1$ Pfaffian state has three types of quasiparticles.

We note that the ground state degeneracy on torus is equal to the number of quasiparticle types. So the $\nu = 1$ Pfaffian state has a three-fold degeneracy on a torus. The charge of quasiparticles can be also calculated from the quasiparticle pattern of zeros:

$$Q_\gamma = \frac{1}{m} \sum_{a=1}^{n} (l_{\gamma;a} - l_a) = \frac{1}{m} (S_{\gamma;n} - S_n). \tag{98}$$

Let us list the number of quasiparticle types calculated from pattern of zeros for various FQH states. For the parafermion states $P_{\nu=n/2;Z_n}$ ($m = 2$),

$P_{2/2;Z_2}$	$P_{3/2;Z_3}$	$P_{4/2;Z_4}$	$P_{5/2;Z_5}$	$P_{6/2;Z_6}$	$P_{7/2;Z_7}$	$P_{8/2;Z_8}$	$P_{9/2;Z_9}$	$P_{10/2;Z_{10}}$
3	4	5	6	7	8	9	10	11

For the parafermion states $P_{\nu=n/(2+2n);Z_n}$ ($m = 2 + 2n$)

$P_{2/6;Z_2}$	$P_{3/8;Z_3}$	$P_{4/10;Z_4}$	$P_{5/12;Z_5}$	$P_{6/14;Z_6}$	$P_{7/16;Z_7}$	$P_{8/18;Z_8}$	$P_{9/20;Z_9}$	$P_{10/22;Z_{10}}$
9	16	25	36	49	64	81	100	121

For the generalized parafermion states $P_{\nu=n/m;Z_n^{(k)}}$

$P_{5/8;Z_5^{(2)}}$	$P_{5/18;Z_5^{(2)}}$	$P_{7/8;Z_7^{(2)}}$	$P_{7/22;Z_7^{(2)}}$	$P_{7/18;Z_7^{(3)}}$	$P_{7/32;Z_7^{(3)}}$	$P_{8/18;Z_8^{(3)}}$	$P_{9/8;Z_9^{(2)}}$
24	54	32	88	72	128	81	40

where k and n are co-prime.

For the composite parafermion states $P_{n_1/m_1;Z_{n_1}^{(k_2)}} P_{n_2/m_2;Z_{n_2}^{(k_2)}}$ obtained as products of two parafermion wave functions

$P_{2/2;Z_2}P_{3/2;Z_3}$	$P_{3/2;Z_3}P_{4/2;Z_4}$	$P_{2/2;Z_2}P_{5/2;Z_5}$	$P_{2/2;Z_2}P_{5/8;Z_5^{(2)}}$
30	70	63	117

where n_1 and n_2 are co-prime. The inverse filling fractions of the above composite states are $\frac{1}{\nu} = \frac{1}{\nu_1} + \frac{1}{\nu_2} = \frac{m_1}{n_1} + \frac{m_2}{n_2}$. More results can be found in [26].

All those results from the pattern of zeros agree with the results from parafermion CFT [27]:

$$\#\text{ of quasiparticles} = \frac{1}{\nu}\prod_i \frac{n_i(n_i+1)}{2} \tag{99}$$

for the generalized composite parafermion state

$$P = \prod_i P_{n_i/m_i;Z_{n_i}^{(k_i)}}, \quad \{n_i\}\text{ co-prime, } (k_i, n_i)\text{ co-prime.} \tag{100}$$

The filling fraction for such generalized composite parafermion state is given by $\nu = (\sum_i \frac{m_i}{n_i})^{-1}$.

4.3.4 Quasiparticle Fusion Algebra: $\gamma_1\gamma_2 = \sum_{\gamma_3} N_{\gamma_1\gamma_2}^{\gamma_3}\gamma_3$

When we fuse quasiparticles γ_1 and γ_2 together, we can get a third quasiparticle γ_3. However, for non-Abelian quasiparticles, the fusion can be more complicated. Fusing γ_1 and γ_2 may produce several kind of quasiparticles. Such kind of fusion is described by quasiparticle fusion algebra (see Fig. 8): $\gamma_1\gamma_2 = \sum_{\gamma_3} N_{\gamma_1\gamma_2}^{\gamma_3}\gamma_3$, where $N_{\gamma_1\gamma_2}^{\gamma_3}$ are non-negative integers.

To calculate the fusion coefficients $N_{\gamma_1\gamma_2}^{\gamma_3}$ from the pattern of zeros, let us put the quasiparticle γ_1 at $z = 0$. Far away from $z = 0$, such a quasiparticle has a pattern of zeros $n_{\gamma_1;l}$ (in the occupation representation). We then insert a quasiparticle γ_2 at $z = R$ for a large R. At $z = r \gg R$, the occupation becomes the occupation of the quasiparticle γ_3: $n_{\gamma_3;l}$. We see that the fusion of γ_2 changes the occupation pattern from $n_{\gamma_1;l}$ to $n_{\gamma_3;l}$:

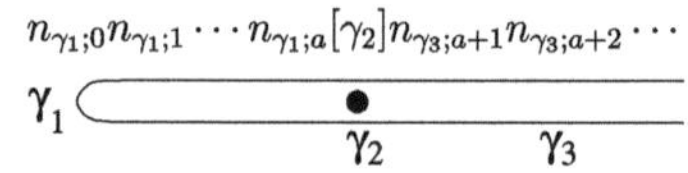

(101)

So the quasiparticle γ_2 becomes a "domain wall" between the γ_1 occupation pattern and the γ_3 occupation pattern [49].

From the above domain wall structure, we can see only $n_{\gamma_1;l}$ and $n_{\gamma_3;l}$, but we cannot see $n_{\gamma_2;l}$. But this is enough for us. We are able to find a condition on $n_{\gamma_2;l}$

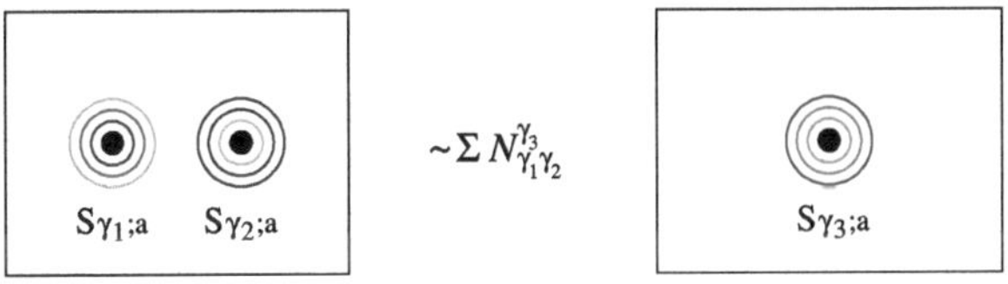

Fig. 8 The graphic picture of the fusion of two quasiparticles. *Each box* represent a many-boson wave function. *In the left box*, we have quasiparticle γ_1 and γ_2 described by patterns of zeros $S_{\gamma_1;a}$ and $S_{\gamma_2;a}$. Far away from the two quasiparticles, the wave function may contain several different patterns of zeros $S_{\gamma_3;a}$ that correspond to several different quasiparticle types γ_3. So we say that γ_1 and γ_2 may fuse into several different types of quasiparticles labeled by γ_3

so that it can induce a domain wall between $n_{\gamma_1;l}$ and $n_{\gamma_3;l}$ [27]:

$$\sum_{j=1}^{b}\left(l^{\mathrm{sc}}_{\gamma_1;j+a}+l^{\mathrm{sc}}_{\gamma_2;j+c}\right)\leq\sum_{j=1}^{b}\left(l^{\mathrm{sc}}_{\gamma_3;j+a+c}+l^{\mathrm{sc}}_{j}\right) \tag{102}$$

for any $a,b,c\in Z_+$, where $l^{\mathrm{sc}}_{\gamma;a}=l_{\gamma;a}-\frac{m(Q_\gamma+a-1)}{n}$.

Solving the above equation allows us to determine when $N^{\gamma_3}_{\gamma_1\gamma_2}$ can be non-zero. If we further assume that $N^{\gamma_3}_{\gamma_1\gamma_2}=0,1$, then the fusion algebra can be determined. Knowing $N^{\gamma_3}_{\gamma_1\gamma_2}$ allows us to determine the ground state degeneracies of FQH state on any closed Riemann surfaces.

We like to mention that for the generalized composite parafermion states which have a CFT description, the pattern-of-zeros approach and the CFT approach give rise to the same fusion algebra. However, the pattern-of-zeros approach applies to other FQH states whose CFT may not be known.

5 The Vertex-Algebra + Pattern-of-Zeros Approach

5.1 Z-Graded Vertex Algebra

The symmetric polynomial $P(\{z_i\})$ and the corresponding derived polynomial $P_{\mathrm{derived}}(\{z_i^{(a_i)}\})$ can be expressed as correlation functions in a vertex algebra:

$$P(\{z_i\})=\left\langle\prod_i V(z_i)\right\rangle,\qquad P_{\mathrm{derived}}(\{z_i^{(a)}\})=\left\langle\prod_{i,a}V_a(z_i^{(a)})\right\rangle$$
$$V_a(z)=V^a,\qquad V_aV_b=V_{a+b}. \tag{103}$$

The vertex algebra is generated by vertex operator $V(z)$ and is described by the following operator product expansion:

$$V_a(z)V_b(w)=\frac{C_{ab}}{(z-w)^{h_a+h_b-h_{a+b}}}V_{a+b}(w)+\cdots \tag{104}$$

where h_a is the scaling dimension of V_a and C_{ab} the structure constant of the vertex algebra. Such a vertex algebra is a Z-graded vertex algebra.

The pattern of zeros S_a discuss before is directly related to h_a:

$$h_{a+b} - h_a - h_b = D_{a,b} = S_{a+b} - S_a - S_b \tag{105}$$

The n-cluster condition implies that $h_a \propto a^2$ if $a = 0 \bmod n$. This allows us to obtain

$$h_a = S_a - \frac{aS_n}{n} + \frac{am}{2} \tag{106}$$

We see that the pattern of zeros S_a only describe the scaling dimensions of the vertex operators. It does not describe the structure constants $C_{a,b}$. So a more complete characterization of FQH wave functions (symmetric polynomials) is given by $(\frac{m}{n}; S_a; C_{ab}, \ldots)$. But $(\frac{m}{n}; S_a; C_{ab}, \ldots)$ may be an overkill. We like to find out what is the minimal set of date that can completely characterize the FQH wave functions (or the symmetric polynomials).

5.2 Z_n-Vertex Algebra

If the above Z-graded vertex algebra satisfies the n-cluster condition, then it can be viewed a Z_n-vertex algebra $\otimes$ a $U(1)$ current algebra:

$$V_a(z) = \psi_a(z) e^{ia\phi(z)\sqrt{m/n}} \tag{107}$$

where $j = \partial\phi$ generates the $U(1)$ current algebra and ψ_a generates the Z_n-vertex algebra:

$$\psi_a(z)\psi_b(w) = \frac{C_{ab}}{(z-w)^{h_a^{\text{sc}} + h_b^{\text{sc}} - h_{a+b}^{\text{sc}}}} \psi_{a+b}(w) + \cdots \tag{108}$$

where $\psi_n = 1$ as the result of the n-cluster condition. The scaling dimension of $\psi_a(z)$ is

$$h_a^{\text{sc}} = h_a - \frac{a^2 m}{2n} = S_a - \frac{aS_n}{n} + \frac{am}{2} - \frac{a^2 m}{2n}, \quad h_a^{\text{sc}} = h_{a+n}^{\text{sc}} \tag{109}$$

The two sets of data $(\frac{m}{n}; S_2, \ldots, S_n)$ and $(\frac{m}{n}; h_1^{\text{sc}}, \ldots, h_{n-1}^{\text{sc}})$ completely determine each other:

$$S_a = h_a^{\text{sc}} - a h_1^{\text{sc}} + \frac{a(a-1)m}{2n}. \tag{110}$$

So we can also use $(\frac{m}{n}; h_1^{\text{sc}}, \ldots, h_{n-1}^{\text{sc}})$ to describe the pattern of zeros.

From the pattern-of-zeros consideration, we find that h_a^{sc} must satisfy

$$S_a = h_a^{\text{sc}} - a h_1^{\text{sc}} + \frac{a(a-1)m}{2n} = \text{integer} \geq 0$$

$$h_{a+b}^{\text{sc}} - h_a^{\text{sc}} - h_b^{\text{sc}} + \frac{abm}{n} = D_{ab} = \text{integer} \geq 0 \tag{111}$$

$$h_{a+b+c}^{\text{sc}} - h_{a+b}^{\text{sc}} - h_{b+c}^{\text{sc}} - h_{a+c}^{\text{sc}} + h_a^{\text{sc}} + h_b^{\text{sc}} + h_c^{\text{sc}}$$
$$= \Delta_3(a,b,c) = \text{even integer} \geq 0 \tag{112}$$

But the above conditions are only on h_a^{sc}. To get the conditions on C_{ab}, we can use the generalized Jacobi identity [50] to obtain a set a non-linear equations for $(h_a^{\text{sc}}, C_{ab}, \ldots)$ [35]. Those conditions may be sufficient and necessary which may lead to a classification of Z_n-vertex algebra.

For some simple pattern of zeros h_a^{sc}, we are able to build a closed set of non-linear equations for $(h_a^{\text{sc}}, C_{ab}, \ldots)$, which lead to a well defined Z_n-vertex algebra. This allows us to calculate quasiparticle scaling dimensions, quasiparticle statistics, central charge (edge spectrum) [35], … We would like to point out that in [32] and [34], a very interesting approach based the pattern of zeros and modular transformation of torus is proposed, that allows us to calculate the fractional statistics of some quasiparticles directly from the pattern-of-zeros data. We also like to point out that finding valid $(h_a^{\text{sc}}, C_{ab}, \ldots)$ corresponds to finding a well defined Z_n vertex algebra. Finding the quasiparticle patterns of zeros corresponds to finding the representations of the Z_n vertex algebra.

But at moment, we cannot handle more general pattern of zeros h_a^{sc}, in the sense that we have some difficulties to obtain a closed set of non-linear algebraic equations for $(h_a^{\text{sc}}, C_{ab}, \ldots)$. We hope that, after some further research, the pattern-of-zeros approach may lead to a classification of Z_n-vertex algebra, which in turn lead to a classification of symmetric polynomials and FQH states.

6 Summary

Although still incomplete, the pattern-of-zeros approach provides quite a powerful way to study symmetric polynomials with infinite variables and FQH states. It connects several very different fields, such as strongly correlated electron systems, topological quantum field theory, CFT (for the edge states), modular tensor category theory (for the quasiparticle statistics), and maybe a new field of infinite-variable symmetric polynomial. This article only reviews the first step in this very exciting direction. More exciting results are yet to come.

Acknowledgements This research is supported by NSF Grant No. DMR-1005541 and NSFC 11074140.

References

1. Tsui, D.C., Stormer, H.L., Gossard, A.C.: Two-dimensional magnetotransport in the extreme quantum limit. Phys. Rev. Lett. **48**, 1559 (1982)
2. Laughlin, R.B.: Anomalous quantum Hall effect: an incompressible quantum fluid with fractionally charged excitations. Phys. Rev. Lett. **50**, 1395 (1983)
3. Chen, X., Gu, Z.-C., Wen, X.-G.: Local unitary transformation, long-range quantum entanglement, wave function renormalization, and topological order. Phys. Rev. B **82**, 155138 (2010). arXiv:1004.3835
4. Landau, L.D.: Theory of phase transformations I. Phys. Z. Sowjetunion **11**, 26 (1937)
5. Ginzburg, V.L., Landau, L.D.: On the theory of superconductivity. Zh. Èksp. Teor. Fiz. **20**, 1064 (1950)
6. Landau, L.D., Lifschitz, E.M.: Statistical Physics—Course of Theoretical Physics, vol. 5. Pergamon, London (1958)
7. Wen, X.-G.: Vacuum degeneracy of chiral spin state in compactified spaces. Phys. Rev. B **40**, 7387 (1989)
8. Wen, X.-G.: Topological orders in rigid states. Int. J. Mod. Phys. B **4**, 239 (1990)
9. Wen, X.-G., Niu, Q.: Ground state degeneracy of the FQH states in presence of random potentials and on high genus Riemann surfaces. Phys. Rev. B **41**, 9377 (1990)
10. Wen, X.-G.: Topological orders and edge excitations in FQH states. Adv. Phys. **44**, 405 (1995)
11. Witten, E.: Quantum field theory and the Jones polynomial. Commun. Math. Phys. **121**, 351 (1989)
12. Zhang, S.C., Hansson, T.H., Kivelson, S.: Effective-field-theory model for the fractional quantum Hall effect. Phys. Rev. Lett. **62**, 82 (1989)
13. Frohlich, J., King, C.: The Chern-Simons theory and knot polynomials. Commun. Math. Phys. **126**, 167 (1989)
14. Blok, B., Wen, X.-G.: Effective theories of fractional quantum Hall effect at generic filling fractions. Phys. Rev. B **42**, 8133 (1990)
15. Blok, B., Wen, X.-G.: Effective theories of fractional quantum Hall effect: hierarchical construction. Phys. Rev. B **42**, 8145 (1990)
16. Read, N.: Excitation structure of the hierarchy scheme in the fractional quantum Hall effect. Phys. Rev. Lett. **65**, 1502 (1990)
17. Frohlich, J., Kerler, T.: Universality in quantum Hall systems. Nucl. Phys. B **354**, 369 (1991)
18. Wen, X.-G., Zee, A.: A classification and matrix formulation of the Abelian FQH states. Phys. Rev. B **46**, 2290 (1992)
19. Frohlich, J., Studer, U.M.: Gauge invariance and current algebra in nonrelativistic manybody theory. Rev. Mod. Phys. **65**, 733 (1993)
20. Frohlich, J., Zee, A.: Large scale physics of the quantum Hall fluid. Nucl. Phys. B **364**, 517 (1991)
21. Moore, G., Read, N.: Nonabelions in the fractional quantum Hall effect. Nucl. Phys. B **360**, 362 (1991)
22. Wen, X.-G.: Non-Abelian statistics in the FQH states. Phys. Rev. Lett. **66**, 802 (1991)
23. Blok, B., Wen, X.-G.: Many-body systems with non-Abelian statistics. Nucl. Phys. B **374**, 615 (1992)
24. Wen, X.-G., Wu, Y.-S.: Chiral operator product algebra hidden in certain FQH states. Nucl. Phys. B **419**, 455 (1994). cond-mat/9310027
25. Wen, X.-G., Wang, Z.: Classification of symmetric polynomials of infinite variables: construction of Abelian and non-Abelian quantum Hall states. Phys. Rev. B **77**, 235108 (2008). arXiv:0801.3291
26. Wen, X.-G., Wang, Z.: Topological properties of Abelian and non-Abelian quantum Hall states from the pattern of zeros. Phys. Rev. B **78**, 155109 (2008). arXiv:0803.1016
27. Barkeshli, M., Wen, X.-G.: Structure of quasiparticles and their fusion algebra in fractional quantum Hall states. Phys. Rev. B **79**, 195132 (2009). arXiv:0807.2789

28. Barkeshli, M., Wen, X.-G.: Classification of Abelian and non-Abelian multilayer fractional quantum Hall states through the pattern of zeros. Phys. Rev. B **82**, 245301 (2010). arXiv:0906.0337
29. Seidel, A., Lee, D.-H.: Abelian and non-Abelian Hall liquids and charge density wave: quantum number fractionalization in one and two dimensions. Phys. Rev. Lett. **97**, 056804 (2006)
30. Bergholtz, E.J., Kailasvuori, J., Wikberg, E., Hansson, T.H., Karlhede, A.: The Pfaffian quantum Hall state made simple–multiple vacua and domain walls on a thin torus. Phys. Rev. B **74**, 081308 (2006)
31. Bernevig, B.A., Haldane, F.D.M.: Generalized clustering conditions of Jack polynomials at negative Jack parameter α. Phys. Rev. B **77**, 184502 (2008). arXiv:0711.3062
32. Seidel, A.: Pfaffian statistics through adiabatic transport in the 1D coherent state. Phys. Rev. Lett. **101**, 196802 (2008). arXiv:0804.4691
33. Seidel, A.: S-duality constraints on 1D patterns associated with fractional quantum Hall states. Phys. Rev. Lett. **105**, 026802 (2010). arXiv:1002.2647
34. Flavin, J., Seidel, A.: Abelian and non-Abelian statistics in the coherent state representation. Phys. Rev. X **1**, 021015 (2011). arXiv:1108.2734
35. Lu, Y.-M., Wen, X.-G., Wang, Z., Wang, Z.: Non-Abelian quantum Hall states and their quasiparticles: from the pattern of zeros to vertex algebra. Phys. Rev. B **81**, 115124 (2010). arXiv:0910.3988
36. Greiter, M., Wen, X.-G., Wilczek, F.: Pairing in the $\nu = 1/2$ FQH state. Phys. Rev. Lett. **66**, 3205 (1991)
37. Greiter, M., Wen, X.-G., Wilczek, F.: Paired Hall states. Nucl. Phys. B **374**, 567 (1992)
38. Leinaas, J.M., Myrheim, J.: On the theory of identical particles. Nuovo Cimento B **37B**, 1 (1977)
39. Wilczek, F.: Quantum mechanics of fractional-spin particles. Phys. Rev. Lett. **49**, 957 (1982)
40. Halperin, B.I.: Statistics of quasiparticles and the hierarchy of fractional quantized Hall states. Phys. Rev. Lett. **52**, 1583 (1984)
41. Arovas, D., Schrieffer, J.R., Wilczek, F.: Fractional statistics and the quantum Hall effect. Phys. Rev. Lett. **53**, 722 (1984)
42. Berry, M.V.: Quantal phase factors accompanying adiabatic changes. Proc. R. Soc. Lond. A **392**, 45 (1984)
43. Nayak, C., Wilczek, F.: $2n$ Quasihole states realize 2^{n-1}-dimensional spinor braiding statistics in paired quantum Hall states Nucl. Phys. B **479**, 529 (1996)
44. Wen, X.-G.: Theory of the edge excitations in FQH effects. Int. J. Mod. Phys. B **6**, 1711 (1992)
45. Wen, X.-G.: Topological order and edge structure of $\nu = 1/2$ quantum Hall state. Phys. Rev. Lett. **70**, 355 (1993)
46. Wen, X.-G.: Chiral Luttinger liquid and the edge excitations in the FQH states. Phys. Rev. B **41**, 12838 (1990)
47. Macdonald, I.G.: Symmetric Functions and Hall Polynomials, 2nd edn. Oxford Mathematical Monographs. Oxford University Press, New York (1995)
48. Hazewinkel, M.: Symmetric functions, noncommutative symmetric functions, and quasisymmetric functions. Acta Appl. Math. **75**, 55 (2003). arXiv:math/0410468
49. Ardonne, E., Bergholtz, E.J., Kailasvuori, J., Wikberg, E.: Degeneracy of non-Abelian quantum Hall states on the torus: domain walls and conformal field theory. J. Stat. Mech. Theory Exp. **2008**, P04016 (2008)
50. Noyvert, B.: Algebraic approach to parafermionic conformal field theories. J. High Energy Phys. **2**, 74 (2007)

Virasoro Central Charges for Nichols Algebras

A.M. Semikhatov

Abstract A Virasoro central charge can be associated with each Nichols algebra with diagonal braiding in a way that is invariant under the Weyl groupoid action. The central charge takes very suggestive values for some items in Heckenberger's list of rank-2 Nichols algebras. In particular, this might be viewed as an indication of the existence of reasonable logarithmic extensions of $W_3 \equiv WA_2$, WB_2, and WG_2 models of conformal field theory. In the W_3 case, the construction of an *octuplet* extended algebra—a counterpart of the triplet $(1, p)$ algebra—is outlined.

1 Introduction

In [1], we described a paradigm treating screening operators in two-dimensional conformal field theory as a braided Hopf algebra, a Nichols algebra [2–9]. This immediately suggests that the inverse relation may also exist. Is any finite-dimensional Nichols algebra with diagonal braiding an algebra of screenings in some conformal model? This is a fascinating problem, especially considering the recent remarkable development in the theory of Nichols algebras—originally a "technicality" in Andruskiewitsch and Schneider's program of classification of pointed Hopf algebras, which has grown into a beautiful theory in and of itself (in addition to the papers cited above and the references therein, also see [8, 10–17]). Diagonal braiding is assumed in what follows.

As many "inverse" problems, that of identifying a conformal field model behind a given Nichols algebra is not necessarily well defined. It is of course well known that screenings can be used to define models of conformal field theory; in particular, defining logarithmic models as kernels of screening [18] turned out to be especially useful. But passing from a Nichols algebras to screenings involves various ambiguities. Nevertheless, the central charges associated with Nichols algebras in what follows have the nice property of being invariant under the Weyl groupoid action—the natural "symmetry" up to which Nichols algebras are classified [10, 19, 20].

To proceed beyond the central charge identification, I restrict myself to Nichols algebras of rank two (already a fairly large number in terms of the possible confor-

A.M. Semikhatov
Lebedev Physics Institute, Moscow 119991, Russia

C. Bai et al. (eds.), *Conformal Field Theories and Tensor Categories*,
Mathematical Lectures from Peking University, DOI 10.1007/978-3-642-39383-9_3,

mal models). All of these were listed by Heckenberger [21, 22] (the general classification, for any rank, was achieved in [7] and was reproduced in a different and independent way in [15–17]). These notes are in fact a compilation of the original Heckenberger's list with explicit results on the presentation of some Nichols algebras (obtained in [15] for the standard type and in [23] in several nonstandard cases), and with several CFT constructions added. The extended algebra of a logarithmic model—the octuplet algebra extending the W_3 algebra—is offered in only one case; the other CFT constructions are merely a starting point for finding extended algebras.[1]

For the Nichols algebra $\mathfrak{B}(X)$ of a θ-dimensional braided linear space (X,Ψ) (a rank-θ Nichols algebra), we fix a basis $(F_i)_{1\leq i\leq\theta}$ in X that diagonalizes the braiding $\Psi : X\otimes X\to X\otimes X$,

$$\Psi(F_i\otimes F_j)=q_{i,j}F_j\otimes F_i, \tag{1}$$

and call the matrix $(q_{i,j})_{1\leq i,j\leq\theta}$ the braiding matrix.

The relation to conformal field theory is based on representing the F_i as screening operators

$$F_i=\oint e^{\alpha_i\cdot\varphi} \tag{2}$$

acting in a space of bosonic fields (or simply "bosons"). Here, $\varphi(z)=(\varphi^1(z),\dots,\varphi^\theta(z))$ is a θ-component boson field with OPEs (16) (Appendix A), the dot denotes Euclidean scalar product, and $\alpha_i\in\mathbb{C}^\theta$ are such that the screenings have the self-braidings and the monodromies coincident with those in (1):[2]

$$\begin{aligned} e^{i\pi\alpha_i.\alpha_i} &= q_{i,i},\\ e^{2i\pi\alpha_i.\alpha_j} &= q_{i,j}q_{j,i}, \quad i\neq j. \end{aligned} \tag{3}$$

The ambiguities inherent in passing from a braiding matrix to screenings realized in terms of free bosons are numerous. Already the "θ-boson space" on which the F_i act can be chosen differently, e.g., including or not including exponentials $e^{\omega.\varphi(z)}$, where ω ranges a lattice in $\mathbb{C}^\theta$. Furthermore, solving relations (3) for $\alpha_i\in\mathbb{C}^\theta$ involves taking logarithms, which introduces arbitrary integer parameters.

And yet the idea to look for a conformal model corresponding to a given Nichols algebra is not altogether meaningless because *the Virasoro central charge is invariant under the Weyl groupoid action.* I go into some detail here because the statement implicitly refers to a *procedure* to deal with the ambiguities such that the invariance is nevertheless ensured.

---

[1]Since the submission of this paper, some progress has been achieved in the cases described in 2.2 and 3.1 in what follows [60].

[2]Notably, the conditions on the braiding matrix elements selecting a Nichols algebra involve only the self-braidings $q_{i,i}$ and the *monodromies* $q_{i,j}q_{j,i}$ for $i\neq j$.

Whenever the α_i are linearly independent, the screenings in (2) uniquely define a Virasoro algebra in their centralizer in the space of differential polynomials in the $\partial\varphi^j(z)$ ($\partial = \partial/\partial z$). The general case is considered in Appendix A, and for $\theta = 2$, for example, the central charge of Virasoro algebra is

$$c = 2 - 3\Big(\Big(\big(4 + (\alpha_1.\alpha_1)(\alpha_2.\alpha_2)\big)(\alpha_1 - \alpha_2).(\alpha_1 - \alpha_2)$$
$$+ 4(\alpha_1 - \alpha_2).\big((\alpha_1.\alpha_1)\alpha_2 - (\alpha_2.\alpha_2)\alpha_1\big)\Big)$$
$$/\big((\alpha_1.\alpha_1)(\alpha_2.\alpha_2) - (\alpha_1.\alpha_2)^2\big)\Big). \tag{4}$$

On the Nichols algebra side, the Weyl groupoid action is defined as follows [10, 19, 20, 24–26]. There exists a generalized Cartan matrix $(a_{i,j})_{1\le i,j\le\theta}$ such that $a_{i,i} = 2$ and

$$q_{i,i}^{a_{i,j}} = q_{i,j}q_{j,i} \quad \text{or} \quad q_{i,i}^{1-a_{i,j}} = 1 \tag{5}$$

holds for each pair $i \neq j$. The Weyl groupoid is generated by pseudoreflections acting on the set of braiding matrices and defined for any k, $1 \le k \le \theta$. The reflected braiding matrix has the entries

$$\mathfrak{R}^{(k)}(q_{i,j}) = q_{i,j}q_{i,k}^{-a_{k,j}}q_{k,j}^{-a_{k,i}}q_{k,k}^{a_{k,i}a_{k,j}}. \tag{6}$$

It may or may not have the same generalized Cartan matrix.[3] The use of this tool has remarkably resulted in the classification of Nichols algebras with diagonal braiding [7].

With the screening momenta $\alpha_i \in \mathbb{C}^\theta$, $1 \le i \le \theta$, defined such that (3) holds, condition (5) is "lifted" to the scalar products as the condition

$$2\alpha_i.\alpha_j = a_{i,j}\alpha_i.\alpha_i \quad \text{or} \quad (1 - a_{i,j})\alpha_i.\alpha_i = 2 \tag{7}$$

to be satisfied for each pair $i \neq j$. Several particular choices have been made in writing this (for example, the 2, not some other even integer, in the second relation).

The Weyl reflections are now lifted to the scalar products by "naively taking the logarithm" of (6) (which amounts to actual pseudoreflections in $\mathbb{C}^\theta$):

$$\mathfrak{R}^{(k)}(\alpha_i.\alpha_j) = \alpha_i.\alpha_j - a_{k,j}\alpha_i.\alpha_k - a_{k,i}\alpha_k.\alpha_j + a_{k,i}a_{k,j}\alpha_k.\alpha_k. \tag{8}$$

Weyl-reflecting the central charge amounts to replacing each $\alpha_i.\alpha_j$ with $\mathfrak{R}^{(k)}(\alpha_i.\alpha_j)$ in the system of equations that defines the central charge (see Appendix A).

[3] If the diagonal braiding is of Cartan type, then Weyl reflections preserve the Cartan matrix. If a generalized Cartan matrix (not of Cartan type) is the same for the entire class of Weyl-reflected braided matrices, then such a generalized Cartan matrix and the braiding matrix are said to belong to the *standard* type. Nonstandard braidings do exist [15, 27].

Theorem *The central charge of the Virasoro algebra centralizing θ screenings in the θ-boson space is invariant under* (8) *if conditions* (7) *hold.*

This is proved in Appendix A; for example, in the rank-2 case (see (4)), the Weyl-reflected central charge is expressed rather explicitly as

$$\begin{aligned}\mathfrak{R}^{(1)}(c) - c = {} & \frac{3}{(\alpha_1.\alpha_1)(\alpha_2.\alpha_2) - (\alpha_1.\alpha_2)^2}\\ & \times (2\alpha_1.\alpha_2 - a_{1,2}\alpha_1.\alpha_1)\big((a_{1,2} - 1)\alpha_1.\alpha_1 + 2\big)\\ & \times \big(a_{1,2}^2\alpha_1.\alpha_1 - a_{1,2}\alpha_1.\alpha_1 - 2a_{1,2}\alpha_1.\alpha_2 + 2\alpha_2.\alpha_2 + 2a_{1,2} - 4\big).\end{aligned}$$

The product $(2\alpha_1.\alpha_2 - a_{1,2}\alpha_1.\alpha_1)((a_{1,2} - 1)\alpha_1.\alpha_1 + 2)$ vanishes whenever (7) holds for $i = 1$ and $j = 2$, and hence c is indeed invariant under $\mathfrak{R}^{(1)}$; the invariance under $\mathfrak{R}^{(2)}$ is verified similarly.

1.1 From Virasoro to Extended Algebras

The central charge value alone does not specify a conformal field theory uniquely. In "good" cases, however—when the central charge found from (4) is a function of a (discrete) parameter—the form of this dependence does suggest what type of operators extend the Virasoro algebra and therefore what the resulting conformal model is; and the centralizer of the screenings then turns out to be sufficiently ample for an interesting conformal field theory to live in it. An exemplary case is the W_3 algebra, which centralizes two screenings associated with a braiding matrix such that $q_{1,1} = q_{2,2} = q$ and $q_{1,2}q_{2,1} = q^{-1}$, with a primitive root of unity q. From the logarithmic perspective, this W_3 algebra is a *nonextended* algebra, playing the same role in relation to an extended algebra as the Virasoro algebra plays in relation to the triplet algebras of $(p, 1)$ [28–32] and (p, p') [33] logarithmic models. Specifically in the W_3 case, the *extended* algebra is the octuplet algebra described in Appendix B. Similar constructions are expected in other good cases; I am optimistic about the fact that the same generalized Dynkin diagram gives rise to a finite-dimensional Nichols algebra *and* to an interesting conformal field theory. The intricate machinery underlying the finite dimensionality of the corresponding Nichols algebra may manifest itself in constructing new logarithmic models.[4]

In what follows, I therefore reproduce Heckenberger's list of rank-2 finite-dimensional Nichols algebras [21, 22], with the only difference that I `enumerate`, not `itemize` the subitems. For several items, I also add the presentations

[4]Recall that *rational* conformal field theories are generally defined as the cohomology of a complex associated with the screenings, whereas *logarithmic* models are defined by the kernel (cf. [18, 32–36]). In particular, this allows interesting logarithmic conformal models to exist in the cases where the rational model is nonexistent (the $(p, 1)$ series) or trivial (the $(2, 3)$ model).

known from [15] and explicitly borrowed from [23], including the case number in [23]. From the Nichols-algebra data, I move toward conformal field theory by analyzing the conditions on the screening momenta. When it is clear what current algebra extends the Virasoro algebra with the central charge obtained from (4), I recall the explicit construction, presenting it in the form that manifestly refers to the corresponding pair of screenings (once again, all extended algebras except the one in Appendix B are not logarithmic extensions, but rather starting points for such extensions).

1.2 Points to Note

1. In conformal field theory, *fermionic* screenings are often interesting. Their Nichols-algebra counterparts are the diagonal entries -1 in braiding matrices. But given a $q_{i,i} = -1$ and trying to reconstruct a screening in general leads to the condition $\alpha_i.\alpha_i = 1 + 2m$ on the screening momentum, with $m \in \mathbb{Z}$. For the corresponding screening current $f(z) = e^{\alpha_i \cdot \varphi(z)}$, it then follows that $f(z)f(w)$ develops a $(1+2m)$th-order zero as $z \to w$. The cases where this zero is actually a pole are somewhat pathological from the CFT standpoint; as regards the cases of a zero of an order ≥ 3, I am unaware of any such examples of screenings. Only $m = 0$ is a "good" value. Remarkably, *solving conditions* (7) *with* $q_{i,i} = -1$ *has the tendency to select the value* $m = 0$, *thus ensuring a true fermionic screening*.
2. Other integers appearing in "taking the logarithms" are not disposed of that easily. There are solutions of (7) where these integers vanish (and the central charge depends on another integer parameter, the *order* of a root of unity); such solutions are referred to as "regular" in what follows. But there also exist "peculiar" solutions of (7) where some of these parasitic integers persist, and which have somewhat reduced chances to correspond to interesting CFT models. In fact, *some* peculiar solutions are eliminated already by the conditions in Heckenberger's list: in some items, the order of the corresponding root of unity must not be too small, and the peculiar solutions do require just one of those excluded values. This might suggest that peculiar solutions should somehow be eliminated altogether, but if so, then I have overlooked the argument.
3. Things get worse with the many items in the list that do not involve a free discrete parameter such as the order of a root of unity. Isolated central charge values are by no means illuminating, and remain entirely unsuggestive when expanded into families by the occurrence of "parasitic integers."

   The unwieldiness of the "peculiar" central charges also thwarted my original intention to provide each item in the list with a central charge. This can be done, but the results are not indicative of anything. The corresponding items in the list are therefore left in their original form given in [21, 22].

4. In the "regular" cases, I choose a primitive pth root of unity as $e^{2i\pi/p}$ with an integer p. This might unnecessarily restrict the generality, but the cases that follow with this choice are already interesting. In "peculiar" cases, by contrast, I try to work out the cases with $e^{2i\pi r/p}$, where r is coprime with p. The r parameter sometimes survives till the central charge, but that's where the story ends, because I do not construct any current algebra generators beyond Virasoro in peculiar cases.

   Mostly, I take the logarithm of relations such as $e^{i\pi x} = e^{2i\pi r/s}$ (where x is typically a linear combination of scalar products) "honestly," as $x = \frac{2r}{s} + 2\ell$, $\ell \in \mathbb{Z}$. In *some* cases, the ensuing dependence on ℓ turns out to be "under control" (something like a shift of the level of an affine Lie algebra with which the corresponding conformal field theory is associated—which interestingly corresponds to a *twist* equivalence of the braiding matrix), and I sometimes omit it.
5. Strictly speaking, identifying a CFT model from its central charge that depends on a parameter is an ill-defined procedure in the sense that given a central charge $c = f(p)$ and redefining the parameterization by an arbitrary function, $p' = g(p)$, changes the "functional form" of c arbitrarily. It is tacitly understood that some "natural" parameterizations are considered and very limited reparameterizations are allowed (typically those that are known to occur in some CFT constructions).

1.3 Notation

The notation

$$R_\ell = \text{the set of primitive } \ell\text{th roots of unity}$$

is copied from [21, 22] as part of the defining conditions in the list items. A braiding matrix (1) is encoded in a generalized Dynkin diagram $\overset{q_{1,1}}{\circ}\!\!\overset{m_{1,2}}{\text{———}}\!\!\overset{q_{2,2}}{\circ}$, where $m_{1,2} = q_{1,2}q_{2,1}$. The A_2, B_2, and G_2 Cartan matrices are $\left(\begin{smallmatrix} 2 & -1 \\ -1 & 2 \end{smallmatrix}\right)$, $\left(\begin{smallmatrix} 2 & -2 \\ -1 & 2 \end{smallmatrix}\right)$, and $\left(\begin{smallmatrix} 2 & -3 \\ -1 & 2 \end{smallmatrix}\right)$. Vectors $\alpha, \beta \in \mathbb{C}^2$ are the momenta of two screenings in what follows (α_1 and α_2 in the nomenclature of (2)).

2 The List, Item 1

The defining conditions are

$$\boxed{q_{12}q_{21} = 1 \quad \text{and} \quad q_{11}, q_{22} \in \bigcup_{a=2}^{\infty} R_a.}$$

This is the "trivial" $A_1 \times A_1$ case. The corresponding CFT model is the product $(p', 1) \times (1, p)$ of two "$(p, 1)$" models [32, 37], or, in the degenerate case where α and β are collinear (and hence only one boson is needed), the (p', p) model [33, 38].

3 The List, Items 2.*

The defining conditions are

$$\boxed{q_{12}q_{21}q_{22} = 1 \quad \text{and} \quad q_{12}q_{21} \neq 1,}$$

plus any of conditions 2.1–2.7. In terms of the momenta $\alpha, \beta \in \mathbb{C}^2$ of the screenings, the common condition for all these cases takes the form

$$2\alpha.\beta + \beta.\beta = 2m \quad (m \in \mathbb{Z}).$$

2.1 (5.7(1)$_{[23]}$) $q_{11}q_{12}q_{21} = 1$, $q_{12}q_{21} \in \bigcup_{a=2}^{\infty} R_a$, Cartan type A_2, $\overset{q}{\circ}\overset{q^{-1}}{\text{———}}\overset{q}{\circ}$.
In terms of scalar products, the conditions are

$$\alpha.\alpha + 2\alpha.\beta = 2n \quad (n \in \mathbb{Z}), \qquad 2\alpha.\beta = -\frac{2}{p} + 2j, \quad |p| \geq 2 \ (j \in \mathbb{Z}).$$

The braiding matrix (which is stable under Weyl reflections) is then parameterized as

$$\begin{pmatrix} e^{2i\pi/p} & (-1)^j e^{-i\pi/p} \\ (-1)^j e^{-i\pi/p} & e^{2i\pi/p} \end{pmatrix}.$$

None of the two screenings is fermionic unless $|p| = 2$.

Conditions (7) are not satisfied for all m, n, j, and p. There are several "peculiar" solutions and a "regular" solution. The peculiar solutions are ($m = 0$, $n = -k - \frac{3}{2}$, $p = 2$, $j = -k - \frac{3}{2}$), ($m = 0$, $n = -k - \frac{3}{2}$, $p = -2$, $j = -k - \frac{5}{2}$), ($m = -k - \frac{3}{2}$, $n = 0$, $p = 2$, $j = -k - \frac{3}{2}$), ($m = -k - \frac{3}{2}$, $n = 0$, $p = -2$, $j = -k - \frac{5}{2}$), ($m = k + \frac{3}{2}$, $n = k + \frac{3}{2}$, $p = 2$, $j = k + \frac{3}{2}$), and ($n = m = k + \frac{3}{2}, k + \frac{3}{2}$, $p = -2$, $j = k + \frac{1}{2}$) with half-integer k in all cases; with this parameterization, the resulting central charge is in each case equal to $\frac{3k}{k+2} - 1$, which is the central charge of the $\widehat{s\ell}(2)_k/\widehat{\mathfrak{h}}$ coset (more on it is to be said below, when it occurs as a "regular" solution).

The regular solution is $m = n = 0$, yielding the central charge

$$c = 50 - \frac{24}{k+3} - 24(k+3), \tag{9}$$

where $k + 3 = \frac{1}{p} - j$ (or, in view of the structure of the formula, $\frac{1}{k+3} = \frac{1}{p} - j$). This is the central charge of the W_3 algebra parameterized in terms of the level k of the $\widehat{s\ell}(3)$ affine Lie algebra from which W_3 can be obtained by Hamiltonian reduction.

The centralizer of the screenings does indeed contain a dimension-3 primary field $W(z)$ (unique up to an overall factor) in the space of differential polynomials in the fields

$$\partial\varphi_\alpha(z) = \alpha.\partial\varphi(z) \quad \text{and} \quad \partial\varphi_\beta(z) = \beta.\partial\varphi(z).$$

Explicitly, setting $j = 0$ for simplicity and omitting the (z) arguments in the right-hand side for brevity,

$$\begin{aligned} W(z) = {} & \partial\varphi_\alpha\partial\varphi_\alpha\partial\varphi_\alpha + \frac{3}{2}\partial\varphi_\alpha\partial\varphi_\alpha\partial\varphi_\beta - \frac{3}{2}\partial\varphi_\alpha\partial\varphi_\beta\partial\varphi_\beta - \partial\varphi_\beta\partial\varphi_\beta\partial\varphi_\beta \\ & - \frac{9(p-1)}{2p}\partial^2\varphi_\alpha\partial\varphi_\alpha - \frac{9(p-1)}{4p}\partial^2\varphi_\alpha\partial\varphi_\beta \\ & + \frac{9(p-1)}{4p}\partial^2\varphi_\beta\partial\varphi_\alpha + \frac{9(p-1)}{2p}\partial^2\varphi_\beta\partial\varphi_\beta \\ & + \frac{9(p-1)^2}{4p^2}\partial^3\varphi_\alpha - \frac{9(p-1)^2}{4p^2}\partial^3\varphi_\beta. \end{aligned} \tag{10}$$

The Nichols algebra $\mathfrak{B}(X)$ (of the two-dimensional braided vector space X with basis F_1 and F_2 with braiding matrix (1)) is in this case the quotient [23]

$$\mathfrak{B}(X) = T(X)/\big([F_1,F_1,F_2],[F_1,F_2,F_2],F_1^p,[F_1,F_2]^p,F_2^p\big) \tag{11}$$

if $p \geq 3$. Here and hereafter, square brackets denote iterated q-commutators constructed in accordance with Lyndon word decomposition. If $p = 2$ (the screenings *are* fermionic!), the triple-bracket generators of the ideal are absent. The dimension is $\dim\mathfrak{B}(X) = p^3$.

The elements F_1^p and F_2^p in the ideal indicate the "positions" of the *long screenings*

$$\begin{aligned} \mathcal{E}_\alpha &= \oint e^{-\alpha^\vee.\varphi} = \oint e^{-p\alpha.\varphi} \quad \text{and} \\ \mathcal{E}_\beta &= \oint e^{-\beta^\vee.\varphi} = \oint e^{-p\beta.\varphi} \quad \left(\alpha^\vee = \frac{2\alpha}{\alpha.\alpha}\right), \end{aligned} \tag{12}$$

i.e., F_1^p and F_2^p "tend to be" the operators "opposite" to the respective long screening. Generally, these long screenings are to produce m-plet structures in logarithmic models, similarly to how the triplet structure of the $(p,1)$ logarithmic models [28–31] is generated by the corresponding long screening [18]. For the current W_3-case, the resulting *octuplet* algebra is outlined in Appendix B.

2.2 (5.7(3)[23]) $q_{11} = -1$, $q_{12}q_{21} \in \bigcup_{a=3}^{\infty} R_a$, Cartan type A_2, $\overset{-1}{\circ}\!\overset{q^{-1}}{\text{———}}\!\overset{q}{\circ}$.

The conditions are restated in terms of the scalar products as

$$\alpha.\alpha = 1 + 2n \quad (n \in \mathbb{Z}), \qquad 2\alpha.\beta = -\frac{2}{p} + 2j, \quad |p| \geq 3 \ (j \in \mathbb{Z}).$$

The braiding matrix is then parameterized as

$$\begin{pmatrix} -1 & (-1)^j e^{-i\pi/p} \\ (-1)^j e^{-i\pi/p} & e^{2i\pi/p} \end{pmatrix},$$

and its Weyl reflections different from the original matrix is

$$\begin{pmatrix} -1 & -(-1)^j e^{i\pi/p} \\ -(-1)^j e^{i\pi/p} & -1 \end{pmatrix}.$$

The Weyl orbit also contains the braiding matrix

$$\begin{pmatrix} e^{2i\pi/p} & (-1)^j e^{-i\pi/p} \\ (-1)^j e^{-i\pi/p} & -1 \end{pmatrix}.$$

The first screening "wants to be fermionic." Remarkably, conditions (7) have a solution only if $m = n = 0$ (would-be solutions with nonzero m or n require $|p| = 2$). Thus the -1 in the braiding matrix does indeed correspond to a fermionic screening in the standard sense—an operator of the form $\oint F(z)$, where $F(z)F(w)$ has a first-order, not a higher-order, zero as $z \to w$.

The solution for the scalar products with $m = n = 0$ yields the central charge

$$c = \frac{3k}{k+2} - 1,$$

where $k + 2 = \frac{1}{p} - j$. This is the central charge of the $\widehat{s\ell}(2)_k/\widehat{\mathfrak{h}}$ coset (where $\widehat{\mathfrak{h}}$ is the Heisenberg subalgebra). The two currents $j^+(z)$ and $j^-(z)$ that are in the centralizer of the screenings and generate the coset algebra can be expressed in terms of the two bosons "in the direction" of each screening as[5]

$$\begin{aligned} j^+(z) &= e^{-(1/k)(2\varphi_\alpha(z)+\varphi_\beta(z))}, \\ j^-(z) &= -\big(\partial\varphi_\alpha(z)\partial\varphi_\beta(z) + \partial\varphi_\alpha(z)\partial\varphi_\alpha(z) + (k+1)\partial^2\varphi_\alpha(z)\big) \\ &\quad \times e^{(1/k)(2\varphi_\alpha(z)+\varphi_\beta(z))}. \end{aligned} \tag{13}$$

Adding a boson $\chi(z)$ associated with the $\widehat{\mathfrak{h}}$ algebra immediately yields the three $\widehat{s\ell}(2)_k$ currents

$$J^\pm(z) = j^\pm(z) e^{\pm\sqrt{(2/k)}\chi(z)} \quad \text{and} \quad J^0(z) = \sqrt{\frac{k}{2}}\partial\chi(z).$$

For $p \geq 3$, the Nichols algebra is the quotient [23]

$$\mathfrak{B}(X) = T(X)/\big([F_1, F_2, F_2], F_1^2, F_2^p\big),$$

with $\dim \mathfrak{B}(X) = 4p$.

---

[5]The exponentials in (13) are assumed to be normal ordered, and the second line involves the (standard) abuse of notation: nested normal ordering from right to left is in fact understood after the expression is expanded.

A long screening here is

$$\mathcal{E}_\beta = \oint e^{-p\beta.\varphi}.$$

2.3 (5.11(3)[23]) $q_{11} \in R_3$, $q_{12}q_{21} \in \bigcup_{a=2}^{\infty} R_a$, $q_{11}q_{12}q_{21} \neq 1$, Cartan type B_2, $\overset{\zeta}{\circ}\overset{q^{-1}}{\text{———}}\overset{q}{\circ}$, $R_3 \ni \zeta \neq q$.

In terms of the screening momenta, the conditions become

$$\alpha.\alpha = \frac{2s}{3}, \qquad 2\alpha.\beta = -\frac{2}{p} + 2j, \quad |p| \geq 2 \; (j \in \mathbb{Z}),$$

where s is coprime with 3. The braiding matrix is parameterized as

$$\begin{pmatrix} e^{2i\pi s/3} & (-1)^j e^{-i\pi/p} \\ (-1)^j e^{-i\pi/p} & e^{2i\pi/p} \end{pmatrix}$$

and its Weyl reflection noncoincident with the original is

$$\begin{pmatrix} e^{2i\pi s/(3)} & (-1)^j e^{-i\pi(4ps-3)/(3p)} \\ (-1)^j e^{-i\pi(4ps-3)/(3p)} & e^{2i\pi(4ps-3)/(3p)} \end{pmatrix}.$$

Conditions (7) can be satisfied only if ($m = 0, p = 3, s = 3\ell - 1, j = 1 - 2\ell$) or ($m = 0, p = -3, s = 3\ell + 1, j = -1 - 2\ell$) (two peculiar solutions), or ($m = 0$, $s = 1$) (the regular solution).

In both peculiar cases, the central charge is $86 - 60(k+3) - \frac{30}{k+3}$, where $k+3 = -\frac{1}{3} + \ell$ (or $\frac{1}{k+3} = -\frac{2}{3} + 2\ell$) in the first case and $k+3 = \frac{1}{3} + \ell$ (or $\frac{1}{k+3} = \frac{2}{3} + 2\ell$) in the second case. The central charge is that of the WB_2 algebra, discussed in more detail below when it appears as a "regular" solution.

In the regular case $m = 0$, the central charge is

$$c = -\frac{21}{2} - 6z - \frac{27}{2(4z-3)},$$

where $\frac{1}{z} = \frac{1}{p} - j$ or $\frac{1}{z} = -\frac{1}{p} + \frac{4}{3} + j$.

The condition $q_{11}q_{12}q_{21} \neq 1$ excludes the value $p = 3$.

If $p \geq 4$, and $p' = \operatorname{ord}(q_{11}q_{22}^{-1}) = \operatorname{ord}(e^{2i\pi/3 - 2i\pi/p})$, then [23]

$$\mathfrak{B}(X) = T(X)/\big([F_1, F_2, F_2], F_1^3, [F_1, F_1, F_2]^{p'}, F_2^p\big),$$

with $\dim \mathfrak{B}(X) = 9pp'$.

2.4 $q_{11} \in \bigcup_{n=4}^{\infty} R_a$, with two subcases listed below. To identify the central charges in what follows, we use the formula [39]

$$c(k) = \ell - 12\frac{|(k+h^\vee)\rho^\vee - \rho|^2}{k+h^\vee} \tag{14}$$

for the central charge of a W-algebra obtained by Hamiltonian reduction of a level-k affine Lie algebra; $h^\vee$ is the dual Coxeter number, ρ is half the sum of positive roots, $\rho^\vee$ half the sum of their duals, and ℓ is the rank of the corresponding finite-dimensional Lie algebra.

2.4.1 (5.11(1)[23]): $q_{12}q_{21} = q_{11}^{-2}$, Cartan type B_2, $\overset{q}{\circ}\!\overset{q^{-2}}{\text{———}}\!\overset{q^2}{\circ}$.
In terms of scalar products, we then have

$$\alpha.\alpha = \frac{2}{p} + 2j, \quad |p| \geq 4 \; (j \in \mathbb{Z}), \qquad 2\alpha.\beta + 2\alpha.\alpha = 2n \quad (n \in \mathbb{Z}).$$

The braiding matrix (stable under Weyl reflections) is

$$\begin{pmatrix} e^{2i\pi/p} & (-1)^n e^{-2i\pi/p} \\ (-1)^n e^{-2i\pi/p} & e^{4i\pi/p} \end{pmatrix}.$$

Conditions (7) hold in two peculiar cases and the regular case. The peculiar cases are $(m = -2j, n = 0, p = 4)$ with $c = -1 - \frac{24}{4j+1} + \frac{24}{4j-1}$ and $(m = 1 - 2j, n = 0, p = -4)$ with $c = -1 - \frac{24}{4j-1} + \frac{24}{4j-3}$.
The regular case is $m = n = 0$, with

$$c = 86 - 60(k+3) - \frac{30}{k+3},$$

where $k + 3 = \frac{1}{p} + j$ (or $\frac{1}{k+3} = \frac{2}{p} + 2j$). This is the central charge of the WB_2 algebra [40–42] (also see [43]) obtained by Hamiltonian reduction of the level-k $B_2^{(1)}$ algebra (by formula (14), with $|\rho|^2 = \frac{5}{2}$, $|\rho^\vee|^2 = 5$, and $\langle \rho, \rho^\vee \rangle = \frac{7}{2}$ for B_2).
The WB_2 algebra contains a unique primary field of dimension 4. Explicitly, it is a rather long (20 terms) differential polynomial in $\partial\varphi_\alpha(z)$ and $\partial\varphi_\beta(z)$,

$$\begin{aligned}
&p(p-3)(27p-32)\partial\varphi_\alpha\partial\varphi_\alpha\partial\varphi_\alpha\partial\varphi_\alpha + 2p(p-3)(27p-32)\partial\varphi_\alpha\partial\varphi_\alpha\partial\varphi_\alpha\partial\varphi_\beta \\
&\quad - 21p\big(p^2-2\big)\partial\varphi_\alpha\partial\varphi_\alpha\partial\varphi_\beta\partial\varphi_\beta - p(3p-2)(16p-27)\partial\varphi_\alpha\partial\varphi_\beta\partial\varphi_\beta\partial\varphi_\beta \\
&\quad - \frac{p}{4}(3p-2)(16p-27)\partial\varphi_\beta\partial\varphi_\beta\partial\varphi_\beta\partial\varphi_\beta \\
&\quad + \cdots \\
&\quad - \frac{(p-3)(3p-4)(30p^3-115p^2+144p-60)}{3p^2}\partial^4\varphi_\alpha \\
&\quad + \frac{(2p-3)(3p-2)(15p^3-72p^2+115p-60)}{3p^2}\partial^4\varphi_\beta
\end{aligned}$$

(*all* coefficients are polynomials in p with integer coefficients after the overall renormalization by $12p^2$).

If $p \geq 5$ is odd, then [23]

$$\mathfrak{B}(X) = T(X)/\big([F_1, F_1, F_1, F_2], [F_1, F_2, F_2], F_1^p,$$
$$[F_1, F_1, F_2]^p, [F_1, F_2]^p, F_2^p\big),$$

with $\dim \mathfrak{B}(X) = p^4$. If $p \geq 4$ is even, then

$$\mathfrak{B}(X) = T(X)/\big([F_1, F_1, F_1, F_2], [F_1, F_2, F_2], F_1^p,$$
$$[F_1, F_1, F_2]^{p/2}, [F_1, F_2]^p, F_2^{p/2}\big)$$

and $\dim \mathfrak{B}(X) = \frac{p^4}{4}$ (the second generator of the ideal is absent for $p = 4$).

2.4.2: $q_{12}q_{21} = q_{11}^{-3}$, Cartan type G_2, $\overset{q}{\circ}\overset{q^{-3}}{\text{———}}\overset{q^3}{\circ}$.
In terms of scalar products, we now have

$$\alpha.\alpha = \frac{2}{p} + 2j, \quad |p| \geq 4 \ (j \in \mathbb{Z}), \qquad 2\alpha.\beta + 3\alpha.\alpha = 2n \quad (n \in \mathbb{Z}).$$

The braiding matrix is parameterized as

$$\begin{pmatrix} e^{2i\pi/p} & (-1)^{j+n} e^{-3i\pi/p} \\ (-1)^{j+n} e^{-3i\pi/p} & e^{6i\pi/p} \end{pmatrix}$$

and is stable under Weyl reflections. Conditions (7) hold in the peculiar cases $(j = 0, m = 0, p = 4)$ with $c = -10 - \frac{54}{4n+1} + \frac{24}{4n-3}$, $(m = -3j, n = 0, p = 6)$ with $c = -\frac{2}{3} + \frac{400}{3(18j-1)} - \frac{36}{6j+1}$, and $(m = 1 - 3j, n = 0, p = -6)$ with $c = -\frac{2}{3} + \frac{400}{3(18j-7)} - \frac{36}{6j-1}$ and in the regular case $m = n = 0$ with the central charge

$$c = 194 - 168(k+4) - \frac{56}{k+4},$$

where $k+4 = \frac{1}{p} + j$ (or $\frac{1}{k+4} = \frac{3}{p} + 3j$). This is the central charge of the WG_2 algebra [42, 44] (also see [43, 45]) obtained by Hamiltonian reduction of the level-k $G_2^{(1)}$ algebra (by formula (14), with $|\rho|^2 = 14$, $|\rho^\vee|^2 = \frac{14}{3}$, and $\langle \rho, \rho^\vee \rangle = 8$ for G_2). The WG_2 algebra contains a unique primary field of dimension 6, which is by far too long to be given here (see [42, 44, 45]).

The remaining subcases of Case 2 may all be considered "peculiar" to some extent. The values of c are equally "peculiar."

2.5 $q_{12}q_{21} \in R_8$, $q_{11} = (q_{12}q_{21})^2$, Cartan type G_2, $\overset{\zeta^2}{\circ}\overset{\zeta}{\text{———}}\overset{\zeta^{-1}}{\circ}$, $\zeta \in R_8$.
The conditions reformulate in terms of scalar products as

$$2\alpha.\beta = \frac{r}{4} + 2j \quad (j \in \mathbb{Z}), \qquad \alpha.\alpha - 4\alpha.\beta = 2n,$$

where r is odd. The braiding matrix is

$$\begin{pmatrix} i^r & (-1)^j e^{i\pi r/8} \\ (-1)^j e^{i\pi r/8} & e^{-(1/4)i\pi r} \end{pmatrix}$$

and its Weyl reflection different from the original matrix is

$$\begin{pmatrix} i^r & (-1)^j e^{(3/8)i\pi r} \\ (-1)^j e^{(3/8)i\pi r} & (-1)^r \end{pmatrix}.$$

Conditions (7) can be satisfied only if $(m=0, r=1-8j-4n)$, yielding the central charge $c=-10-\frac{48}{4n-1}+\frac{108}{4n-9}$. For $n=0$, this gives the celebrated central charge value

$$c=26.$$

2.6 $q_{12}q_{21}\in R_{24}, q_{11}=(q_{12}q_{21})^6$, $\overset{\zeta^6}{\circ}\overset{\zeta}{\text{———}}\overset{\zeta^{-1}}{\circ}$, $\zeta\in R_{24}$.
The conditions for the scalar products are

$$2\alpha.\beta=\frac{r}{12}+2j \quad (j\in\mathbb{Z}), \qquad \alpha.\alpha-12\alpha.\beta=2n,$$

where r is coprime with 2 and 3. The braiding matrix

$$\begin{pmatrix} i^r & (-1)^j e^{i\pi r/24} \\ (-1)^j e^{i\pi r/24} & e^{-i\pi r/12} \end{pmatrix}$$

has the G_3 Cartan matrix, but its nontrivial Weyl reflection is

$$\begin{pmatrix} i^r & (-1)^j e^{(11/24)i\pi r} \\ (-1)^j e^{(11/24)i\pi r} & e^{2i\pi r/3} \end{pmatrix},$$

with the associated generalized Cartan matrix $\left(\begin{smallmatrix} 2 & -3 \\ -2 & 2 \end{smallmatrix}\right)$; various other generalized Cartan matrices are produced under further Weyl reflections.

Conditions (7) are satisfied only if $(m=0, r=1-24j-4n)$, with $c=-10-\frac{144}{4n-1}+\frac{324}{4n-25}$ (that r be coprime with 2 and 3 selects the values $n=2+3\ell$ or $n=3+3\ell$, $\ell\in\mathbb{Z}$).

2.7 $q_{12}q_{21}\in R_{30}, q_{11}=(q_{12}q_{21})^{12}$, $\overset{\zeta^{12}}{\circ}\overset{\zeta}{\text{———}}\overset{\zeta^{-1}}{\circ}$, $\zeta\in R_{30}$.
In terms of scalar products,

$$2\alpha.\beta=\frac{r}{15}+2j \quad (j\in\mathbb{Z}), \qquad \alpha.\alpha-24\alpha.\beta=2n,$$

where r is coprime with 30. The braiding matrix, parameterized as

$$\begin{pmatrix} e^{4i\pi r/5} & (-1)^j e^{i\pi r/30} \\ (-1)^j e^{i\pi r/30} & e^{-i\pi r/15} \end{pmatrix},$$

has the associated generalized Cartan matrix $\left(\begin{smallmatrix}2 & -4\\ -1 & 2\end{smallmatrix}\right)$. The nontrivial Weyl reflection is

$$\begin{pmatrix} e^{4i\pi r/5} & (-1)^{j+r}e^{-(7/30)i\pi r}\\ (-1)^{j+r}e^{-(7/30)i\pi r} & (-1)^r \end{pmatrix},$$

with the same generalized Cartan matrix, but other (generalized) Cartan matrices are produced under further Weyl reflections. Conditions (7) are solved by $(m = 0, n = 1 + 2\ell, r = -2 - 5\ell - 30j)$, where $\ell = 1 + 6u$ or $\ell = 3 + 6u$, $u \in \mathbb{Z}$, respectively yielding the incomprehensible $c = -\frac{62}{5} + \frac{2916}{5(30u-17)} - \frac{180}{30u+7}$ and $c = -\frac{62}{5} + \frac{2916}{5(30u-7)} - \frac{180}{30u+17}$.

4 The List, Items 3.*

The defining conditions are

$$\boxed{q_{12}q_{21} \neq 1, \qquad q_{11}q_{12}q_{21} \neq 1, \qquad q_{12}q_{21}q_{22} \neq 1,}$$

$$\boxed{q_{22} = -1, \qquad q_{11} \in R_2 \cup R_3,}$$

plus any of conditions 3.1–3.7. In terms of the screening momenta, the common conditions are

$$\beta.\beta = 1 + 2m \quad (m \in \mathbb{Z}), \qquad \alpha.\alpha = 1 \text{ or } \frac{2s}{3},$$

where s is coprime with 3.

3.1 (5.7(4)[23]) $q_{11} = -1$, $q_{12}q_{21} \in \bigcup_{a=3}^{\infty} R_a$, Cartan type A_2, $\overset{-1}{\circ}\!\overset{q}{\text{———}}\!\overset{-1}{\circ}$.
In terms of scalar products of the screening momenta, these conditions are

$$\alpha.\alpha = 1 + 2n \quad (n \in \mathbb{Z}), \qquad 2\alpha.\beta = \frac{2}{p} + 2j, \quad |p| \geq 3 \ (j \in \mathbb{Z}).$$

Both screenings "want to be fermionic." The braiding matrix is

$$\begin{pmatrix} -1 & (-1)^j e^{i\pi/p}\\ (-1)^j e^{i\pi/p} & -1 \end{pmatrix}$$

and its Weyl reflections are

$$\begin{pmatrix} -1 & -(-1)^j e^{-i\pi/p}\\ -(-1)^j e^{-i\pi/p} & e^{2i\pi/p} \end{pmatrix} \quad \text{and} \quad \begin{pmatrix} e^{2i\pi/p} & -(-1)^j e^{-i\pi/p}\\ -(-1)^j e^{-i\pi/p} & -1 \end{pmatrix}.$$

Remarkably, conditions (7) are satisfied (with $|p| \geq 3$) only for $m = n = 0$ (no "peculiar" solutions!), yielding the $\widehat{s\ell}(2)_k/\widehat{\mathfrak{h}}$ central charge

$$c = \frac{3k}{k+2} - 1,$$

where $k + 1 = \frac{1}{p} + j$. For $j = 0$, in particular, this relation between k and p takes the form

$$\frac{1}{p+1} + \frac{1}{k+2} = 1.$$

This "duality" between two levels, k and $p - 1$, was extensively used in [46, 47] (see also the references therein); in particular,

$$\widehat{s\ell}(2)_k/\widehat{\mathfrak{h}} = \widehat{s\ell}(2|1)_{p-1}/\widehat{g\ell}(2)_{p-1},$$

offering another view on what the CFT counterpart of the Nichols algebra is.[6]

The currents generating the $\widehat{s\ell}(2)_k/\widehat{\mathfrak{h}}$ coset algebra are given by

$$\begin{aligned} j^+(z) &= \partial\varphi_\beta(z)e^{(1/k)(\varphi_\alpha(z)-\varphi_\beta(z))}, \\ j^-(z) &= \partial\varphi_\alpha(z)e^{-(1/k)(\varphi_\alpha(z)-\varphi_\beta(z))} \end{aligned} \tag{15}$$

(as before, $\varphi_\alpha(z) = \alpha.\varphi(z)$ and $\varphi_\beta(z) = \beta.\varphi(z)$ are the boson fields "in the direction" of the corresponding screening). With an extra boson $\chi(z)$ added to account for the missing $\widehat{\mathfrak{h}}$, the $\widehat{s\ell}(2)_k$ algebra currents are reconstructed as

$$J^\pm(z) = j^\pm(z)e^{\pm\sqrt{(2/k)}\chi(z)}, \qquad J^0(z) = \sqrt{\frac{k}{2}}\partial\chi(z).$$

The $\widehat{s\ell}(2)$ algebra is well known, since the "old" studies of the Wakimoto bosonization, to be described as a centralizer of two fermionic screenings "at an angle" to each other.[7] In this item in the list, we see again that imposing relations (7) *implies* that both -1 in matrix translate exactly into true fermionic screenings.

For $p \geq 3$, the Nichols algebra is given by [23]

$$\mathfrak{B}(X) = T(X)/\big(F_1^2, [F_1, F_2]^p, F_2^2\big)$$

with $\dim \mathfrak{B}(X) = 4p$.

---

[6]This coset equivalence belongs to a vast subject discussed in [48, 49].

[7]The Wakimoto bosonization [50] yields two essentially different three-boson realizations of $\widehat{s\ell}(2)$—the "symmetric" and the "nonsymmetric" ones, respectively centralizing two fermionic screenings and one bosonic plus one fermionic screening. The names refer to the "$j^+ \leftrightarrow j^-$ symmetric" structure of (15) and the "asymmetric" structure of (13). Somewhat broader, the "variously symmetric" realizations are discussed in [47].

3.2 There are two subcases.

3.2.1: $q_{11} \in R_3$, $q_{12}q_{21} = q_{11}$, Cartan type B_2, $\overset{\zeta}{\circ}\!\!\overset{\zeta}{\text{———}}\!\!\overset{-1}{\circ}$, $\zeta \in R_3$.
This reformulates in terms of scalar products of the screening momenta as

$$\alpha.\alpha = \frac{2s}{3} + 2\ell \quad (\ell \in \mathbb{Z}), \qquad 2\alpha.\beta = \alpha.\alpha + 2n \quad (n \in \mathbb{Z}),$$

where s is coprime with 3. The braiding matrix, which is then parameterized as

$$\begin{pmatrix} e^{2i\pi s/3} & (-1)^{n+\ell} e^{i\pi s/3} \\ (-1)^{n+\ell} e^{i\pi s/3} & -1 \end{pmatrix},$$

has the Cartan type B_2. Its nontrivial Weyl reflection is

$$\begin{pmatrix} -e^{4i\pi s/3} & -(-1)^{n+\ell} e^{-(1/3)i\pi s} \\ -(-1)^{n+\ell} e^{-(1/3)i\pi s} & -1 \end{pmatrix}.$$

Once again, conditions (7) ensure that the tentative fermionic screening is such indeed, i.e., $m = 0$: the conditions can be solved only if $(m = 0, s = 1 - 3\ell)$ or $(m = 0, s = -n - 3\ell)$. These cases respectively yield the unilluminating central charges $2 - \frac{6(12n-7)}{9n^2+6n-5}$ and $-1 - \frac{36}{2n+3} + \frac{18}{n}$.

3.2.2 (5.11(4)[23]): $q_{11} \in R_3$, $q_{12}q_{21} = -q_{11}$, Cartan type B_2, $\overset{\zeta}{\circ}\!\!\overset{-\zeta}{\text{———}}\!\!\overset{-1}{\circ}$, $\zeta \in R_3$.
Then

$$\alpha.\alpha = \frac{2s}{3}, \qquad 2\alpha.\beta - \alpha.\alpha = 1 + 2n,$$

where s is coprime with 3. The braiding matrix

$$\begin{pmatrix} e^{2i\pi s/3} & (-1)^{n+\ell} i e^{i\pi s/3} \\ (-1)^{n+\ell} i e^{i\pi s/3} & -1 \end{pmatrix}$$

is of Cartan type B_2. Its Weyl reflections are

$$\begin{pmatrix} e^{2i\pi s/3} & -(-1)^{n+\ell} i e^{i\pi s/3} \\ -(-1)^{n+\ell} i e^{i\pi s/3} & -1 \end{pmatrix} \quad \text{and}$$

$$\begin{pmatrix} e^{4i\pi s/3} & -(-1)^{n+\ell} e^{-i\pi(s/3+1/2)} \\ -(-1)^{n+\ell} e^{-i\pi(s/3+1/2)} & -1 \end{pmatrix}.$$

Conditions (7) are solved only if $(m = 0, s = 1 - 3\ell)$, which leaves us with another incomprehensible $c = 2 - \frac{24(12n-1)}{36n^2+60n+1}$ (which is $c = 26$ at $n = 0$, however). The Nichols algebra is given by the quotient [23]

$$\mathfrak{B}(X) = T(X)/\big([F_1, F_1, F_2, F_1, F_2], F_1^3, F_2^2\big)$$

with $\dim \mathfrak{B}(X) = 36$.

The remaining subcases are equally unsuggestive, and the details are omitted.

3.3 $q_0 := q_{11}q_{12}q_{21} \in R_{12}$, $q_{11} = q_0^4$. This translates into

$$\alpha.\alpha + 2\alpha.\beta = \frac{2r}{12} + 2j, \qquad \alpha.\alpha = 4\alpha.\alpha + 8\alpha.\beta + 2n.$$

3.4 $q_{12}q_{21} \in R_{12}$, $q_{11} = -(q_{12}q_{21})^2$, or

$$2\alpha.\beta = \frac{2r}{12} + 2j, \qquad \alpha.\alpha = 4\alpha.\beta + 1 + 2n.$$

3.5 $q_{12}q_{21} \in R_9$, $q_{11} = (q_{12}q_{21})^{-3}$, or

$$2\alpha.\beta = \frac{2r}{9} + 2j, \qquad \alpha.\alpha = -6\alpha.\beta + 2n.$$

3.6 $q_{12}q_{21} \in R_{24}$, $q_{11} = -(q_{12}q_{21})^4$, or

$$2\alpha.\beta = \frac{2r}{24} + 2j, \qquad \alpha.\alpha = 8\alpha.\beta + 2n.$$

3.7 $q_{12}q_{21} \in R_{30}$, $q_{11} = -(q_{12}q_{21})^5$, or

$$2\alpha.\beta = \frac{2r}{30} + 2j, \qquad \alpha.\alpha = 10\alpha.\beta + 1 + 2n.$$

5 The List, Items 4.*

The conditions are

$$\boxed{q_{12}q_{21} \neq 1, \qquad q_{11}q_{12}q_{21} \neq 1, \qquad q_{12}q_{21}q_{22} \neq 1,}$$

$$\boxed{q_{22} = -1, \qquad q_{11} \notin R_2 \cup R_3,}$$

plus any of the conditions in cases 4.1.–4.8.

In terms of the screening momenta, the common condition is

$$\beta.\beta = 1 + 2m,$$

showing that F_β is a candidate for a fermionic screening.

4.1 (5.11(2)[23]) $q_{11} \in \bigcup_{a=5}^{\infty} R_a$, $q_{12}q_{21} = q_{11}^{-2}$, Cartan type B_2, 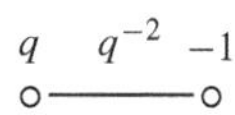

.

In terms of screenings,

$$\alpha.\alpha = \frac{2}{p} + 2j, \quad |p| \geq 5 \; (j \in \mathbb{Z}), \qquad 2\alpha.\alpha + 2\alpha.\beta = 2n.$$

Then the braiding matrix is parameterized as

$$\begin{pmatrix} e^{2i\pi/p} & (-1)^n e^{-2i\pi/p} \\ (-1)^n e^{-2i\pi/p} & -1 \end{pmatrix}.$$

Its Weyl reflection noncoincident with the original matrix is

$$\begin{pmatrix} -e^{-2i\pi/p} & -(-1)^n e^{2i\pi/p} \\ -(-1)^n e^{2i\pi/p} & -1 \end{pmatrix}.$$

Remarkably, once again, conditions (7) are satisfied *only* for $m = n = 0$ (tentative "peculiar" solutions all have $|p| \le 4$). Hence, F_β is indeed a standard fermionic screening. The central charge is then given by

$$c = -25 + \frac{24}{k+3} + 6(k+3),$$

where $\frac{1}{p} + j = -\frac{1}{k+1}$ (or $\frac{1}{p} + j = \frac{1}{2} + \frac{1}{k+1}$). Somewhat mysteriously, this is *minus* the central charge

$$c_{\mathcal{W}_3^{(2)}} = 25 - \frac{24}{k+3} - 6(k+3)$$

of the $\mathcal{W}_3^{(2)}$ algebra, which can be obtained by a "partial" Hamiltonian reduction of $\widehat{s\ell}(3)_k$ and which has a *three*-boson realization [47, 51, 52].[8]

The Nichols algebra is the quotient [23]

$$\mathfrak{B}(X) = T(X)/\big([F_1, F_1, F_1, F_2], F_1^p, [F_1, F_2]^{p'}, F_2^2\big),$$

where $p' = \mathrm{ord}(-e^{2i\pi/p})$, with $\dim \mathfrak{B}(X) = 4pp'$.

None of the remaining cases currently seems illuminating in any respect.

4.2 $q_{11} \in R_5 \cup R_8 \cup R_{12} \cup R_{14} \cup R_{20}$, $q_{12}q_{21} = q_{11}^{-3}$.

4.3 $q_{11} \in R_{10} \cup R_{18}$, $q_{12}q_{21} = q_{11}^{-4}$.

4.4 $q_{11} \in R_{14} \cup R_{24}$, $q_{12}q_{21} = q_{11}^{-5}$.

4.5 $q_{12}q_{21} \in R_8$, $q_{11} = (q_{12}q_{21})^{-2}$.

4.6 $q_{12}q_{21} \in R_{12}$, $q_{11} = (q_{12}q_{21})^{-3}$.

4.7 $q_{12}q_{21} \in R_{20}$, $q_{11} = (q_{12}q_{21})^{-4}$.

---

[8]T. Creutzig has suggested that this a quotient of some CFT that has central charge zero and contains the $\mathcal{W}_3^{(2)}$ subalgebra. The $\mathcal{W}_n^{(2)}$ algebras can be rather versatile [53–55].

4.8 $q_{12}q_{21} \in R_{30}, q_{11} = (q_{12}q_{21})^{-6}$.

6 The List, Items 5.*

I merely reproduce the original items from the list of Nichols algebras. The basic conditions are

$$\boxed{q_{12}q_{21} \neq 1, \qquad q_{11}q_{12}q_{21} \neq 1, \qquad q_{12}q_{21}q_{22} \neq 1,}$$

$$\boxed{q_{11} \neq -1, \qquad q_{22} \in R_3,}$$

to which further conditions in 5.1.–5.5. are to be added one by one.

5.1 $q_0 := q_{11}q_{12}q_{21} \in R_{12}, q_{11} = q_0^4, q_{22} = -q_0^2$.

5.2 $q_{12}q_{21} \in R_{12}, q_{11} = q_{22} = -(q_{12}q_{21})^2$.

5.3 $q_{12}q_{21} \in R_{24}, q_{11} = (q_{12}q_{21})^{-6}, q_{22} = (q_{12}q_{21})^{-8}$.

5.4 $q_{11} \in R_{18}, q_{12}q_{21} = q_{11}^{-2}, q_{22} = -q_{11}^3$.

5.5 $q_{11} \in R_{30}, q_{12}q_{21} = q_{11}^{-3}, q_{22} = -q_{11}^5$.

7 Conclusions

Some additions to the above list (including the currently uninteresting items?!) might hopefully follow in the future. A variety of isolated central charge values for lower-rank W-algebras can be found in [56, 57] (also see [58]), with interesting possibilities of an overlap with the isolated values which I deemed uninteresting. I know nothing about a CFT counterpart of one "regular" case, 2.3 (a G(3) reduction?). More presentations of nonstandard type appeared recently in [59].

As a more specific result, the construction of an octuplet algebra—a W_3-counterpart of the $(1, p)$ triplet algebra—should be noted.

Acknowledgements It is a pleasure to acknowledge that these notes were compiled partly as a result of the Conference on Conformal Field Theory and Tensor Categories in Beijing. Very special thanks go to N. Andruskiewitsch, J. Fjelstad, J. Fuchs, A. Gainutdinov, M. Mombelli, I. Runkel, C. Schweigert, and A. Virelizier for the very useful discussions and, of course, to Yi-Zhi Huang and the other organizers for their hospitality in Beijing. I also thank T. Creutzig, I. Tipunin, and S. Wood for useful discussions and I. Angiono for comments, and the referee for suggestions. This paper was supported in part by the RFBR grant 11-01-00830 and the RFBR–CNRS grant 09-01-93105.

Appendix A: Virasoro Algebra

In CFT, the Virasoro algebra

$$[L_m, L_n] = (m-n)L_{m+n} + \frac{c}{12}(m-1)m(m+1), \qquad m, n \in \mathbb{Z}$$

standardly appears in the guise of an energy–momentum tensor $T(z) = \sum_{n\in\mathbb{Z}} L_n \times z^{-n-2}$—a (chiral) field on the complex plane that satisfies the OPEs

$$T(z)T(w) = \frac{c/2}{(z-w)^4} + \frac{2T(z)}{(z-w)^2} + \frac{\partial T(z)}{z-w}.$$

The c parameter (understood to be multiplied by the unit operator whenever necessary) is called the central charge.

For θ bosonic fields $\varphi(z) = (\varphi^1(z), \dots, \varphi^\theta(z))$ with the OPEs

$$\varphi^i(z)\varphi^j(w) = \delta^{ij}\log(z-w), \tag{16}$$

which are also frequently used in calculations in the form

$$\partial\varphi^i(z)\partial\varphi^j(w) = \frac{\delta^{ij}}{(z-w)^2},$$

the energy–momentum tensors are parameterized by $\xi \in \mathbb{C}^\theta$,

$$T_\xi(z) = \frac{1}{2}\partial\varphi(z).\partial\varphi(z) + \xi.\partial^2\varphi(z). \tag{17}$$

The corresponding central charge is

$$c_\xi = \theta - 12\xi.\xi. \tag{18}$$

The OPE of $T_\xi(z)$ with a vertex operator $e^{\mu.\varphi(z)}$ is

$$T_\xi(z)e^{\mu.\varphi(w)} = \frac{\Delta e^{\mu.\varphi(w)}}{(z-w)^2} + \frac{\partial e^{\mu.\varphi(w)}}{z-w}, \qquad \Delta = \frac{1}{2}\mu.\mu - \xi.\mu$$

A screening operator is, by definition, any expression $\oint V(\cdot)$, where $V(z)$ is a field of dimension $\Delta = 1$ (and the contour integration is essentially equivalent to taking a residue "after the action of $V(z)$ is evaluated"). For $\theta = 2$, any two exponentials $e^{\alpha.\varphi(z)}$ and $e^{\beta.\varphi(z)}$ with noncollinear $\alpha, \beta \in \mathbb{C}^2$ define screening operators with respect to the energy–momentum tensor

$$T(z) = \frac{1}{2}\partial\varphi(z).\partial\varphi(z)$$
$$- \frac{(2+\alpha.\beta - \alpha.\alpha)\beta.\beta - 2\alpha.\beta}{2\delta^2}\partial^2\varphi_\alpha(z)$$

$$-\frac{(2+\alpha.\beta-\beta.\beta)\alpha.\alpha-2\alpha.\beta}{2\delta^2}\partial^2\varphi_\beta(z),$$

where $\partial\varphi_\alpha(z)=\alpha.\partial\varphi(z)$ and $\partial\varphi_\beta(z)=\beta.\partial\varphi(z)$, and $\delta^2=(\alpha.\alpha)(\beta.\beta)-(\alpha.\beta)^2$. This gives formula (4) for the central charge.

Next, I show that the central charge of the θ-boson energy–momentum tensor that centralizers θ screenings $\oint e^{\alpha_i.\varphi(z)}$, $1\le i\le\theta$, with linearly independent momenta is invariant under Weyl reflections (8) if Eqs. (7) hold.

Given the α_i, $1\le i\le\theta$, the condition that all the exponentials $e^{\alpha_i.\varphi(z)}$ have dimension 1 is expressed by the system of equations for ξ

$$\frac{1}{2}\alpha_i.\alpha_i-\xi.\alpha_i=1,\quad 1\le i\le\theta.$$

With ξ written as $\xi=\sum_{j=1}^{\theta}x_j\alpha_j$, this becomes a system for the x_j,

$$\frac{1}{2}\alpha_i.\alpha_i-\sum_{j=1}^{\theta}x_j\alpha_j.\alpha_i=1,\quad 1\le i\le\theta,\tag{19}$$

uniquely solvable if the α_i are linearly independent.

Under a Weyl groupoid operation $\mathfrak{R}^{(k)}$ in (8), the scalar products change and the solution (x_j) also changes. The "old" and "new" central charges are

$$c=\theta-12\sum_{\ell,j=1}^{\theta}x_\ell x_j\alpha_\ell.\alpha_j\quad\text{and}\quad\mathfrak{R}^{(k)}(c)=\theta-12\sum_{\ell,j=1}^{\theta}\tilde{x}_\ell\tilde{x}_j\mathfrak{R}^{(k)}(\alpha_\ell.\alpha_j),$$

where the $\tilde{x}_j$ solve the system "$\mathfrak{R}^{(k)}((19))$." With $\tilde{x}_j=x_j+y_j$, this system becomes

$$\begin{aligned}&\frac{1}{2}\alpha_i.\alpha_i-a_{k,i}\alpha_i.\alpha_k+\frac{1}{2}a_{k,i}^2\alpha_k.\alpha_k\\&\qquad-\sum_{j=1}^{\theta}(x_j+y_j)(\alpha_j.\alpha_i-a_{k,i}\alpha_j.\alpha_k-a_{k,j}\alpha_k.\alpha_i+a_{k,j}a_{k,i}\alpha_k.\alpha_k)\\&\quad=1,\quad 1\le i\le\theta\end{aligned}$$

(for a chosen k). The claim is that this system of equations for the "deformation" of the original solution is solved by the ansatz $y_j=\delta_{j,k}y$. To see this, substitute such y_j and use (19) in the resulting equations, which then become

$$a_{k,i}\left(\frac{1}{2}\alpha_k.\alpha_k(a_{k,i}+1)-\alpha_i.\alpha_k-1\right)+\left(y+\sum_{j=1}^{\theta}x_ja_{k,j}\right)(\alpha_k.\alpha_i-a_{k,i}\alpha_k.\alpha_k)=0,$$

where $1 \leq i \leq \theta$. Remarkably, if (7) holds, then the above equations are indeed solved by

$$y = 1 - \frac{2}{\alpha_k.\alpha_k} - \sum_{j=1}^{\theta} x_j a_{k,j}. \tag{20}$$

It remains to find the new central charge. With $\tilde{x}_j = x_j + \delta_{j,k} y$,

$$\sum_{\ell,j=1}^{\theta} \tilde{x}_\ell \tilde{x}_j \mathfrak{R}^{(k)}(\alpha_\ell.\alpha_j) = \sum_{\ell,j=1}^{\theta} x_\ell x_j (\alpha_\ell.\alpha_j - 2a_{k,\ell}\alpha_j.\alpha_k + a_{k,l}a_{k,j}\alpha_k.\alpha_k)$$
$$+ 2\sum_{j=1}^{\theta} y x_j (a_{k,j}\alpha_k.\alpha_k - \alpha_k.\alpha_j) + y^2 \alpha_k.\alpha_k$$

and yet another use of (19) shows that this is

$$= \sum_{\ell,j=1}^{\theta} x_\ell x_j \alpha_\ell.\alpha_j + \Bigl(y + \sum_{j=1}^{\theta} x_j a_{k,j}\Bigr)\Bigl(y\alpha_k.\alpha_k + 2 - \alpha_k.\alpha_k + \sum_{\ell=1}^{\theta} x_\ell a_{k,\ell}\alpha_k.\alpha_k\Bigr),$$

where the last factor vanishes by virtue of (20). The central charge is invariant.

Appendix B: W_3 Logarithmic Octuplet Algebras

With the two screenings as in case 2.1 (the "regular" solution there, with central charge (9) and the $W(z)$ field (10)), I propose a W_3 counterpart of the $(1, p)$ algebra [28–31] by closely following the constructions in [18].

An *octuplet* of primary fields is generated from the field $e^{\gamma.\varphi(z)}$ with $\gamma \in \mathbb{C}^2$ such that $\gamma.\alpha^\vee = p$ and $\gamma.\beta^\vee = p$, i.e., from the field

$$\mathcal{W}(z) = e^{\gamma.\varphi(z)}, \quad \gamma = \alpha^\vee + \beta^\vee$$

(which *is* in the kernel of the two screenings $F_\alpha = \oint e^{\alpha.\varphi}$ and $F_\beta = \oint e^{\beta.\varphi}$). This is a Virasoro primary field of dimension $\Delta = 3p - 2$, that is,

$$L_n \mathcal{W}(z) = 0, \quad n \geq 1,$$
$$L_0 \mathcal{W}(z) = \Delta \mathcal{W}(z), \quad \Delta = 3p - 2,$$

and, moreover, a W_3 primary: as is easy to verify, the modes of the dimension-3 field $W(z) = \sum_{n\in\mathbb{Z}} W_n z^{-n-3}$ in (10) act on $\mathcal{W}(z)$ such that

$$W_n \mathcal{W}(z) = 0, \quad n \geq 0.$$

Then the long screenings (12) generate the octuplet

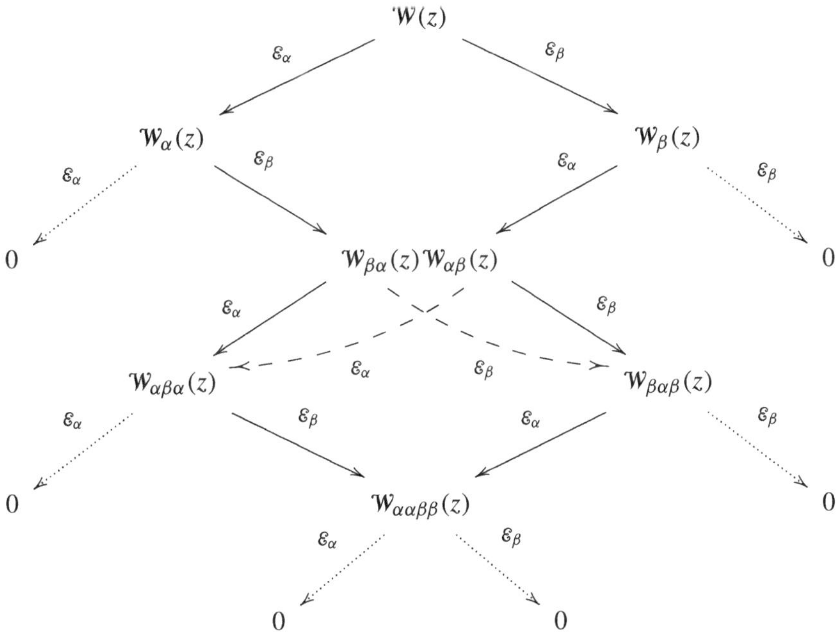

Here, $\mathcal{W}_\alpha(z) = \mathcal{E}_\alpha \mathcal{W}(z)$, $\mathcal{W}_{\beta\alpha}(z) = \mathcal{E}_\beta \mathcal{W}_\alpha(z)$, and so on, and $\mathcal{W}_{\alpha\alpha\beta\beta}(z) = \mathcal{E}_\beta \times \mathcal{W}_{\alpha\beta\alpha}(z) = \mathcal{E}_\alpha \mathcal{W}_{\beta\alpha\beta}(z)$; the dashed arrows represent maps to the target field up to a nonzero overall factor $(\frac{(-1)^p}{2})$. All the fields in the diagram are W_3-algebra primaries, with the same Virasoro dimension. All fields below the top are of the form $\mathcal{W}_\bullet(z) = \mathcal{P}_\bullet(\partial\varphi(z))e^{\mu_\bullet\cdot\varphi(z)}$, where the momenta $\mu_\bullet$ are immediately read off from the diagram as $\mu_\alpha = \gamma - \alpha^\vee$, $\mu_{\alpha\beta} = \mu_{\beta\alpha} = \gamma - \alpha^\vee - \beta^\vee = 0$, and so on, and the $\mathcal{P}_\bullet(\partial\varphi(z))$ are differential polynomials in $\partial\varphi_\alpha(z)$ and $\partial\varphi_\beta(z)$, of the orders $\mathrm{ord}(\mathcal{P}_\alpha) = \mathrm{ord}(\mathcal{P}_\beta) = p-1$, $\mathrm{ord}(\mathcal{P}_{\alpha\beta}) = \mathrm{ord}(\mathcal{P}_{\beta\alpha}) = 3p-2$, $\mathrm{ord}(\mathcal{P}_{\alpha\beta\alpha}) = \mathrm{ord}(\mathcal{P}_{\beta\alpha\beta}) = 3p-3$, and $\mathrm{ord}(\mathcal{P}_{\alpha\alpha\beta\beta}) = 4p-4$.

Calculations in particular examples show the OPE

$$\mathcal{W}(z)\mathcal{W}_{\alpha\alpha\beta\beta}(w) = \frac{c_1 \cdot 1}{(z-w)^{6p-4}} + \frac{c_2 T(w)}{(z-w)^{6p-6}} + \frac{c_2/2\partial T(w)}{(z-w)^{6p-7}} + \cdots$$

with *nonzero* coefficients (and no dimension-3 $W(w)$ field), and the OPEs $\mathcal{W}_\alpha(z) \times \mathcal{W}_{\beta\alpha\beta}(w)$ and $\mathcal{W}_\beta(z)\mathcal{W}_{\alpha\beta\alpha}(w)$ that start very similarly. The adjoint-$s\ell(3)$ nature of the octuplet manifests itself in the OPEs such as

$$\mathcal{W}_\alpha(z)\mathcal{W}_\beta(w) = \frac{c_3 \mathcal{W}(w)}{(z-w)^{3p-2}} + \cdots,$$

$$\mathcal{W}_\alpha(z)\mathcal{W}_{\alpha\beta\alpha}(w) = \mathcal{O}(z-w),$$

$$\mathcal{W}_{\beta}(z)\mathcal{W}_{\beta\alpha\beta}(w) = \mathcal{O}(z-w),$$
$$\mathcal{W}_{\alpha\beta\alpha}(z)\mathcal{W}_{\beta\alpha\beta}(w) = \frac{c_3'\mathcal{W}_{\alpha\alpha\beta\beta}(w)}{(z-w)^{3p-2}} + \cdots .$$

References

1. Semikhatov, A.M., Tipunin, I.Y.: The Nichols algebra of screenings. Commun. Contemp. Math. **14**, 1250029 (2012). arXiv:1101.5810
2. Nichols, W.D.: Bialgebras of type one. Commun. Algebra **6**, 1521–1552 (1978)
3. Andruskiewitsch, N., Graña, M.: Braided Hopf algebras over non abelian finite groups. Bol. Acad. Nacional de Ciencias (Cordoba) **63**, 45–78 (1999). arXiv:math/9802074
4. Andruskiewitsch, N., Schneider, H.-J.: Pointed Hopf algebras. In: New Directions in Hopf Algebras. MSRI Publications, vol. 43, pp. 1–68. Cambridge University Press, Cambridge (2002)
5. Andruskiewitsch, N.: Some remarks on Nichols algebras. In: Bergen, J., Catoiu, S., Chin, W. (eds.) Hopf Algebras, pp. 25–45. Dekker, New York (2004)
6. Andruskiewitsch, N., Schneider, H.-J.: On the classification of finite-dimensional pointed Hopf algebras. Ann. Math. **171**, 375–417 (2010). arXiv:math/0502157
7. Heckenberger, I.: Classification of arithmetic root systems. Adv. Math. **220**, 59–124 (2009). math/0605795
8. Andruskiewitsch, N., Heckenberger, I., Schneider, H.-J.: The Nichols algebra of a semisimple Yetter-Drinfeld module. Amer. J. Math. **132**, 1493–1547 (2010). arXiv:0803.2430
9. Andruskiewitsch, N., Radford, D., Schneider, H.-J.: Complete reducibility theorems for modules over pointed Hopf algebras. J. Algebra **324**, 2932–2970 (2010). arXiv:1001.3977
10. Heckenberger, I.: The Weyl groupoid of a Nichols algebra of diagonal type. Invent. Math. **164**, 175–188 (2006)
11. Graña, M., Heckenberger, I., Vendramin, L.: Nichols algebras of group type with many quadratic relations. arXiv:1004.3723
12. Graña, M., Heckenberger, I.: On a factorization of graded Hopf algebras using Lyndon words. J. Algebra **314**, 324–343 (2007)
13. Andruskiewitsch, N., Fantino, F., Garcia, G.A., Vendramin, L.: On Nichols algebras associated to simple racks. In: Groups, Algebras and Applications. Contemp. Math., vol. 537, pp. 31–56. Am. Math. Soc., Providence (2011). arXiv:1006.5727
14. Andruskiewitsch, N., Angiono, I., Yamane, H.: On pointed Hopf superalgebras. In: New Developments in Lie Theory and Its Applications. Contemp. Math., vol. 544. Am. Math. Soc., Providence (2011). arXiv:1009.5148
15. Angiono, I.: On Nichols algebras with standard braiding. Algebra Number Theory **3**, 35–106 (2009). arXiv:0804.0816
16. Angiono, I.E.: A presentation by generators and relations of Nichols algebras of diagonal type and convex orders on root systems. arXiv:1008.4144
17. Angiono, I.: On Nichols algebras of diagonal type. arXiv:1104.0268
18. Fuchs, J., Hwang, S., Semikhatov, A.M., Tipunin, I.Y.: Nonsemisimple fusion algebras and the Verlinde formula. Commun. Math. Phys. **247**, 713–742 (2004). hep-th/0306274
19. Heckenberger, I., Yamane, H.: A generalization of Coxeter groups, root systems, and Matsumoto's theorem. Math. Z. **259**, 255–276 (2008)
20. Heckenberger, I., Schneider, H.-J.: Right coideal subalgebras of Nichols algebras and the Duflo order on the Weyl groupoid. arXiv:0909.0293
21. Heckenberger, I.: Finite dimensional rank 2 Nichols algebras of diagonal type I: examples. arXiv:math/0402350

22. Heckenberger, I.: Finite dimensional rank 2 Nichols algebras of diagonal type II: classification. arXiv:math/0404008
23. Helbig, M.: On the lifting of Nichols algebras. arXiv:1003.5882
24. Cuntz, M., Heckenberger, I.: Weyl groupoids with at most three objects. J. Pure Appl. Algebra **213**, 1112–1128 (2009). arXiv:0805.1810
25. Cuntz, M., Heckenberger, I.: Finite Weyl groupoids of rank three. arXiv:0912.0212
26. Cuntz, M., Heckenberger, I.: Finite Weyl groupoids. arXiv:1008.5291
27. Andruskiewitsch, N., Angiono, I.E.: On Nichols algebras with generic braiding. In: Brzezinski, T., Gómez Pardo, J.L., Shestakov, I., Smith, P.F. (eds.) Modules and Comodules. Trends in Mathematics, pp. 47–64 (2008). arXiv:math/0703924
28. Kausch, H.G.: Extended conformal algebras generated by a multiplet of primary fields. Phys. Lett. B **259**, 448 (1991)
29. Gaberdiel, M.R., Kausch, H.G.: Indecomposable fusion products. Nucl. Phys. B **477**, 293–318 (1996). hep-th/9604026
30. Gaberdiel, M.R., Kausch, H.G.: A rational logarithmic conformal field theory. Phys. Lett. B **386**, 131–137 (1996). hep-th/9606050
31. Gaberdiel, M.R., Kausch, H.G.: A local logarithmic conformal field theory. Nucl. Phys. B **538**, 631–658 (1999). hep-th/9807091
32. Feigin, B.L., Gainutdinov, A.M., Semikhatov, A.M., Tipunin, I.Y.: Modular group representations and fusion in logarithmic conformal field theories and in the quantum group center. Commun. Math. Phys. **265**, 47–93 (2006). arXiv:hep-th/0504093
33. Feigin, B.L., Gainutdinov, A.M., Semikhatov, A.M., Tipunin, I.Y.: Logarithmic extensions of minimal models: characters and modular transformations. Nucl. Phys. B **757**, 303–343 (2006). arXiv:hep-th/0606196
34. Adamović, D., Milas, A.: On the triplet vertex algebra $W(p)$. Adv. Math. **217**, 2664–2699 (2008). arXiv:0707.1857v2
35. Adamović, D., Milas, A.: The $N = 1$ triplet vertex operator superalgebras. Commun. Math. Phys. **288**, 225–270 (2009). arXiv:0712.0379
36. Adamović, D., Milas, A.: Lattice construction of logarithmic modules for certain vertex algebras. arXiv:0902.3417
37. Feigin, B.L., Gainutdinov, A.M., Semikhatov, A.M., Tipunin, I.Y.: Kazhdan-Lusztig correspondence for the representation category of the triplet W-algebra in logarithmic CFT. Theor. Math. Phys. **148**, 1210–1235 (2006). arXiv:math/0512621
38. Feigin, B.L., Gainutdinov, A.M., Semikhatov, A.M., Tipunin, I.Y.: Kazhdan-Lusztig-dual quantum group for logarithmic extensions of Virasoro minimal models. J. Math. Phys. **48**, 032303 (2007). arXiv:math/0606506
39. Frenkel, E., Kac, V., Wakimoto, M.: Characters and fusion rules for W-algebras via quantized Drinfeld-Sokolov reduction. Commun. Math. Phys. **147**, 295–328 (1992)
40. Hamada, K.-j., Takao, M.: Spin-4 current algebra. Phys. Lett. B **209**, 247–251 (1988)
41. Figueroa-O'Farrill, J.M., Schrans, S., Thielemans, K.: On the Casimir algebra of B_2. Phys. Lett. B **263**, 378–384 (1991)
42. Ito, K., Terashima, S.: Free field realization of WBC_n and WG_2 algebras. Phys. Lett. B **354**, 220–231 (1995). hep-th/9503165
43. Kausch, H.G., Watts, G.M.T.: Duality in quantum Toda theory and W-algebras. Nucl. Phys. B **386**, 166–192 (1992). hep-th/9202070
44. Figueroa-O'Farrill, J.M., Schrans, S.: The spin 6 extended conformal algebra. Phys. Lett. B **245**, 471–476 (1990)
45. Zhu, C.-J.: The complete structure of the WG_2 algebra and its BRST quantization. hep-th/9508126
46. Bowcock, P., Feigin, B.L., Semikhatov, A.M., Taormina, A.: $\widehat{s\ell}(2|1)$ and $\widehat{D}(2|1;\alpha)$ as vertex operator extensions of dual affine $s\ell(2)$ algebras. Commun. Math. Phys. **214**, 495–545 (2000). hep-th/9907171
47. Feigin, B.L., Semikhatov, A.M.: $\mathcal{W}_n^{(2)}$ algebras. Nucl. Phys. B **698**, 409–449 (2004). math/0401164

48. Blumenhagen, R., Eholzer, W., Honecker, A., Hornfeck, K., Huebel, R.: Unifying W-algebras. Phys. Lett. B **332**, 51–60 (1994). hep-th/9404113
49. Blumenhagen, R., Eholzer, W., Honecker, A., Hornfeck, K., Huebel, R.: Coset realization of unifying W-algebras. Int. J. Mod. Phys. A **10**, 2367–2430 (1995). hep-th/9406203
50. Wakimoto, M.: Fock representations of the affine Lie algebra $A_1^{(1)}$. Commun. Math. Phys. **104**, 605–609 (1986)
51. Polyakov, A.M.: Gauge transformations and diffeomorphisms. Int. J. Mod. Phys. A **5**, 833 (1990)
52. Bershadsky, M.: Conformal field theories via Hamiltonian reduction. Commun. Math. Phys. **139**, 71 (1991)
53. Creutzig, T., Gao, P., Linshaw, A.R.: Fermionic coset, critical level $W_4^{(2)}$-algebra and higher spins. arXiv:1111.6603
54. Creutzig, T., Gao, P., Linshaw, A.R.: A commutant realization of $W_n^{(2)}$ at critical level. arXiv: 1109.4065
55. Creutzig, T., Ridout, D.: W-Algebras extending affine $gl(1|1)$. arXiv:1111.5049
56. Blumenhagen, R., Flohr, M., Kliem, A., Nahm, W., Recknagel, A., Varnhagen, R.: W-algebras with two and three generators. Nucl. Phys. B **361**, 255–289 (1991)
57. Blumenhagen, R., Eholzer, W., Honecker, A., Huebel, R.: New $N = 1$ extended superconformal algebras with two and three generators. Int. J. Mod. Phys. A **7**, 7841–7871 (1992). hep-th/9207072
58. Eholzer, W., Honecker, A., Huebel, R.: How complete is the classification of W-symmetries? Phys. Lett. B **308**, 42–50 (1993). hep-th/9302124
59. Angiono, I.: Nichols algebras of unidentified diagonal type. arXiv:1108.5157
60. Semikhatov, A.M., Tipunin, I.Y.: Logarithmic $\widehat{s\ell}(2)$ CFT models from Nichols algebras. 1. J. Phys. A, Math. Theor. (to appear). arXiv:1301.2235

Logarithmic Bulk and Boundary Conformal Field Theory and the Full Centre Construction

Ingo Runkel, Matthias R. Gaberdiel, and Simon Wood

Abstract We review the definition of bulk and boundary conformal field theory in a way suited for logarithmic conformal field theory. The notion of a maximal bulk theory which can be non-degenerately joined to a boundary theory is defined. The purpose of this construction is to obtain the more complicated bulk theories from simpler boundary theories. We then describe the algebraic counterpart of the maximal bulk theory, namely the so-called full centre of an algebra in an abelian braided monoidal category. Finally, we illustrate the previous discussion in the example of the $W_{2,3}$-model with central charge 0.

1 Introduction

In two-dimensional conformal field theory, one usually considers correlation functions where the fields have power law singularities as they approach each other, for example $\langle\sigma(z)\sigma(w)\rangle = |z-w|^{1/4}$ for the correlator of two spin fields in the critical Ising model. Power law behaviour occurs if the two fields approaching each other are eigenvectors of the generator of infinitesimal scale transformations $\Delta = L_0 + \overline{L}_0$. In unitary theories one has $\Delta^\dagger = \Delta$, so that Δ can be diagonalised. In non-unitary theories, however, there is no a priori reason to impose diagonalisability of Δ, and in this case additional logarithmic singularities can occur. For example, in the symplectic fermion model of [22] the two-point correlator of the partner of the vacuum state reads $\langle\omega(z)\omega(w)\rangle = 4\log|z-w|$.

I. Runkel (✉)
Fachbereich Mathematik, Universität Hamburg, Bundesstraße 55, 20146 Hamburg, Germany
e-mail: ingo.runkel@uni-hamburg.de

M.R. Gaberdiel
Institute for Theoretical Physics, ETH Zürich, 8093 Zürich, Switzerland
e-mail: gaberdiel@itp.phys.ethz.ch

S. Wood
IPMU, The University of Tokyo, 5-1-5 Kashiwanoha, Kashiwa, 277-8583, Japan
e-mail: simon.wood@ipmu.jp

C. Bai et al. (eds.), *Conformal Field Theories and Tensor Categories*,
Mathematical Lectures from Peking University, DOI 10.1007/978-3-642-39383-9_4,

From the point of view of representation theory, in unitary theories the state spaces are direct sums of irreducible representations of (two copies of) the Virasoro algebra, while in non-unitary theories the indecomposable summands may or may not be irreducible. In fact, if Δ is not diagonalisable one necessarily finds such non-semi-simple behaviour.

A general recipe for constructing examples of non-logarithmic rational conformal field theories, known as the 'Cardy case', is as follows. Take all irreducible representations R_i of the chiral symmetry, that is, of the algebra formed by all modes of all holomorphic fields, or, more formally, of a vertex operator algebra $\mathcal{V}$. The state space relevant to describe the theory on a cylinder, i.e. the space of states on the circle, is $\mathcal{H}_{\text{bulk}} = \bigoplus_i R_i \otimes_{\mathbb{C}} R_i^*$, where the sum runs over all irreducibles and R_i^* is the conjugate representation. This theory can be placed on a strip, in which case we have to fix the state space $\mathcal{H}_{\text{bnd}}$ on an interval with prescribed boundary conditions. If the two boundary conditions coincide one may take $\mathcal{H}_{\text{bnd}} = R \otimes_f R^*$, where R is an arbitrary representation of the chiral symmetry $\mathcal{V}$ (not necessarily irreducible) and '$\otimes_f$' denotes the fusion tensor product. In particular, if we take $R = \mathcal{V}$ then $\mathcal{H}_{\text{bnd}} = \mathcal{V} \otimes_f \mathcal{V}^* \cong \mathcal{V}$. That is, the space of boundary states consists of a single irreducible representation, namely the vacuum representation itself. This leads us to the first theme to keep in mind:

For an appropriate choice of boundary condition, the boundary theory is much simpler than the bulk theory.

It turns out that in all rational conformal field theories which can be defined on surfaces with or without boundary and which have a unique bulk vacuum, the boundary theory determines the bulk theory uniquely [13, 18, 33]. The bulk theory is characterised as the 'largest possible one' which can be matched to the given boundary theory. This principle has also been checked for some logarithmic models [23, 24, 26]. The second theme to keep in mind can be phrased as:

For a given boundary theory, one may find a largest possible bulk theory that can be consistently and non-degenerately joined to the boundary theory. This bulk theory, if it exists, is unique.

This principle has also been established in the operator algebraic approach to unitary conformal field theory on the half-plane with Minkowski signature [36]. (Logarithmic models are not accessible in this setting as the formalism requires unitarity.)

On the representation theoretic side, the construction of the bulk theory as the largest one which fits to a specific boundary theory corresponds to starting from an algebra in an abelian braided monoidal category $\mathcal{C}$ and finding its 'full centre', a commutative algebra in the product category $\mathcal{C} \boxtimes \mathcal{C}^{\text{rev}}$.

This paper consists of three parts. In part one, which is Sect. 2, the definition of bulk and boundary conformal field theory in terms of its correlation functions is reviewed. Using this definition, the characterisation of the bulk theory as the 'largest one' fitting to a given boundary theory is made precise.

Part two (Sect. 3) provides the algebraic counterparts of the conformal field theory notions in Sect. 2 in the setting of abelian braided monoidal categories. This part contains a fairly detailed review of the Deligne product of abelian categories, as this will play an important role. The main notion in part two is that of the full centre of an algebra. We will recall its definition, derive some of its properties, and link its definition to the maximality condition of the bulk theory associated to a boundary theory from Sect. 2.

Part three (Sect. 4) investigates one specific example of a logarithmic conformal field theory, namely the $W_{2,3}$-model of central charge zero. We chose this model because on the one hand it is still relatively simple, for example it only involves 13 distinct irreducible representations, but on the other hand each 'nice' property from non-logarithmic rational theories which is currently known to be violated in logarithmic models with a finite number of irreducibles is already violated in the $W_{2,3}$-model. We discuss properties of a tentative bulk theory for the $W_{2,3}$-model which can be interpreted as a 'logarithmic extension' of the underlying unitary minimal model at $c = 0$, i.e. the trivial theory with a one-dimensional state space. We consider the Virasoro action on states of generalised weight $(0, 0)$ and $(2, 0)$ and discuss the operator product expansion of some of these fields. We also find that an analogue of the indecomposability parameter b is equal to -5. This value has recently appeared in the discussion of bulk theories with $c = 0$ [51].

This paper grew out of two talks given by the first author which were based on the joint works [23–26]. We have tried to make this paper to some extent self-contained. In consequence it became slightly lengthy and contains a large amount of review material. Nonetheless, there are also some new results which we briefly list: the discussion of ideals for homomorphisms of conformal field theories in Sect. 2.2; the reformulation of the computation of the maximal bulk theory in purely categorical language in Sect. 3.5 and Table 2; the treatment in Sect. 3 of a class of abelian monoidal categories more general than finite tensor categories (as defined in [11]); the existence proof of the full centre in this setting in Theorem 3.24; the calculation of the analogue of the indecomposability parameter $b = -5$ in the $W_{2,3}$-bulk theory $R(\mathbf{1}^*)$ and the operator product expansions in this model in Sect. 4.5.

2 Bulk and Boundary Correlators

In this section we give a definition of conformal field theory on the complex plane and on the upper half plane in terms of correlation functions. The presentation is tailored to be self-contained and to make the relation to the algebraic concepts in Sect. 3 apparent.

Bibliographical Note This section is mostly a review. The characterisation of CFT on the complex plane in terms of correlators and operator product expansion is used in [2]. Axiomatic formulations close in spirit to the one presented below are [21, 28] (other approaches can be found in [16, 27, 32, 48, 49]). The point that the requirement of modular invariance poses severe constraints on a CFT was stressed in [3, 5]. CFT on the upper half plane as presented below was developed in [4, 6, 35]; an axiomatic formulation can be found in [33]. The idea to obtain the CFT on the complex plane from correlators of boundary fields was first implemented in [44, 45] and further developed in the context of non-logarithmic rational CFT in [13, 18, 34]. The first application of this principle to logarithmic models can be found in [23, 24].

2.1 Consistency Conditions for CFT on the Complex Plane

We will take the point of view that a two-dimensional conformal field theory (or any statistical or quantum field theory in any dimension, for that matter) is defined in terms of its correlation functions. That is, we are given a *space of fields* F, which is a $\mathbb{C}$-vector space whose elements we call *fields*. The space F is typically infinite dimensional, because with each field it contains all its derivatives (see Remark 2.5(iii) below for a precise statement). In addition we have a *collection of correlators* $(C_n)_{n\in\mathbb{Z}_{>0}}$. We call C_n an *n-point correlator*. It assigns a complex number to n fields and n mutually distinct complex numbers, i.e.

$$C_n : (\mathbb{C}^n \setminus \mathrm{diag}) \times F^n \longrightarrow \mathbb{C}, \tag{2.1}$$

where $\mathbb{C}^n \setminus \mathrm{diag}$ stands for points $(z_1, \dots, z_n) \in \mathbb{C}^n$ such that $z_i \neq z_j$ for $i \neq j$. The collection $(C_n)_{n\in\mathbb{Z}_{>0}}$ must obey conditions (C1)–(C5) which we discuss in the following.

(C1) Each C_n is smooth in each argument from $\mathbb{C}$ and linear in each argument from F.

Of course, in practice the correlators may satisfy stronger properties than smoothness. For example they could be holomorphic, or finite sums of holomorphic times anti-holomorphic functions (as in the example in Sect. 4). 'Smooth' is the minimal condition compatible with the physical requirement that the correlators should be continuous, and with condition (C4) below which states that the derivative of a correlator is again a correlator

(C2) Each C_n is invariant under joint permutation of the arguments in $\mathbb{C}^n$ and F^n, i.e. for each permutation $\sigma \in S_n$,

$$C_n(z_1, \dots, z_n, \phi_1, \dots, \phi_n) = C_n(z_{\sigma(1)}, \dots, z_{\sigma(n)}, \phi_{\sigma(1)}, \dots, \phi_{\sigma(n)}).$$

The customary notation for a correlator is

$$C_n(z_1, \dots, z_n, \phi_1, \dots, \phi_n) \equiv \langle \phi_1(z_1) \cdots \phi_n(z_n) \rangle, \tag{2.2}$$

where for us the right hand side is just a notational device. In particular, we do not assign an independent meaning to $\phi(z)$ as an operator. Still, we will say 'the field ϕ is inserted at position z' if the pair ϕ, z is an argument of a correlator.

We now turn to the notion of a 'short distance expansion' or 'operator product expansion'.[1] The OPE links the $n+1$-point and n-point correlators. Namely, if two fields, say ϕ_1 and ϕ_2, are 'close together' in the sense that z_1 is closer to z_2 than any other insertion point, then

$$\langle \phi_1(z_1)\phi_2(z_2)\phi_3(z_3)\cdots \rangle = \sum_\alpha f^\alpha_{\phi_1,\phi_2}(z_1 - z_2) \cdot \langle \varphi_\alpha(z_2)\phi_3(z_3)\cdots \rangle. \tag{2.3}$$

Here φ_α is some basis of F and $f^\alpha_{\phi_1,\phi_2}(x)$ are functions which do not depend on how many or which fields are part of the correlator, apart from ϕ_1 and ϕ_2.

More formally, we demand that F is a direct sum $F = \bigoplus_{\Delta\in\mathbb{R}} F^{(\Delta)}$ where $F^{(\Delta)}$ are the fields of 'generalised scaling dimension' Δ. The set of such scaling dimensions must be bounded below and discrete,[2] that is, for any Δ_0 there are only finitely many $\Delta \le \Delta_0$ with $F^{(\Delta)} \ne 0$. We define $\overline{F}$ to be the algebraic completion, i.e. the direct product $\overline{F} = \prod_{\Delta\in\mathbb{R}} F^{(\Delta)}$. The OPE is a map

$$M : \mathbb{C}^\times \times F \otimes_\mathbb{C} F \longrightarrow \overline{F}, \qquad (z, v) \mapsto M_z(v), \tag{2.4}$$

which is linear in $F \otimes_\mathbb{C} F$. In the notation of (2.3), this amounts to writing $M_x(\phi_1 \otimes \phi_2) = \sum_\alpha f^\alpha_{\phi_1,\phi_2}(x) \cdot \varphi_\alpha$, where the sum is typically infinite, hence the need for a completion. Note that $\overline{F}$ comes with canonical projections to 'states with scaling dimension Δ or less', $P_\Delta : \overline{F} \to \bigoplus_{d\le\Delta} F^{(d)}$. With the help of these, we formulate the OPE condition:

(C3) For $n \ge 1$, $\phi_1, \dots, \phi_{n+1} \in F$, and $(z_1, \dots, z_{n+1}) \in \mathbb{C}^{n+1}\setminus\mathrm{diag}$ such that $|z_1 - z_2| < |z_k - z_2|$ for $k > 2$, we have

$$\begin{aligned} &C_{n+1}(z_1, z_2, z_3, \dots, z_{n+1}, \phi_1, \phi_2, \phi_3, \dots, \phi_{n+1}) \\ &\quad= \lim_{\Delta\to\infty} C_n\big(z_2, z_3, \dots, z_{n+1}, \\ &\qquad P_\Delta \circ M_{z_1-z_2}(\phi_1 \otimes \phi_2), \phi_3, \dots, \phi_{n+1}\big). \end{aligned} \tag{2.5}$$

The limiting procedure in (2.5) is necessary because C_n is defined only on F, not on $\overline{F}$. The existence of the limit is a non-trivial requirement. In fact, if $|z_1 - z_2| \ge |z_k - z_2|$ for some $k > 2$, the expression on the right will typically diverge for $\Delta \to \infty$. That (C3) is only formulated for the first two arguments of C_{n+1} is not a restriction due to the permutation invariance imposed in (C2).

---

[1] In our setting only the first term makes sense literally, but it is customary to use the second term and abbreviate it as OPE, so we will do the same.

[2] That the set of scaling dimensions is bounded below amounts to the physical requirement that the energy should be bounded from below. It is used for example in (C5) to ensure that the sums there are finite. The discreteness condition will be needed to formulate (C3), and in particular to define the projector P_Δ there. The discreteness condition is imposed for simplicity; it rules out all examples with continuous spectrum of scaling dimensions, such as Liouville theory.

Remark 2.1 In addition to (C3), one often requires the existence of a translation invariant vacuum vector, that is, a vector $\Omega \in F^{(0)}$ such that $\langle \Omega(\zeta)\phi_1(z_1)\cdots \phi_n(z_n)\rangle = \langle \phi_1(z_1)\cdots\phi_n(z_n)\rangle$ for $n \geq 1$. We prefer not to include this as an axiom because our example in Sect. 4 below (conjecturally) satisfies (C1)–(C3), as well as (C4) and (C5′) to be discussed below, while not having a vacuum vector.

Finally, let us describe the coinvariance conditions. Denote by Vir the Virasoro algebra. We demand the following properties of F:

- F is equipped with the structure of a Vir ⊕ Vir-module. The generators of the first copy of Vir are denoted by L_n and C, and those of the second copy by $\overline{L}_n$ and $\overline{C}$.
- F has a direct sum decomposition into spaces $F^{(\Delta)}$ of generalised $(L_0 + \overline{L}_0)$-eigenvalue Δ; this decomposition satisfies that for any Δ_0 there are only finitely many $\Delta \leq \Delta_0$ with $F^{(\Delta)} \neq 0$. (This was already imposed above.)
- F is locally finite as a $\mathbb{C}L_0 \oplus \mathbb{C}\overline{L}_0$ module. This means that acting with L_0 and $\overline{L}_0$ on any vector $v \in F$ generates a finite-dimensional subspace.

The last condition guarantees in particular that the exponentials $\exp(\lambda L_0)$ and $\exp(\lambda \overline{L}_0)$, for $\lambda \in \mathbb{C}$, are well defined operators on F. The condition holds automatically if all $F^{(\Delta)}$ are finite-dimensional.

There are two types of coinvariance conditions. The first one is easy to formulate and allows one to move insertion points:

(C4) For $n \geq 1$, $\phi_1, \ldots, \phi_n \in F$, and $z_1, \ldots, z_n \in \mathbb{C}^n \setminus \text{diag}$,

$$\begin{aligned} \frac{d}{dz_1} C_n(z_1, \ldots, z_n, \phi_1, \ldots, \phi_n) &= C_n(z_1, \ldots, z_n, L_{-1}\phi_1, \ldots, \phi_n), \\ \frac{d}{d\bar{z}_1} C_n(z_1, \ldots, z_n, \phi_1, \ldots, \phi_n) &= C_n(z_1, \ldots, z_n, \overline{L}_{-1}\phi_1, \ldots, \phi_n). \end{aligned} \tag{2.6}$$

By permutation invariance, the fact that (C4) is formulated only for the first argument only is not a restriction.

The second type of coinvariance condition is a bit more involved. Let f be a meromorphic function on $\mathbb{C} \cup \{\infty\}$ (i.e. a rational function) which has poles at most at the points $z_1, \ldots, z_n$ and ∞, and which satisfies the growth condition $\lim_{\zeta\to\infty} \zeta^{-3} f(\zeta) = 0$. Denote the expansion parameters around each of the z_k as $f(\zeta) = \sum_{m=-\infty}^{\infty} f_m^k \cdot (\zeta - z_k)^{m+1}$.

(C5) For $n \geq 1$, $\phi_1, \ldots, \phi_n \in F$, and $z_1, \ldots, z_n \in \mathbb{C}^n \setminus \text{diag}$, and for all f as above,

$$\sum_{k=1}^{n} \sum_{m=-\infty}^{\infty} f_m^k \cdot C_n(z_1, \ldots, z_n, \phi_1, \ldots, L_m\phi_k, \ldots, \phi_n) = 0, \tag{2.7}$$

and the same condition with $\overline{L}_m$ in place of L_m.

The sum over m in (2.7) is actually finite: Since f is meromorphic, $f_m^k = 0$ for $m \ll 0$, and since the grading by generalised scaling dimensions on F is bounded from below, $L_m\phi_k = 0$ for $m \gg 0$.

Remark 2.2 In place of (C5) one could put the stronger requirement of the existence of a stress tensor. This would be a pair of fields $T, \overline{T} \in F^{(2)}$ (called the holomorphic and anti-holomorphic component of the stress tensor) such that $\overline{L}_{-1}T = 0$ and $L_{-1}\overline{T} = 0$, and

$$M_z(T \otimes \phi) = \sum_{m=-\infty}^{\infty} z^{-m-2} L_m \phi, \qquad M_z(\overline{T} \otimes \phi) = \sum_{m=-\infty}^{\infty} \bar{z}^{-m-2} \overline{L}_m \phi. \quad (2.8)$$

Note that $M_z(T \otimes \phi)$ and $M_z(\overline{T} \otimes \phi)$ are elements of $\overline{F}$, as they should be. Furthermore, one requires that the limit $\lim_{\zeta \to \infty} |\zeta|^4 \langle T(\zeta)\phi_1(z_1) \cdots \phi_n(z_n)\rangle$ exists for all n and all z_i, ϕ_i, and similar for $\overline{T}$ (this is $sl(2, \mathbb{C})$-invariance of the out vacuum). The conditions (2.7) arise from the contour integral

$$\frac{1}{2\pi i} \oint f(\zeta) \cdot \big\langle T(\zeta)\phi_1(z_1) \cdots \phi_n(z_n)\big\rangle d\zeta = 0 \quad (2.9)$$

where the contour is a big circle enclosing $z_1, \ldots, z_n$. Deforming the contour to a union of small circles, one around each z_i, and applying the OPE (2.8) results in (2.7). Not requiring the existence of a stress tenser is akin to not requiring the existence of a vacuum vector (cf. Remark 2.1): the correlators in the example in Sect. 4 conjecturally satisfy the convariance condition (C5) without having a stress tensor.

Remark 2.3

(i) One consequence of (C5) is that all correlators are translation invariant,

$$\big\langle \phi_1(z_1 + s) \cdots \phi_n(z_n + s)\big\rangle = \big\langle \phi_1(z_1) \cdots \phi_n(z_n)\big\rangle \quad \text{for all } s \in \mathbb{C}. \quad (2.10)$$

To see this apply (C5) to the constant function $f = 1$, in which case $f^k_m = \delta_{m,-1}$ for $k = 1, \ldots, n$ and so $\sum_{k=1}^n C_n(z_1, \ldots, z_n, \phi_1, \ldots, L_{-1}\phi_k, \ldots, \phi_n) = 0$. Combining the corresponding relation for $\overline{L}_{-1}$ with (C4) yields translation invariance. Along the same lines one shows covariance (not invariance) under Möbius transformations which map none of the points $z_1, \ldots, z_n$ to infinity.

(ii) A CFT on $\mathbb{C}$ can be used to define n-point correlators on the Riemann sphere. This is done by choosing an isomorphism from the Riemann sphere to $\mathbb{C} \cup \{\infty\}$ such that no field insertion gets mapped to infinity and by then evaluating C_n on the resulting configuration. (One also needs to include local coordinates around the insertions, we skip the details.)

Since the OPE allows one to reduce[3] C_{n+1} to C_n, all correlators are uniquely determined by the OPE and C_1. By translation invariance, $C_1(z, \phi)$ is independent

[3]Of course, the OPE can only be applied if the condition on the distances of insertion points in (C3) is met. But one can always choose a pair z_i, z_j of distinct points such that $|z_i - z_j|$ is minimal among all distances between pairs of insertion points. If necessary, one can then pick a point z_i' arbitrarily close to z_i such that $|z_i' - z_j|$ is strictly smaller than all other distances. The OPE (C3) applies to the pair z_i', z_j and the value of the correlator at z_i is determined by continuity.

of z and thus yields a function $\Omega^* : F \to \mathbb{C}$. It follows from (C5) with $f(\zeta) = (\zeta - z)^{m+1}$ that

$$C_1(z, L_m\phi) = 0 = C_1(z, \overline{L}_m\phi) \quad \text{for all } m \leq 1. \tag{2.11}$$

If $\phi \in F^{(\Delta)}$, then by definition $(L_0 + \overline{L}_0 - \Delta)^N\phi = 0$ for some $N > 0$. This gives $0 = C_1(z, (L_0 + \overline{L}_0 - \Delta)^N\phi) = (-\Delta)^N C_1(z, \phi)$, because, as we just saw, $C_1(z, (L_0 + \overline{L}_0)^k\phi) = 0$ for all $k > 0$. It follows that $C_1(z, \phi)$ can be non-zero only if $\Delta = 0$.

Let us collect the discussion so far into a definition.

Definition 2.4 A *conformal field theory on the complex plane* is a triple (F, M, Ω^*), where

- F (the *space of fields*) is a $\mathrm{Vir} \oplus \mathrm{Vir}$-module which is a direct sum of generalised $(L_0 + \overline{L}_0)$-eigenspaces $F^{(\Delta)}$ whose generalised eigenvalues are bounded from below and discrete, and which is locally finite as a $\mathbb{C}L_0 \oplus \mathbb{C}\overline{L}_0$ module,
- M (the *operator product expansion*) is a function $\mathbb{C}^\times \times F \otimes_{\mathbb{C}} F \to \overline{F}$ which is linear in $F \otimes_{\mathbb{C}} F$,
- Ω^* (the *out-vacuum*) is a linear map $F^{(0)} \to \mathbb{C}$,

such that there exists a collection of correlators $(C_n)_{n\in\mathbb{Z}_{>0}}$ which satisfy (C1)–(C5) and the normalisation condition $C_1(z, \phi) = \langle \Omega^*, \phi \rangle$.

Remark 2.5

(i) The definition shows that a CFT contains only a relatively small amount of data which has to satisfy an infinite number of intricate linear and differential equations. It is in fact very hard to prove that a triple (F, M, Ω^*) gives a CFT. To some extent, the formalism of vertex operator algebras, its representations and intertwining operators was developed with this aim. The VOA formalism allows one to prove that non-logarithmic rational CFTs provide examples of Definition 2.4, see [28]. We are not aware of a full proof of the existence of a logarithmic CFT in the above sense, e.g. using the formalism [30]. (This is merely to indicate that logarithmic CFTs are more difficult, not that we doubt their existence.)

(ii) We have deliberately not included non-degeneracy of the 2-point correlator $\langle \phi(z)\psi(w) \rangle$ into Definition 2.4; this will be discussed in the next subsection.

(iii) If the space F is finite-dimensional, then the $\mathrm{Vir} \oplus \mathrm{Vir}$-action on F has to be trivial,[4] and so in particular L_{-1} and $\overline{L}_{-1}$ would act trivially on F. By (C5)

[4] All finite dimensional Vir-modules M are trivial. The proof is easy. The Jordan normal form of L_0 splits M into generalised L_0-eigenspaces. Let $\Lambda \geq 0$ be such that all generalised L_0-eigenvalues have real parts of absolute value less or equal to Λ. Since L_m changes the generalised L_0-eigenvalue by $-m$, all L_m with $|m| > 2\Lambda$ must act as zero. For $m \neq 0$ and $2N + m \neq 0$ we can write $L_m = [L_{N+m}, L_{-N}]/(2N + m)$. For N large enough, both L_N and L_{m-N} act trivially on M, and so all L_m with $m \neq 0$ must act trivially. Therefore, also $L_0 = \frac{1}{2}[L_1, L_{-1}]$ and $C = 2[L_2, L_{-2}] - 4[L_1, L_{-1}]$ act trivially.

this implies that all correlators are independent of the insertion points. Such a conformal field theory is called a *topological field theory*.

(iv) As mentioned in the initial bibliographical remark, the above axiomatic approach is closest in spirit to [2, 21, 28]. All of these require that the generator of scale transformations $L_0 + \overline{L}_0$ acts diagonally and with eigenvalue Δ on $F^{(\Delta)}$, and so they do not accommodate logarithmic theories. Some similarities and further differences are:

[2]: There, (C1), (C2) are implicitly assumed, and also (C3) is implicit when requiring the existence of an OPE. The existence of a stress tensor is demanded, which then entails (C4) and (C5).

[21]: The emphasis is on constructing a (meromorphic) CFT from a generating set of fields and their correlators. (C2) is assumed, and (C1) is replaced by the much stronger condition that all correlators should be analytic in the insertion points. The OPE and (C3) are a consequence of the construction (see Theorem 3 there). The covariance conditions (C4) and (C5) are imposed only for the Möbius group (that is, for L_0, $L_{\pm 1}$). Accordingly, existence of a stress tensor is not required (but the consequences of the existence of a stress tensor are investigated in Sect. 7 there).

[28]: There, the notion of a 'conformal full field algebra' is introduced (see Definitions 1.1 and 1.19). (C1) and (C2) are stated in Definition 1.1, and (C3) is replaced by the stronger 'convergence property' which involves multiple simultaneous OPEs. In Definition 1.19, the existence of a stress tensor is imposed, from which (C4) and (C5) follow.

The present set of axioms (C1)–(C5) is intended to be a minimal set of conditions which one would want to require from a conformally invariant theory.

Suppose (F, M, Ω^*) is a conformal field theory. By assumption there exists a collection of correlators $(C_n)_{n\in\mathbb{Z}_{>0}}$ satisfying (C1)–(C5) and we have seen above that this determines the C_n uniquely. As a small example computation with the above axioms, let us look at $\langle\phi(z)\psi(w)\rangle$. By translation invariance, we may assume $w = 0$. By (C3),

$$\begin{aligned}\big\langle\phi(z)\psi(0)\big\rangle &= \lim_{\Delta\to\infty} C_1\big(0, P_\Delta \circ M_z(\phi\otimes\psi)\big)\\ &= \big\langle\Omega^*, M_z(\phi\otimes\psi)\big\rangle.\end{aligned} \tag{2.12}$$

The limit can be dropped because Ω^* is non-vanishing only on $F^{(0)}$. Next, by (C5) with $f(\zeta) = \zeta$ we know that $C_2(z, 0, (L_0 + zL_{-1})\phi, \psi) + C_2(z, 0, \phi, L_0\psi) = 0$, together with (C4) we find

$$-z\frac{d}{dz}\big\langle\Omega^*, M_z(\phi\otimes\psi)\big\rangle = \big\langle\Omega^*, M_z(L_0\phi\otimes\psi)\big\rangle + \big\langle\Omega^*, M_z(\phi\otimes L_0\psi)\big\rangle \tag{2.13}$$

and a corresponding equation with $\frac{d}{d\bar{z}}$ and $\overline{L}_0$. The solution to these first order differential equations reads

$$\begin{aligned}\big\langle \phi(z)\psi(0)\big\rangle = \big\langle \Omega^*, M_1 \circ \exp\big\{ &- \ln(z)(L_0 \otimes id_F + id_F \otimes L_0) \\ &- \ln(\bar{z})(\overline{L}_0 \otimes id_F + id_F \otimes \overline{L}_0)\big\}\phi \otimes \psi\big\rangle.\end{aligned} \tag{2.14}$$

From this we see two things: when evaluated on Ω^*, M_z is uniquely fixed by M_1; and if the action of L_0 or $\overline{L}_0$ has a nilpotent part, the two-point correlators may contain logarithms. In sub-representations of F which are irreducible, L_0 acts diagonalisably since $\exp(2\pi i L_0)$ commutes with all Virasoro modes, and hence by Schur's Lemma has to be a multiple of the identity. In this sense, the appearance of logarithms is linked to (but not equivalent to) the presence of non-semi-simple Vir-modules.

2.2 Background States, Non-degeneracy, and Ideals

We would like to allow more general out-states—or background states—than the out-vacuum Ω^*, namely, we would like to be able to place an arbitrary state from the graded dual of F 'at infinity'. The graded dual of F is defined as

$$F' = \big\{ u : F \to \mathbb{C} \text{ linear} \mid \exists \Delta_{\max}(u) : u\big(F^{(\Delta)}\big) = 0 \text{ for } \Delta > \Delta_{\max}(u)\big\}. \tag{2.15}$$

The graded dual is again a Vir $\oplus$ Vir-module via $(L_m u)(v) := u(L_{-m}v)$ and $(\overline{L}_m u)(v) := u(\overline{L}_{-m}v)$. With this definition, the generalised $(L_0 + \overline{L}_0)$-eigenvalues of F' are the same as those of F and each element $u \in F'$ is annihilated by L_m and $\overline{L}_m$ for large enough $m > 0$.

Define a *CFT on* $\mathbb{C}$ *with background states* as a pair (F, M), where F and M are as in Definition 2.4. However, for each $u \in F'$ we now demand the existence of functions $C_n(u|z_1, \dots, z_n, \phi_1, \dots, \phi_n)$, which we will also write as

$$^u\big\langle \phi_1(z_1) \cdots \phi_n(z_n)\big\rangle, \tag{2.16}$$

and which have to satisfy the following conditions:

- $^u\langle\phi(0)\rangle = u(\phi)$ for all $u \in F'$, $\phi \in F$.
- (C1)–(C4) from before, but with $C_n(u|\cdots)$ in place of $C_n(\cdots)$.
- (C5′), which is a modified version of (C5) to be described now.

Let f be a rational function on $\mathbb{C} \cup \{\infty\}$ as for (C5), but without imposing the growth condition at infinity. Define the f_m^k as for (C5) and define f_m^∞ via the expansion around infinity: $f(\zeta) = \sum_{m=-\infty}^{\infty} f_m^\infty \zeta^{m+1}$ for $|\zeta|$ larger than all of the $|z_i|$.

(C5′) For $n \geq 1$, $u \in F'$, $\phi_1, \ldots, \phi_n \in F$, and $z_1, \ldots, z_n \in \mathbb{C}^n \setminus \text{diag}$, and for all f as above,

$$\sum_{k=1}^{n} \sum_{m=-\infty}^{\infty} f_m^k \cdot C_n(u|z_1, \ldots, z_n, \phi_1, \ldots, L_m\phi_k, \ldots, \phi_n)$$
$$= \sum_{m=-\infty}^{\infty} f_{-m}^{\infty} \cdot C_n(L_m u|z_1, \ldots, z_n, \phi_1, \ldots, \phi_n) \tag{2.17}$$

and the same condition with $\overline{L}_m$ in place of L_m.

As for (C5), the sums in (2.17) only involve a finite number of non-zero terms.

Let (F, M) be a CFT on $\mathbb{C}$ with background states. Let $\Omega^* \in F'$ be a primary $sl(2, \mathbb{C})$-invariant state, that is, $L_m\Omega^* = 0 = \overline{L}_m\Omega^*$ for all $m \geq -1$. Then (F, M, Ω^*) is a CFT on $\mathbb{C}$ in the sense of Definition 2.4, with correlators $C_n(\Omega^*|\cdots)$. Indeed, (C5′) reduces to (C5) if we fix u to be Ω^* and impose the growth condition $\lim_{\zeta\to\infty} \zeta^{-3} f(\zeta) = 0$.

Remark 2.6 As was noted in Remark 2.5(iii), when F is a trivial Vir $\oplus$ Vir-module, (F, M) is a topological field theory. One can easily convince oneself that then the pair (F, M) is just a commutative, associative algebra with multiplication $M : F \otimes F \to F$ (F is concentrated in grade 0, so $\overline{F} = F$, and M is position independent). Indeed, a useful way to think about a conformal field theory on the complex plane is as a generalisation of a commutative, associative algebra where the product depends on a non-zero complex parameter.

Continuing the analogy with algebra, let us define a *homomorphism of CFTs* (F, M) *and* (G, N) to be a Vir $\oplus$ Vir-intertwiner $f : F \to G$ such that $f \circ M_x = N_x \circ (f \otimes f)$. Since $f(F^\Delta) \subset G^{(\Delta)}$, the map f is well-defined as a map $\overline{F} \to \overline{G}$. By an *ideal* in F we mean a Vir $\oplus$ Vir-submodule I of F such that for all $\iota \in I$, $\phi \in F$ and $x \in \mathbb{C}^\times$ we have $M_x(\iota \otimes \phi) \in \overline{I}$ and $M_x(\phi \otimes \iota) \in \overline{I}$ (actually one of the two conditions implies the other). The kernel of a homomorphism is an ideal. Given an ideal $I \subset F$, we obtain a CFT on the quotient F/I such that the canonical projection $\pi : F \to F/I$ is a homomorphism of CFTs.

Another class of examples of ideals is the following. Let (F, M, Ω^*) be a CFT on $\mathbb{C}$. Let F_0 be the kernel of the 2-point correlator, i.e. fix $z \neq w$ and define

$$F_0 = \left\{\eta \in F \middle| \left\langle \phi(z)\eta(w) \right\rangle = 0 \text{ for all } \phi \in F\right\}. \tag{2.18}$$

From (2.14) one concludes that F_0 is independent of z, w. It follows from (C5) that F_0 is a Vir $\oplus$ Vir-submodule of F. Let $\eta \in F_0$ and $\phi, \psi \in F$. By expressing the 3-point correlator $\langle \eta(x)\phi(y)\psi(z)\rangle$ as a limit of two-point correlators via (C3) in two ways, one involving $M_{x-y}(\eta \otimes \phi)$ and one $M_{y-z}(\phi \otimes \psi)$, one sees that F_0 is an ideal in F. Again because of (C3), a correlator $\langle \phi_1(z_1) \cdots \phi_n(z_n)\rangle$ is zero if at least one of the ϕ_i is from F_0.

Definition 2.7 A conformal field theory on the complex plane $(F; M, \Omega^*)$ is *non-degenerate* if F_0 as defined in (2.18) is $\{0\}$.

Remark 2.8

(i) If $\Omega^*(F_0) = \{0\}$, the CFT on the quotient F/F_0 has an out-vacuum induced by Ω^* and is non-degenerate. On the level of correlators, one cannot tell the difference between F and F/F_0 and hence it is common to restrict one's attention to non-degenerate CFTs on $\mathbb{C}$. However, the device of background states allows one to obtain interesting correlators also for degenerate CFTs.

(ii) Let $f : F \to G$ be a homomorphism of the CFTs (F, M) and (G, N). Let $\Gamma^* \in G'$ be a primary $sl(2, \mathbb{C})$-invariant state. Then $\Omega^* := \Gamma^* \circ f$ is a primary $sl(2, \mathbb{C})$-invariant state in F'. Because of $f \circ M_x = N_x \circ (f \otimes f)$, the correlators of (F, M, Ω^*) and (G, N, Γ^*) are related by

$$\big\langle \phi_1(z_1) \cdots \phi_n(z_n) \big\rangle_F = \big\langle \phi_1'(z_1) \cdots \phi_n'(z_n) \big\rangle_G, \quad \text{where } \phi_i' = f(\phi_i). \tag{2.19}$$

(iii) With the notation of (ii), if $\ker(f) \neq \{0\}$, it follows from (2.19) that the CFT (F, M, Ω^*) is necessarily degenerate. Explicitly, $\langle \phi(z)\psi(w) \rangle_F = \langle \Omega^*, M_{z-w}(\phi \otimes \psi) \rangle = \langle \Gamma^*, f \circ M_{z-w}(\phi \otimes \psi) \rangle = \langle \Gamma^*, N_{z-w}(f(\phi) \otimes f(\psi)) \rangle = \langle \phi'(z)\psi'(w) \rangle_G$, so that $\ker(f) \subset F_0$.

(iv) If there is an isomorphism $f : F \to F'$ of Vir$\oplus$Vir-modules, one can define the non-degenerate pairing $(u, v) = {}^{f(u)}\langle v(0) \rangle = \langle f(u), v \rangle$ on $F \times F$. This pairing is invariant in the sense that $(L_m u, v) = (u, L_{-m} v)$ and $(\overline{L}_m u, v) = (u, \overline{L}_{-m} v)$ for all $u, v \in F$ and $m \in \mathbb{Z}$. In this situation one can also ask if the inversion $z \mapsto 1/z$ is a symmetry of the theory, i.e. if, for all $\phi_i, \psi_i \in F$,

$$^{f(\psi_1)}\big\langle \phi_1(z_1) \cdots \phi_n(z_n) \psi_2(0) \big\rangle = {}^{f(\psi_2)}\big\langle \phi_1'(1/z_1) \cdots \phi_n'(1/z_n) \psi_1(0) \big\rangle, \tag{2.20}$$

where (see, e.g., Sect. 3.2 in [20])

$$\phi_i' = \exp\big(\ln(-z_i^{-2}) L_0 + \ln(-\bar{z}_i^{-2}) \overline{L}_0\big) \exp\big(-z_i^{-1} L_1 - \bar{z}_i^{-1} \overline{L}_1\big) \phi_i. \tag{2.21}$$

An inversion-covariant CFT with background states provides us with an alternative way to define correlators on the Riemann sphere as compared to Remark 2.3(ii). For a Riemann sphere with two or more insertions, choose an isomorphism with $\mathbb{C} \cup \{\infty\}$ which maps one of the insertion points to infinity and evaluate the resulting configuration with $C_n(f(\cdot)|\cdots)$. Different such choices are related by a Möbius transformation which maps one of the insertion points (including infinity) to infinity. (As in Remark 2.3(ii) one needs to choose local coordinates around the insertions, we skip the details.)

In Sect. 4.5, we will encounter the special situation of a (conjectural) CFT on $\mathbb{C}$ with background states (F, M) which has a surjective homomorphism $\pi : F \to \mathbb{C}$ to the trivial CFT $(\mathbb{C}, \cdot)$ with one-dimensional state space, where '$\cdot$' stands for the product on $\mathbb{C}$. Because π is a Vir $\oplus$ Vir-intertwiner, this situation can only occur for $c = 0$. If we take $\Omega^* = \pi$ (which is a primary $sl(2,\mathbb{C})$-invariant state in F') as out-vacuum, the correlators of (F, M, Ω^*) satisfy

$$\langle \phi_1(z_1)\phi_2(z_2)\cdots\phi_n(z_n)\rangle_F = \pi(\phi_1)\cdot\pi(\phi_2)\cdots\pi(\phi_n). \tag{2.22}$$

Thus, if we want to tell the theory (F, M) apart from the trivial theory we must consider correlators with background states other than Ω^*. This small observation is the reason for including this subsection.

2.3 Modular Invariant Partition Functions

Given a Vir $\oplus$ Vir-module F as in the Definition 2.4, the *graded trace* of F is

$$Z(F;\tau) = \mathrm{tr}_F\big(q^{L_0 - C/24}\bar{q}^{\overline{L}_0 - \overline{C}/24}\big), \tag{2.23}$$

where $q = e^{2\pi i\tau}$ and τ is a complex number with $\mathrm{Im}(\tau) > 0$. Suppose for the moment that C and $\overline{C}$ both act on F by multiplication with a number c. Then we can rewrite Z as

$$Z(F;\tau) = \sum_{\Delta} e^{-2\pi\,\mathrm{Im}(\tau)\cdot(\Delta - c/12)}\,\mathrm{tr}_{F(\Delta)}\big(e^{2\pi i\,\mathrm{Re}(\tau)(L_0 - \overline{L}_0)}\big). \tag{2.24}$$

The graded trace may be ill-defined, for example L_0 and $\overline{L}_0$ might have infinite common eigenspaces or the sum over Δ may not converge. If $Z(F;\tau)$ is well-defined, it is a generating function sorting states in F by their scaling dimension (or energy)—with dual parameter $\mathrm{Im}(\tau)$—and by their spin with dual parameter $\mathrm{Re}(\tau)$.

Given a conformal field theory on the complex plane (F, M, Ω^*), one may ask if the set of correlators C_n determined by it is part of a larger family of correlators which allow Riemann surfaces other than the complex plane. The simplest additional surface would be a torus of complex modulus τ,

$$T_\tau = \mathbb{C}/(\mathbb{Z} + \tau\mathbb{Z}). \tag{2.25}$$

A correlator of n fields on T_τ is then required to be related to a sum of correlators of $n+2$ fields on the Riemann sphere by 'inserting a sum over intermediate states'. Schematically, this is shown in Fig. 1. We will not go into any detail, but we point out that the sum is over a basis $\{\varphi_\alpha\}$ of F and a basis $\{\varphi'_\alpha\}$ dual to the first basis with respect to the 2-point correlator on the Riemann sphere. For this procedure to make sense, the 2-point correlator has to be non-degenerate. Such correlators on the Riemann sphere can be obtained from a non-degenerate CFT on $\mathbb{C}$ via Remark 2.3(ii) or from a CFT with background states and an isomorphism $F \to F'$ as in Remark 2.8(iv).

If the system of correlators on the Riemann sphere form part of a larger collection defined on other Riemann surfaces including the torus, then the amplitude for the

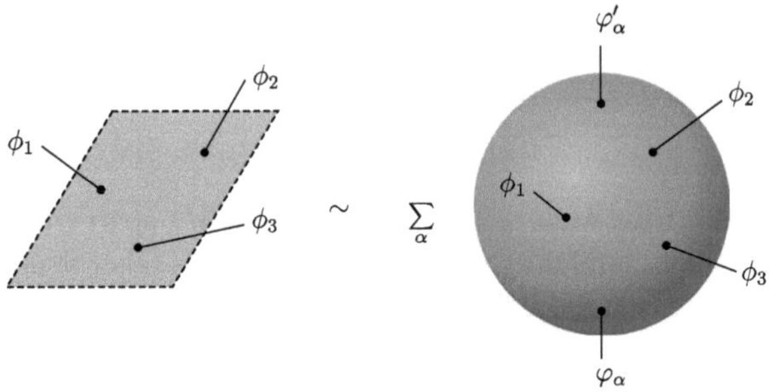

Fig. 1 A correlator of n bulk fields $\phi_i \in F$ on the torus can be expressed as a sum of correlators of $n+2$ fields on the Riemann sphere, where the two additional bulk fields are taken from a basis $\{\varphi_\alpha\}$ of F and its dual basis $\{\varphi'_\alpha\}$

torus T_τ is described by the function $Z(F;\tau)$. It must therefore only depend on the conformal equivalence class of T_τ, that is, it must be *modular invariant*,

$$Z(F;-1/\tau) = Z(F;\tau+1) = Z(F;\tau). \tag{2.26}$$

The function $Z(F;\tau)$ is called the *partition function* of the CFT.

Remark 2.9

(i) Physically, if the CFT arises as a continuum limit of a two-dimensional critical lattice model, one would expect its partition function to be modular invariant since the lattice model could equally be evaluated in a finite geometry with periodic boundary conditions.

(ii) For non-logarithmic rational conformal field theories, modular invariance of the partition function proved to be very constraining. Understanding which $\mathrm{Vir} \oplus \mathrm{Vir}$-modules F (or $\mathcal{V} \otimes_{\mathbb{C}} \mathcal{V}$-modules for a vertex operator algebra $\mathcal{V}$) give rise to a modular invariant graded trace is an important step in attacking classification questions. A typical behaviour in non-logarithmic rational examples is that if $\mathcal{V}$ has order N distinct irreducible representations, then a modular invariant F splits into order N^2 irreducible direct summands. In this sense, modular invariant CFTs for a fixed $\mathcal{V}$ are all 'equally complicated'.[5]

2.4 Consistency Conditions for CFT on the Upper Half Plane

The description of conformal field theory on the upper half plane is very similar to that on the complex plane. The main difference is that there are now two spaces of fields: *bulk fields*, which are the ones already discussed in Sect. 2.1 and are inserted in the interior of the upper half plane, and *boundary fields*, which must be inserted

[5]This statement can be made more precise: a non-logarithmic rational CFT with symmetry $\mathcal{V} \otimes_{\mathbb{C}} \mathcal{V}$, which has a unique vacuum and is modular invariant, has the property that the categorical dimension of F is equal to the global dimension of $\mathrm{Rep}(\mathcal{V})$, see Theorem 3.4 in [34] for details. In particular all such F have the same categorical dimension.

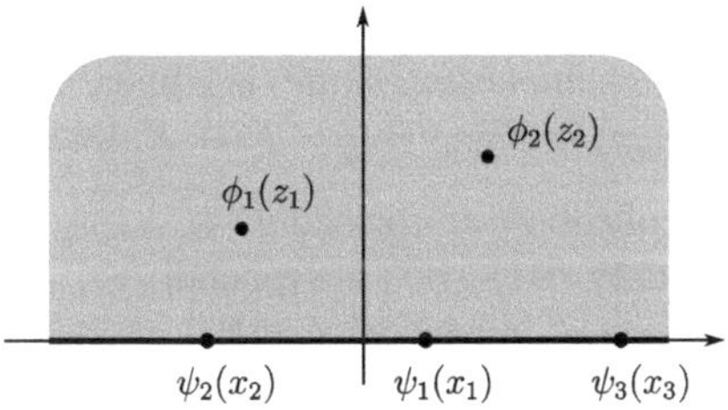

Fig. 2 Bulk and boundary field insertions on the upper half plane. Here $\phi_i \in F$, $\psi_i \in B$ and $\mathrm{Im}(z_i) > 0$, $x_i \in \mathbb{R}$. The figure describes the correlator $\langle \psi_1(x_1)\psi_2(x_2)\psi_3(x_3)\phi_1(z_1)\phi_2(z_2)\rangle = U_{3,2}(x_1, x_2, x_3, z_1, z_2, \psi_1, \psi_2, \psi_3, \phi_1, \phi_2)$

on the real axis, cf. Fig. 2. Correspondingly, the collection of correlators $(U_{m,n})_{m,n}$ now depends on two integers, m counting the number of boundary fields and n counting the number of bulk fields. Let $\mathbb{H} := \{z \in \mathbb{C} \,|\, \mathrm{Im}(z) > 0\}$ be the open upper half plane. Then

$$U_{m,n} : \big(\mathbb{R}^m \backslash \mathrm{diag}\big) \times \big(\mathbb{H}^n \backslash \mathrm{diag}\big) \times B^m \times F^n \longrightarrow \mathbb{C}, \tag{2.27}$$

where $\mathbb{H}^n \backslash \mathrm{diag}$ and $\mathbb{R}^n \backslash \mathrm{diag}$ refer the set of n mutually distinct points. The customary notation is, for $\phi_i \in F$, $\psi_i \in B$, $z_i \in \mathbb{H}$ and $x_i \in \mathbb{R}$,

$$\begin{aligned} &U_{m,n}(x_1, \ldots, x_m, z_1, \ldots, z_n, \psi_1, \ldots, \psi_m, \phi_1, \ldots, \phi_n) \\ &\quad = \big\langle \psi_1(x_1) \cdots \psi_m(x_m)\phi_1(z_1) \cdots \phi_n(z_n)\big\rangle. \end{aligned} \tag{2.28}$$

Remark 2.10 In the discussion of correlators on the upper half plane above we are implicitly assuming that the entire real axis carries the same boundary condition. In more generality one would allow different intervals to carry different boundary conditions. We will not treat this case explicitly, but we note that it is included in the present formalism: One can always think of the real line with several boundary conditions as a real line with a single boundary condition given by their superposition, together with appropriate boundary field insertions that project to the individual constituents.

Definition 2.11 A *conformal field theory on the upper half plane* is a tuple

$$\big(F, M, \Omega^*; B, m, \omega^*; b\big),$$

where

- (F, M, Ω^*) is a CFT on the complex plane,
- B (the *space of boundary fields*) is a Vir-module which is a direct sum of generalised L_0-eigenspaces $B^{(h)}$, whose generalised L_0-eigenvalues h are bounded below and discrete,[6]

[6]The space B is automatically locally finite as a $\mathbb{C}L_0$ module (cf. Sect. 2.1 for the definition of 'locally finite'). This is so because any vector v in B can be written as a finite sum of vectors

- m (the *boundary OPE*) is a map $\mathbb{R}_{>0} \times B \otimes_{\mathbb{C}} B \to \overline{B}$, linear in $B \otimes_{\mathbb{C}} B$,
- ω^* (the *out-vacuum on the upper half plane*) is a linear function $B^{(0)} \to \mathbb{C}$,
- b (the *bulk-boundary map*) is a map $\mathbb{R}_{>0} \times F \to \overline{B}$, linear in F,

such that there exists a collection of correlators $(U_{m,n})_{m,n}$, with $m, n \in \mathbb{Z}_{\geq 0}$ and $(m, n) \neq (0, 0)$, which satisfy (B1)–(B5) in Appendix A, as well as the normalisation condition $U_{1,0}(0, \psi) = \langle \omega^*, \psi \rangle$ for all $\psi \in B$.

Conditions (B1)–(B5) are the same in spirit as (C1)–(C5), just more tiresome to write down, and they have been moved to Appendix A for this reason. Here we merely note that there are now three different types of short distance expansions. The OPE of two bulk fields as in (C3), the expansion of a bulk field ϕ close to the boundary in terms of boundary fields via $b_y(\phi) \in \overline{B}$, and the OPE of two boundary fields $(\psi, \psi') \mapsto m_x(\psi \otimes \psi') \in \overline{B}$.

The basic class of examples is provided by the Virasoro minimal models with A-series modular invariant. In this case the central charge is $c = 1 - 6(p-q)^2/pq$ with $p, q \geq 2$ and coprime. Denote by i a Kac-label for that central charge and by R_i the corresponding irreducible representation[7] of Vir. Then $F = \bigoplus_i R_i \otimes_{\mathbb{C}} R_i$, where i runs over all Kac-labels (modulo their $\mathbb{Z}/2$-identification) and for B we can take the vacuum representation $B = R_{(1,1)}$ of L_0-weight 0. There are many more possible spaces of boundary fields for this bulk theory, namely $B = U \otimes_f U^*$, where U is any direct sum of the R_i and $\otimes_f$ denotes the fusion product (resulting again in a direct sum of the R_i according to the fusion rules).

Let us stress again the point made in the introduction and in Remark 2.9(ii). The space of bulk fields in a modular invariant CFT tends to be 'big' in the sense that it involves many different irreducible representations (in logarithmic CFT this should be taken as a statement about the composition series or about the character). On the other hand, there often exists a CFT on the upper half plane with bulk fields F and a much simpler set of boundary fields B involving only very few irreducible representations. One may thus attempt to first gain control over the boundary theory and then try to construct a fitting bulk theory. This is the topic of the next subsection.

2.5 From Boundary to Bulk

In this subsection we make precise the following idea: Given a boundary theory, i.e. a space of boundary fields and their correlators on the upper half plane, try to build the 'biggest bulk theory' that can be made to fit to this boundary theory. We will find

---

$v_h \in B^{(h)}$, and on each v_h we have $(L_0 - h)^N v_h = 0$ for some large enough N. For F, the same argument only gives local finiteness as a $\mathbb{C}(L_0 + \overline{L}_0)$ module, which is why local finiteness as a $\mathbb{C}L_0 \oplus \mathbb{C}\overline{L}_0$ module was included as a separate condition.

[7] Of course there are many more irreducible representations of the Virasoro algebra with this value of the central charge, but only those corresponding to entries in the Kac table are also representations of the simple Virasoro vertex operator algebra with this central charge.

that this bulk theory, if it exists, is unique. The algebraic version of the question of existence and the description of the data F, M, and b will be addressed in Sect. 3.

Remark 2.12 In Sects. 2.1–2.4 we have discussed CFTs with Virasoro symmetry. This can be generalised to other vertex operator algebras $\mathcal{V}$ as underlying symmetry of the CFT. The space of bulk fields F is then a representation of $\mathcal{V} \otimes_{\mathbb{C}} \mathcal{V}$ and the space of boundary fields B a representation of $\mathcal{V}$. The coinvariance conditions become those of $\mathcal{V}$ and will contain the Virasoro conditions in (C5) and (B5) as a subset. It is important for us to allow this generalisation, even if we avoided spelling out the formalism for general $\mathcal{V}$. The reason is that in the examples we study, we want the category of representations Rep($\mathcal{V}$) to have certain finiteness properties (in particular a finite number of irreducibles, more details will follow in condition (PF) in Sect. 3.2 below). If we were only to allow $\mathcal{V}$ to be the simple Virasoro vertex operator algebra at a given central charge, the finiteness conditions would limit us to (non-logarithmic) minimal models. Therefore, we allow for more general $\mathcal{V}$, in particular the vertex operator algebra $\mathcal{W}$ at $c = 0$ for the $W_{2,3}$-model discussed in Sect. 4. It is not currently clear to us to which extent the construction below is the right ansatz if we were to drop these finiteness conditions.

Definition 2.13 A *boundary theory* is a triple (B, m, ω^*) with B, m, ω^* as in Definition 2.11, such that there exists a collection of correlators on the upper half plane $(U_{m,0})_{m\in\mathbb{Z}_{>0}}$ involving only boundary fields, and which satisfy (B1)–(B5) restricted to $U_{m,0}$, as well as $U_{1,0}(0, \psi) = \langle\omega^*, \psi\rangle$ for all $\psi \in B$.

Thus, a CFT on the upper half plane consists of a CFT on the complex plane, a boundary theory, and a consistent interaction between them via the bulk-boundary map. In analogy with Definition 2.7 we say

Definition 2.14 A boundary theory (B, m, ω^*) is called *non-degenerate* if for all $x, y \in \mathbb{R}$ with $x \neq y$ and $\psi \in B$ there is a $\psi' \in B$ such that $\langle\psi(x)\psi'(y)\rangle \neq 0$.

Remark 2.15

(i) Continuing from Remark 2.6, it is again helpful to briefly consider the much simpler special case of topological field theory. One checks that a non-degenerate boundary theory (B, m, ω^*) with trivial Vir-action on B is the same as an associative but not necessarily commutative algebra B, together with a map $\omega^* : B \to \mathbb{C}$ such that $(a, b) \mapsto \langle\omega^*, a \cdot b\rangle$ is a non-degenerate pairing on B.
(ii) As in Sect. 2.2 one can introduce boundary theories with background states (B, m) which involve a modified version of condition (B5). We have chosen not to discuss boundary theories with background states in detail. The construction of the 'biggest bulk theory' below is therefore formulated in terms of a non-degenerate boundary theory, but one could alternatively use a boundary theory with background states.

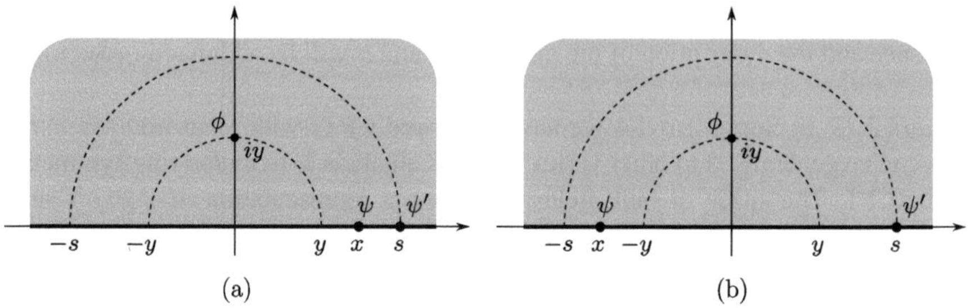

Fig. 3 Geometric setting for the centrality condition. The limit (2.30) defining $\tilde{U}_{2,1}(s, x, iy, \psi', \psi, \phi)$ is assumed to exist for $y < |x|$. This gives rise to two functions $U_\pm(x)$ on the open interval $(-s, s)$: $U^+(x)$ equals $\tilde{U}_{2,1}$ for $x \in (y, s)$ as shown in **a**, while $U^-(x)$ equals $\tilde{U}_{2,1}$ for $x \in (-s, -y)$ as shown in **b**

Let us now fix a non-degenerate boundary theory (B, m, ω^*). The 'biggest bulk theory' will be characterised as a terminal object in a category of pairs $\mathcal{P}$, which we proceed to define. An object of $\mathcal{P}$ is a pair $(\tilde{F}, \tilde{b})$, where

- $\tilde{F}$ is a 'candidate space of bulk fields'. Namely it is a $\text{Vir} \oplus \text{Vir}$-module with boundedness condition as in Definition 2.4 (or more generally a $\mathcal{V} \otimes_{\mathbb{C}} \mathcal{V}$-module).
- $\tilde{b}$ is a 'candidate bulk-boundary map'. By this we mean that $\tilde{b} : \mathbb{R}_{>0} \times \tilde{F} \to \overline{B}$ as in Definition 2.11, such that there exists a function $\tilde{U}_{1,1} : \mathbb{R} \times \mathbb{H} \times B \times \tilde{F} \to \mathbb{C}$ (a 'candidate correlator' of one bulk field and one boundary field) which satisfies the derivative property (B4), the coinvariance condition (B5), and which for $|x| > y$ can be expressed through the candidate bulk-boundary map and the boundary 2-point correlator as

$$\tilde{U}_{1,1}(x, iy, \psi, \phi) = \lim_{h\to\infty} U_{2,0}\big(x, 0, \psi, P_h \circ \tilde{b}_y(\phi)\big); \quad \psi \in B, \phi \in \tilde{F}. \quad (2.29)$$

  Here $U_{2,0}$ is a boundary correlator from Definition 2.13 which is uniquely fixed by (B, m, ω^*), and P_h is the canonical projection $\overline{B} \to \bigoplus_{d\leq h} B^{(d)}$, analogous to P_Δ in (C3).
- $\tilde{b}$ has to be *central*, a condition which we will detail momentarily.

To formulate the centrality condition, we define a candidate correlator $\tilde{U}_{2,1}$ of two boundary fields $\psi, \psi' \in B$ and one bulk field $\phi \in \tilde{F}$ via

$$\tilde{U}_{2,1}\big(s, x, iy, \psi', \psi, \phi\big) = \lim_{h\to\infty} U_{3,0}\big(s, x, 0, \psi', \psi, P_h \circ \tilde{b}_y(\phi)\big), \quad (2.30)$$

at least for $y < |x| < s$ (we take $s > 0$); we assume (as part of the centrality condition) that the limit exists. There are then two disconnected domains for x: it can be in (y, s) or in $(-s, -y)$, see Fig. 3 for an illustration. We now try to use the derivative property (B5) in the form

$$\frac{d}{dx}\tilde{U}_{2,1}\big(s, x, iy, \psi', \psi, \phi\big) = \tilde{U}_{2,1}\big(s, x, iy, \psi', L_{-1}\psi, \phi\big) \quad (2.31)$$

to extend the function $\tilde{U}_{2,1}$ to all of $(-s,s)$. Depending on whether we start from (y,s) or in $(-s,-y)$, we a priori obtain two different functions $U^+(x)$ and $U^-(x)$ on $(-s,s)$. We call $\tilde{b}$ *central* if these two extensions coincide: $U^+(x) = U^-(x)$ for all $x \in (-s,s)$.

The centrality condition holds automatically in a CFT on the upper half plane (because the correlator $U_{2,1}$ is a smooth function and satisfies the expansion conditions (B3)). The point here, of course, is to impose only a small subset of the conditions a CFT has to satisfy. For example, to define the pairs $(\tilde{F},\tilde{b})$ we are only ever looking at candidate correlators with one bulk field and one or two boundary fields.

But back to the category of pairs $\mathcal{P}$. Now that we have defined its objects, it is easy to give the space of morphisms from $(\tilde{F},\tilde{b})$ to $(\tilde{G},\tilde{c})$. It consists of all Vir $\oplus$ Vir-intertwiners $f : \tilde{F} \to \tilde{G}$ (or more generally $\mathcal{V} \otimes_{\mathbb{C}} \mathcal{V}$-intertwiners) such that the diagram of maps

$$\begin{array}{ccc} \tilde{F} & \xrightarrow{\;\;f\;\;} & \tilde{G} \\ & {\scriptstyle \tilde{b}_y} \searrow \quad \swarrow {\scriptstyle \tilde{c}_y} & \\ & \overline{B} & \end{array} \tag{2.32}$$

commutes for all $y > 0$.

An object T in a category $\mathcal{C}$ is called *terminal* if for every object $U \in \mathcal{C}$ there exists a unique morphism $U \to T$. A category $\mathcal{C}$ may or may not have a terminal object, but if one exists, it is unique up to unique isomorphism (take two terminal objects T and T' and play with maps between them). We have now gathered all ingredients to define:

$$\big(F(B), b(B)\big) \text{ is a terminal object in } \mathcal{P}. \tag{2.33}$$

We want to think of $F(B)$ as the maximal space of bulk fields which can be consistently joined to our prescribed boundary theory (B, m, ω^*), and of course we take $b(B)$ as the bulk-boundary map. The next lemma makes this interpretation precise.

Lemma 2.16 *Let B be a non-degenerate boundary theory. Let $(F(B), b(B))$ be a terminal object in $\mathcal{P}$ and let $(\tilde{F},\tilde{b})$ be an arbitrary object in $\mathcal{P}$.*

(i) *The kernel of $\tilde{b}_y : \tilde{F} \to \overline{B}$ is independent of y.*
(ii) *The map $b(B)_y : F(B) \to \overline{B}$ is injective for each $y > 0$.*
(iii) *If $\tilde{b}_y : \tilde{F} \to \overline{B}$ is injective for $y > 0$, then there is an injective* Vir $\oplus$ Vir-*intertwiner $\iota : \tilde{F} \to F(B)$ such that $\tilde{b}_y = b(B)_y \circ \iota$ for all $y > 0$.*

Proof For part (i), let $K(y)$ be the kernel of $\tilde{b}_y : \tilde{F} \to \overline{B}$. By non-degeneracy of the boundary theory, the kernel of $\tilde{b}_y$ is determined by $\tilde{U}_{1,1}$.

- $K(y)$ is a Vir $\oplus$ Vir-module: Use the coinvariance property to show that $\tilde{U}_{1,1}(x, iy, \psi, \phi) = 0$ for all ψ implies $\tilde{U}_{1,1}(x, iy, \psi, L_m\phi) = 0$ and $\tilde{U}_{1,1}(x, iy, \psi, \overline{L}_m\phi) = 0$ for all ψ.

- $K(y) = K(y')$ for all $y, y' > 0$: There exists a global conformal transformation $\mathbb{H} \to \mathbb{H}$ which leaves a point $x \in \mathbb{R}$ invariant and maps y to y'. The coinvariance property can be integrated to give $\tilde{U}_{1,1}(x, iy, \psi, \phi) = \tilde{U}_{1,1}(x, iy', \psi', \phi')$, where ψ' and ϕ' are obtained from ψ and ϕ by an appropriate exponential of modes L_0, L_1, $\overline{L}_0$, $\overline{L}_1$. Using the previous point we see that $\phi \in K(y')$ implies $\phi' \in K(y')$ and thus $\phi \in K(y)$. Together with the inverse transformation one finds $K(y) = K(y')$.

To see (ii), let K be the kernel of $b(B)_y$ and let $e : K \to F(B)$ be the embedding map. The triangle

$$\begin{array}{ccc} K & \xrightarrow{\quad e \quad} & F(B) \\ & {\scriptstyle 0}\searrow \qquad \swarrow{\scriptstyle b(B)_y} & \\ & \overline{B} & \end{array} \tag{2.34}$$

commutes for all $y > 0$ (since the kernel is independent of y). By the terminal object property, the map $K \to F(B)$ which makes the above triangle commute is unique, and therefore $e = 0$. Hence also $K = \{0\}$.

Part (iii) is now trivial. The existence of ι follows from the terminal object property. Since $\tilde{b}_y = b(B)_y \circ \iota$ with $\tilde{b}_y$ and $b(B)_y$ injective, also ι must be injective. □

Remark 2.17 That the candidate bulk-boundary map $\tilde{b}$ in a pair $(\tilde{F}, \tilde{b})$ is injective has the physical interpretation that all bulk fields can be distinguished in upper half plane correlators. If a bulk field ϕ from the kernel of the bulk-boundary map is inserted in a correlator on the upper half plane, this correlator vanishes, irrespective of the other field insertions. Thus by the above lemma, the space $F(B)$ is maximal in the sense that any candidate space of bulk fields $(\tilde{F}, \tilde{b})$, for which all bulk fields can be distinguished in upper half plane correlators, can be embedded in $F(B)$. This embedding is compatible with the candidate bulk-boundary map.

It remains to address the question of existence of the terminal object $(F(B), b(B))$, to see how the OPE of bulk fields in $F(B)$ is determined, to verify its associativity and commutativity, and to investigate the compatibility of bulk and boundary OPE with the bulk-boundary map $b(B)$. To do so, it is best to leave behind the infinite dimensional vector spaces underlying $F(B)$ and B and the infinite set of coinvariance conditions on the correlators, and to take a fresh look at the problem from the more abstract viewpoint of algebras in braided monoidal categories.[8]

[8]We should also address the non-degeneracy of the 2-point correlator and verify modular invariance. Unfortunately, we currently do not know how to do this at the level of generality used in Sect. 3. We can only point to non-logarithmic rational CFTs, where everything works as it should [13, 34], and to the $W_{1,p}$-series and the $W_{2,3}$-model [24, 26], which give modular invariant torus amplitudes and have a self-contragredient space of bulk fields, $F(B) \cong F(B)'$. The latter condition is necessary for the existence of a non-degenerate 2-point correlator.

3 Algebraic Reformulation

Some aspects of the consistency conditions for a CFT are analytic in nature, such as the convergence condition (C3) for the OPE and the differential equations (C4) to be satisfied by correlation functions. Other aspects have a combinatorial counterpart which can be described using the language of algebras in braided monoidal categories. In this section we present these counterparts, and we point out the corresponding concepts from Sect. 2.

The translation is made by fixing a vertex operator algebra $\mathcal{V}$ as chiral symmetry of the CFT and considering the category Rep($\mathcal{V}$) of representations of $\mathcal{V}$. This category is by definition $\mathbb{C}$-linear and abelian. Under certain conditions on $\mathcal{V}$, one obtains in addition a tensor product and a braiding on Rep($\mathcal{V}$) [29, 30].

In this section, k denotes a field of characteristic 0. We will use the notation $\mathcal{C}(U, V)$ to denote the set of morphisms from an object U to an object V in a category $\mathcal{C}$. The categories $\mathcal{C}$ we will consider have the following properties:

- $\mathcal{C}$ is k-linear, abelian, and satisfies finiteness conditions detailed in Sect. 3.2.
- If $\mathcal{C}$ is monoidal, the tensor product functor is k-linear and right exact in both arguments.[9]

For the algebraic constructions presented in this section, it is irrelevant whether $\mathcal{C}$ is realised as representations of some vertex operator algebra $\mathcal{V}$ or not.

We assume some familiarity with abelian categories, exact functors, monoidal categories, monoidal functors, and braidings; the standard reference is [37]. Other notions, such as conjugates, the Deligne product and related constructions, and algebras in monoidal categories, are reviewed in Sects. 3.1–3.4. The main point of this section is the notion of the 'full centre', introduced in Sect. 3.5, which is the algebraic implementation of the construction of a bulk theory form a boundary theory described in Sect. 2.5. Some properties related to the full centre are discussed in Sect. 3.7.

3.1 Conjugates

In many of the constructions below we will need that every object $U \in \mathcal{C}$ has a *conjugate object* U^*. The extra structure we demand to come along with this conjugation is summarised in

> **Condition (C):** The category $\mathcal{C}$ is equipped with an involutive contragredient k-linear functor $(-)^* : \mathcal{C} \to \mathcal{C}$, together with a natural family of isomorphisms $\delta_U : U \to U^{**}$ which satisfy $(\delta_U)^* = (\delta_{U^*})^{-1} : U^{***} \to U^*$ for all $U \in \mathcal{C}$. Furthermore, $\mathcal{C}$ is equipped with a family of isomorphisms $\pi_{U,V} : \mathcal{C}(U, V^*) \to \mathcal{C}(U \otimes V, \mathbf{1}^*)$, natural in U and V.

[9]For monoidal $\mathcal{C}$, we do not require the tensor unit $\mathbf{1}$ to be simple. Neither do we require it to be absolutely simple, that is, we do not impose that the space of endomorphisms of $\mathbf{1}$ is $k \cdot id_{\mathbf{1}}$.

We do *not* demand that there be maps $\mathrm{ev}_U : U^* \otimes U \to \mathbf{1}$ and $\mathrm{coev}_U : \mathbf{1} \to U \otimes U^*$ which satisfy the properties of a categorical dual. Indeed, this property fails in the $W_{2,3}$-example, see Sect. 4.1 below. We do also not demand the (weaker) property that $(U \otimes V)^*$ be isomorphic to $V^* \otimes U^*$ (which also fails in the $W_{2,3}$-example).

Remark 3.1 Condition (C) was introduced in Sect. 3.1 in [25] (there, the condition $(\delta_U)^* = (\delta_{U^*})^{-1}$ was not spelled out). It is motivated by the relation of Hom-spaces and spaces of conformal blocks on the sphere in the case $\mathcal{C} = \mathrm{Rep}(\mathcal{V})$ for a suitable vertex operator algebra $\mathcal{V}$. Then R^* is the contragredient representation R' of R (see Notation I:2.36 in [30] and (2.15)) and δ_R is the natural isomorphism from a graded vector space with finite-dimensional homogeneous components to its graded double dual, which indeed satisfies $(\delta_R)^* = (\delta_{R^*})^{-1}$. Denote by $\otimes_f$ the fusion-tensor product in $\mathrm{Rep}(\mathcal{V})$. The definition of π is motivated by the observation that $\mathrm{Hom}_{\mathcal{V}}(R \otimes_f S, T)$ is isomorphic to the space of three-point conformal blocks on the Riemann sphere $\mathbb{C} \cup \{\infty\}$ with insertions of R and S at x and y, say, and of the *contragredient* T^* of T at a point z. Since the position of the insertion points are arbitrary, this space of conformal blocks is also isomorphic to $\mathrm{Hom}_{\mathcal{V}}(R \otimes_f T^*, S^*)$. Furthermore, the space of blocks does not change by inserting the vertex operator algebra $\mathcal{V}$ itself. Thus, with $S = \mathcal{V}$, $\mathrm{Hom}_{\mathcal{V}}(R, T^*) \cong \mathrm{Hom}_{\mathcal{V}}(R \otimes_f \mathcal{V}, T^*) \cong \mathrm{Hom}_{\mathcal{V}}(R \otimes_f T, \mathcal{V}^*)$. In the setting of [30], the above reasoning amounts to Proposition II:3.46.

Definition 3.2 A pairing $p : U \otimes V \to \mathbf{1}^*$ is called *non-degenerate* if the map $\pi^{-1}_{U,V}(p) : U \to V^*$ is an isomorphism.

An alternative characterisation of non-degeneracy of p is that $p \circ (f \otimes id_V) = 0$ implies $f = 0$ for all $f : X \to U$, and $p \circ (id_U \otimes g) = 0$ implies $g = 0$ for all $g : Y \to V$ (see Lemma B.7 in [25] for a proof). This justifies the name 'non-degenerate'. There is a canonical non-degenerate pairing

$$\beta_U := \pi_{U,U^*}(\delta_U) : U \otimes U^* \to \mathbf{1}^*, \tag{3.1}$$

which in particular has the property that $\beta_V \circ (h \otimes id_{V^*}) = \beta_U \circ (id_U \otimes h^*)$ for all $h : U \to V$, see Lemma B.3.

3.2 Deligne Product

The point of this subsection is to gain some familiarity with the Deligne product of k-linear abelian categories which will be used extensively below. A reader who deems this too technical (or boring) could maybe have a quick glance at Definition 3.3, condition (PF) and Corollary 3.7, and then continue with Sect. 3.3.

Let A, B be two k-algebras. Denote by A-mod and B-mod the k-linear abelian categories of finitely generated modules over these algebras. We can now ask if we can construct $(A \otimes_k B)$-mod directly from the categories A-mod and B-mod rather than using the algebras A and B. The problem one faces is that in general not every $A \otimes_k B$-module is a direct sum of tensor products of A-modules and B-modules.

For example, if $A = B = k[x]/\langle x^2 \rangle$, we have $A \otimes_k B \cong k[x, y]/\langle x^2, y^2 \rangle$. The $A \otimes_k B$-module $M = k[x, y]/\langle x^2, y^2, x - y \rangle$ has dimension 2 and both x and y act non-trivially. Since up to dimension two, the only A- (or B-) module with non-trivial action is $k[x]/\langle x^2 \rangle$, the module M does not arise as a direct sum of tensor products.

The passage from A-mod $\times$ B-mod (the category of pairs of objects and morphisms) to $(A \otimes_k B)$-mod is a special case of the Deligne product of abelian categories (Sect. 5.1 in [9]). Given two k-linear abelian categories $\mathcal{A}$, $\mathcal{B}$, denote by $\mathcal{F}un_{k,\text{r.ex.}}(\mathcal{A}, \mathcal{B})$ the category of k-linear right exact functors from $\mathcal{A}$ to $\mathcal{B}$ and natural transformations between them.

Definition 3.3 Let $\{\mathcal{A}_\sigma\}_{\sigma \in S}$ be a family of k-linear abelian categories. The *Deligne product* of the $\{\mathcal{A}_\sigma\}_{\sigma \in S}$ is a pair $(\mathcal{A}_S, \boxtimes_S)$, such that

(i) $\mathcal{A}_S$ is a k-linear abelian category, and $\boxtimes_S : \prod_{\sigma \in S} \mathcal{A}_\sigma \to \mathcal{A}_S$ is a functor which is k-linear and right exact in each $\mathcal{A}_\sigma$,

(ii) Let $\mathcal{B}$ be a k-linear abelian category and denote by $\mathcal{F}un_{\text{mult,r.ex.}}(\prod_{\sigma \in S} \mathcal{A}_\sigma, \mathcal{B})$ the category of all functors which are k-linear and right exact in each $\mathcal{A}_\sigma$. Then for all $\mathcal{B}$, the functor

$$(-) \circ \boxtimes_S : \mathcal{F}un_{k,\text{r.ex.}}(\mathcal{A}_S, \mathcal{B}) \longrightarrow \mathcal{F}un_{\text{mult,r.ex.}}\Big(\prod_{\sigma \in S} \mathcal{A}_\sigma, \mathcal{B}\Big), \qquad F \mapsto F \circ \boxtimes_S, \tag{3.2}$$

is an equivalence of categories.

We will also write the Deligne product as $\boxtimes_{\sigma \in S} \mathcal{A}_\sigma$, or, in case there are only a finite number of factors with index set $S = \{1, 2, \ldots, n\}$, as $\mathcal{A}_1 \boxtimes \mathcal{A}_2 \boxtimes \cdots \boxtimes \mathcal{A}_n$. The triangle one would like to draw for the universal property in condition (ii) is

$$\begin{array}{ccc} \prod_{\sigma \in S} \mathcal{A}_\sigma & \xrightarrow{\ \boxtimes_S\ } & \boxtimes_{\sigma \in S} \mathcal{A}_\sigma \\ & {\scriptstyle f} \searrow \quad \swarrow {\scriptstyle \exists! F} & \\ & \mathcal{B} & \end{array}, \tag{3.3}$$

and it should be read as follows: for each $f \in \mathcal{F}un_{\text{mult,r.ex.}}(\prod_{\sigma \in S} \mathcal{A}_\sigma, \mathcal{B})$ there exists an $F \in \mathcal{F}un_{k,\text{r.ex.}}(\mathcal{A}_S, \mathcal{B})$ such that f is naturally isomorphic to $F \circ \boxtimes_S$. Any other F' with this property is naturally isomorphic to F. However, this captures the equivalence of functor categories required in condition (ii) only on the level of objects.

Remark 3.4 In the algebraic reformulation of the construction in Sect. 2.5, the Deligne product appears as follows. Let $\mathcal{V}$ be a suitable vertex operator algebra. The space of boundary fields will be a representation of $\mathcal{V}$, in other words, an object in $\mathcal{C} = \mathrm{Rep}(\mathcal{V})$. The space of bulk fields will be a representation of $\mathcal{V} \otimes_{\mathbb{C}} \mathcal{V}$, that is, an object in $\mathrm{Rep}(\mathcal{V} \otimes_{\mathbb{C}} \mathcal{V})$. In the algebraic description, we will replace[10] $\mathrm{Rep}(\mathcal{V} \otimes_{\mathbb{C}} \mathcal{V})$ by $\mathcal{C} \boxtimes \mathcal{C}$ (or rather by $\mathcal{C} \boxtimes \mathcal{C}^{\mathrm{rev}}$, were 'rev' refers to the inverse braiding, see Sect. 3.3 below).

If it exists, the Deligne product is unique up to an equivalence: Let $(\mathcal{A}'_S, \boxtimes'_S)$ be another Deligne product and set $\mathcal{B} = \mathcal{A}'_S$ and $f = \boxtimes'_S$ in the above triangle. This results in a functor $F : \mathcal{A}_S \to \mathcal{A}'_S$. The converse procedure gives $G : \mathcal{A}'_S \to \mathcal{A}$ and their compositions have to be equivalent to the identity. To make general existence statements, we will need the following finiteness condition (cf. Sect. 2.12.1 in [9]):

> **Condition (F):** The category is k-linear and abelian, each object is of finite length,[11] and all Hom-spaces are finite-dimensional over k.

By Proposition 5.13 in [9], if each $\mathcal{A}_\sigma$ satisfies condition (F) then the Deligne product $\mathcal{A}_S \equiv \boxtimes_{\sigma \in S} \mathcal{A}_\sigma$ exists and equally satisfies condition (F); for each $X_\sigma, Y_\sigma \in \mathcal{A}_\sigma$, the functor $\boxtimes_S$ gives an isomorphism

$$\bigotimes_{k,\sigma \in S} \mathcal{A}_\sigma(X_\sigma, Y_\sigma) \xrightarrow{\sim} \mathcal{A}_S(\boxtimes_{\sigma \in S} X_\sigma, \boxtimes_{\sigma \in S} Y_\sigma). \tag{3.4}$$

A stronger condition than (F) is

> **Condition (PF):** The category is k-linear and abelian, and it has a projective generator P whose endomorphism space is finite-dimensional over k.

That P is a projective generator means that P is projective and for every $U \in \mathcal{A}$ there is an $m \in \mathbb{N}$ and a surjection $P^{\oplus m} \to U$, i.e. every object in $\mathcal{A}$ is a quotient of some m-fold direct sum of P's. Since $\mathcal{A}(P, P)$ is finite-dimensional, so are all other morphism spaces in $\mathcal{A}$ (pick projective resolutions). Since P has finite composition series (or $\mathcal{A}(P, P)$ would have infinite dimension since P would have non-zero maps to every subobject in the descending chain), so does every object in $\mathcal{A}$. Thus (PF) $\Rightarrow$ (F). Categories satisfying (PF) have the following convenient description:

Theorem 3.5 (Corollary 2.17 in [9]) *$\mathcal{A}$ satisfies condition* (PF) *if and only if there exists a unital finite-dimensional k-algebra A such that $\mathcal{A}$ is equivalent, as a k-linear category, to the category* $\mathrm{Rep}_{f.d.}(A)$ *of finite dimensional* (*over k*) *right A-modules.*

---

[10] We are not aware of a statement in the vertex operator algebra literature that says $\mathrm{Rep}(\mathcal{V} \otimes_{\mathbb{C}} \mathcal{W}) = \mathrm{Rep}(\mathcal{V}) \boxtimes \mathrm{Rep}(\mathcal{W})$, but it seems very natural to us that this property should hold, at least for 'sufficiently nice' $\mathcal{V}$ and $\mathcal{W}$, e.g. when their representation categories satisfy condition (PF) below.

[11] An object A has *finite length* if there is a chain of subobjects $0 = A_0 \subset A_1 \subset A_2 \subset \cdots \subset A_{n-1} \subset A_n = A$ such that each $S_i = A_i / A_{i-1}$ is non-zero and simple. The S_i are called *composition factors* and n is the *composition length*.

The proof is maybe instructive to gain some intuition for the finiteness condition (PF); for the convenience of the reader we include it in Appendix B.1. The next theorem confirms the motivation for studying Deligne products which was stated in the beginning of this subsection. It is proved (in greater generality) in Proposition 5.3 in [9]; we sketch a proof in our simpler situation.

Theorem 3.6 *Let A, B be finite-dimensional unital k-algebras. Then*

$$\mathrm{Rep}_{f.d.}(A) \boxtimes \mathrm{Rep}_{f.d.}(B) = \mathrm{Rep}_{f.d.}(A \otimes_k B). \tag{3.5}$$

Sketch of proof Write $\mathcal{A} = \mathrm{Rep}_{f.d.}(A)$, $\mathcal{B} = \mathrm{Rep}_{f.d.}(B)$, $\mathcal{D} = \mathrm{Rep}_{f.d.}(A \otimes_k B)$. The functor $\boxtimes : \mathcal{A} \times \mathcal{B} \to \mathcal{D}$ is $(M, N) \mapsto M \otimes_k N$, seen as an $A \otimes_k B$ right module, and $(f, g) \mapsto f \otimes_k g$ for module maps f, g. (Since k is a field, $\boxtimes$ is actually *exact* in each argument, not only right exact, cf. Corollary 5.4 in [9].)

Let $\mathcal{E}$ be a k-linear abelian category. We need to show that $(-) \circ \boxtimes$ gives an equivalence of functor categories $\mathcal{F}un_{k,\mathrm{r.ex.}}(\mathcal{D}, \mathcal{E}) \to \mathcal{F}un_{\mathrm{mult,r.ex.}}(\mathcal{A} \times \mathcal{B}, \mathcal{E})$, see (3.2). The point is that a k-linear, right exact functor $F : \mathcal{D} \to \mathcal{E}$ is fixed by $F(A \otimes_k B)$, and by $F(f)$ for all right module endomorphisms of $A \otimes_k B$. To see this, just express an arbitrary finite-dimensional $A \otimes_k B$ right module M via the first two terms in a free resolution, $(A \otimes_k B)^{\oplus n} \to (A \otimes_k B)^{\oplus m} \to M \to 0$ for appropriate $m, n \in \mathbb{Z}_{\geq 0}$. Similarly, a functor $G : \mathcal{A} \times \mathcal{B} \to \mathcal{E}$ which is k-linear and right exact in each argument is fixed by $G(A, B)$ and $G(f, g)$ for all right module endomorphisms f of A and g of B. From this one derives that $(-) \circ \boxtimes$ is essentially surjective. Natural transformations are equally determined by evaluating them on $A \otimes_k B$, respectively on (A, B), and from this one can deduce that $(-) \circ \boxtimes$ is full and faithful. □

Corollary 3.7 *If $\mathcal{A}$ and $\mathcal{B}$ satisfy property* (PF), *then so does $\mathcal{A} \boxtimes \mathcal{B}$. If P and Q are projective generators of $\mathcal{A}$ and $\mathcal{B}$, respectively, then $P \boxtimes Q$ is a projective generator of $\mathcal{A} \boxtimes \mathcal{B}$.*

Proof By the explicit construction in Appendix B.1 we have $\mathcal{A} \cong \mathrm{Rep}_{f.d.}(A)$ as k-linear abelian categories for the choice $A = \mathcal{A}(P, P)$, and also $\mathcal{B} \cong \mathrm{Rep}_{f.d.}(B)$ for $B = \mathcal{B}(Q, Q)$. Then by Theorem 3.6 we may take $\mathcal{A} \boxtimes \mathcal{B} \equiv \mathrm{Rep}_{f.d.}(A \otimes_k B)$. With this choice, $P \boxtimes Q = A \otimes_k B$, which indeed is a projective generator. □

Natural transformations of right exact functors whose domain is a Deligne product are determined by their action on 'product objects'. We will use this a number of times, so let us give a short proof (the statement holds for $\boxtimes_{\sigma \in S} \mathcal{A}_\sigma$, but for notational simplicity we only give the case with two factors).

Lemma 3.8 *Let $\mathcal{A}$, $\mathcal{B}$ satisfy property* (F). *Let $\mathcal{C}$ be a k-linear abelian category, let $F, G \in \mathcal{F}un_{k,\mathrm{r.ex.}}(\mathcal{A} \boxtimes \mathcal{B}, \mathcal{C})$ and let $\alpha, \beta : F \Rightarrow G$ be natural transformations. The following are equivalent*:

(i) $\alpha_X = \beta_X$ *for all* $X \in \mathcal{A} \boxtimes \mathcal{B}$,
(ii) $\alpha_{A \boxtimes B} = \beta_{A \boxtimes B}$ *for all* $A \in \mathcal{A}$, $B \in \mathcal{B}$.

Proof We need to check (ii) ⇒ (i). Write $\hat{F} = F \circ \boxtimes$ and $\hat{G} = G \circ \boxtimes$. The functor $(-) \circ \boxtimes$ maps natural transformations $F \Rightarrow G$ to natural transformations $\hat{F} \Rightarrow \hat{G}$ via

$$\{\eta_X\}_{X \in \mathcal{A} \boxtimes \mathcal{B}} \longmapsto \{\eta_{A,B}\}_{(A,B) \in \mathcal{A} \times \mathcal{B}}, \quad \text{where } \eta_{A,B} := \eta_{A \boxtimes B}. \tag{3.6}$$

By condition (ii) in Definition 3.3, the map (3.6) is an isomorphism and hence β is uniquely determined by its values on all $A \boxtimes B$. □

We will be interested in the case that a category $\mathcal{C}$ satisfies property (F) and is in addition monoidal with k-linear right exact tensor product. Then the tensor product $\otimes_{\mathcal{C}} : \mathcal{C} \times \mathcal{C} \to \mathcal{C}$ gives us a right exact functor

$$T_{\mathcal{C}} : \mathcal{C} \boxtimes \mathcal{C} \longrightarrow \mathcal{C}, \tag{3.7}$$

such that $A \otimes_{\mathcal{C}} B = T_{\mathcal{C}}(A \boxtimes B)$ and analogously for morphisms. Let now $\mathcal{D}$ be another such category. Then $\mathcal{C} \boxtimes \mathcal{D}$ is monoidal with right exact tensor product given by

$$\begin{aligned} \otimes_{\mathcal{C}\boxtimes\mathcal{D}} = \big[&(\mathcal{C} \boxtimes \mathcal{D}) \times (\mathcal{C} \boxtimes \mathcal{D}) \xrightarrow{\boxtimes} \mathcal{C} \boxtimes \mathcal{D} \boxtimes \mathcal{C} \boxtimes \mathcal{D} \\ &\xrightarrow{\sim} \mathcal{C} \boxtimes \mathcal{C} \boxtimes \mathcal{D} \boxtimes \mathcal{D} \xrightarrow{T_{\mathcal{C}} \boxtimes T_{\mathcal{D}}} \mathcal{C} \boxtimes \mathcal{D} \big], \end{aligned} \tag{3.8}$$

for details see Sects. 5.16–5.17 in [9]. The unnamed isomorphism is induced by the functor $\mathcal{C} \times \mathcal{D} \times \mathcal{C} \times \mathcal{D} \to \mathcal{C} \times \mathcal{C} \times \mathcal{D} \times \mathcal{D}$ which exchanges the middle two factors. In particular, for $A, B \in \mathcal{C}$ and $U, V \in \mathcal{D}$,

$$(A \boxtimes U) \otimes_{\mathcal{C}\boxtimes\mathcal{D}} (B \boxtimes V) = (A \otimes_{\mathcal{C}} B) \boxtimes (U \otimes_{\mathcal{D}} V). \tag{3.9}$$

3.3 Braiding

For this subsection we fix a braided monoidal k-linear abelian category $\mathcal{C}$ which satisfies property (F), and which has a k-linear right exact tensor product. In the previous subsection we saw that $\mathcal{C} \boxtimes \mathcal{C}$ is again monoidal with right exact tensor product. We will use the braiding on $\mathcal{C}$ for three related constructions:

- turn the functor $T_{\mathcal{C}} : \mathcal{C} \boxtimes \mathcal{C} \to \mathcal{C}$ from above into a tensor functor,
- equip the category $\mathcal{C} \boxtimes \mathcal{C}$ with a braiding,
- define a 'mixed braiding' with one object from $\mathcal{C} \boxtimes \mathcal{C}$ and one object from $\mathcal{C}$.

Let us start with the monoidal structure on $T \equiv T_{\mathcal{C}}$. The tensor product of $\mathcal{C} \boxtimes \mathcal{C}$ will be denoted by $\otimes_{\mathcal{C}^2}$. We have to give isomorphisms

$$T_{2;X,Y} : T(X) \otimes_{\mathcal{C}} T(Y) \to T(X \otimes_{\mathcal{C}^2} Y), \qquad T_0 : \mathbf{1} \to T(\mathbf{1} \boxtimes \mathbf{1}) \equiv \mathbf{1} \otimes_{\mathcal{C}} \mathbf{1}, \tag{3.10}$$

where $T_{2;X,Y}$ is natural in $X, Y \in \mathcal{C} \boxtimes \mathcal{C}$. T_2 and T_0 are required to satisfy the hexagon and triangle identity (given explicitly in (3.28) and (3.29) below for a lax monoidal functor). For T_0 one takes the inverse unit isomorphism of $\mathcal{C}$. For T_2, consider first the two functors from $\mathcal{C}^{\times 4}$ to $\mathcal{C}$ given by

$$\begin{aligned}(A,B,U,V) &\mapsto T(A\boxtimes B)\otimes_{\mathcal{C}} T(U\boxtimes V)\\ &\equiv (A\otimes_{\mathcal{C}} B)\otimes_{\mathcal{C}}(U\otimes_{\mathcal{C}} V) \quad \text{and} \\ (A,B,U,V) &\mapsto T\big((A\boxtimes B)\otimes_{\mathcal{C}^2}(U\boxtimes V)\big)\equiv (A\otimes_{\mathcal{C}} U)\otimes_{\mathcal{C}}(B\otimes_{\mathcal{C}} V).\end{aligned} \tag{3.11}$$

These are linked by the natural isomorphism (not writing out $\otimes_{\mathcal{C}}$ between objects)[12]

$$\begin{aligned}\tilde{T}_{2;(A,B),(U,V)} := \big[(AB)(UV) &\xrightarrow{\text{assoc.}} \big(A(BU)\big)V \\ &\xrightarrow{(id_A\otimes c^{-1}_{U,B})\otimes id_V} \big(A(UB)\big)V \xrightarrow{\text{assoc.}} (AU)(BV)\big].\end{aligned} \tag{3.12}$$

The defining isomorphism of the Deligne product between functor categories transports $\tilde{T}_2$ to the desired natural isomorphism T_2 in (3.10). In particular, T_2 obeys $T_{2;U\boxtimes V,A\boxtimes B} = \tilde{T}_{2;(U,V),(A,B)}$. The hexagon identity for T_2 follows if it holds on product objects (Lemma 3.8), and for these it reduces to the hexagon of the braiding c of $\mathcal{C}$, cf. Proposition 5.2 in [31].

Next we turn to the braiding on $\mathcal{C}\boxtimes\mathcal{C}$ that we wish to use. This will again be defined by transporting a natural isomorphism, this time between two functors $\mathcal{C}^{\times 4}\to\mathcal{C}\boxtimes\mathcal{C}$

$$\begin{aligned}(A,B,U,V) &\mapsto (A\boxtimes B)\otimes_{\mathcal{C}^2}(U\boxtimes V)\equiv(A\otimes_{\mathcal{C}} U)\boxtimes(B\otimes_{\mathcal{C}} V) \quad \text{and}\\ (A,B,U,V) &\mapsto (U\boxtimes V)\otimes_{\mathcal{C}^2}(A\boxtimes B)\equiv(U\otimes_{\mathcal{C}} A)\boxtimes(V\otimes_{\mathcal{C}} B).\end{aligned} \tag{3.13}$$

The natural isomorphism we choose is $\tilde{c}_{(A,B),(U,V)} = c_{A,U}\boxtimes c^{-1}_{V,B}$. The defining property of the Deligne product provides a natural isomorphism $c_{X,Y}: X\otimes_{\mathcal{C}^2} Y\to Y\otimes_{\mathcal{C}^2} X$ which satisfies

$$c_{A\boxtimes B,U\boxtimes V} = \big[(A\boxtimes B)\otimes_{\mathcal{C}^2}(U\boxtimes V)\xrightarrow{c_{A,U}\boxtimes c^{-1}_{V,B}}(U\boxtimes V)\otimes_{\mathcal{C}^2}(A\boxtimes B)\big]. \tag{3.14}$$

One verifies that the hexagon condition for the braiding on $\mathcal{C}$ implies the hexagon for c in $\mathcal{C}\boxtimes\mathcal{C}$ on product objects; by Lemma 3.8 it then holds on all of $\mathcal{C}\boxtimes\mathcal{C}$. We will denote the category $\mathcal{C}\boxtimes\mathcal{C}$ with tensor product (3.9) and braiding (3.14) by

$$\mathcal{C}\boxtimes\mathcal{C}^{\text{rev}}. \tag{3.15}$$

Finally, we turn to the mixed braiding between $\mathcal{C}\boxtimes\mathcal{C}^{\text{rev}}$ and $\mathcal{C}$. The relevant functors $\mathcal{C}^{\times 3}\to\mathcal{C}$ are $\tilde{L}(A,B,U) = (A\otimes_{\mathcal{C}} B)\otimes_{\mathcal{C}} U$ and $\tilde{R}(A,B,U) = U\otimes_{\mathcal{C}}$

[12]The convention to use c^{-1} and not c for T_2 agrees with Sect. 2.4 in [34] but it is opposite to Sect. 7 in [7]. This should be taken into account when referring to proofs in [7]. We use the c^{-1} convention to make Lemma 3.9 true in the form given below. In the context of CFT, the inverse braiding convention means that in the graphical notation 'lines corresponding to holomorphic insertions go on top'.

$(A \otimes_{\mathcal{C}} B)$. Between these we have the natural isomorphism $\tilde{\varphi}_{A,B,U} : \tilde{L} \Rightarrow \tilde{R}$ given by the string diagram (to be read the optimistic way, i.e. upwards from bottom to top)

$$\tilde{\varphi}_{A,B,U} =$$

. (3.16)

In terms of formulas, this translates as[13]

$$\tilde{\varphi}_{A,B,U} = \alpha^{-1}_{U,A,B} \circ (c_{A,U} \otimes_{\mathcal{C}} id_B) \circ \alpha_{A,U,B} \circ \big(id_A \otimes_{\mathcal{C}} c^{-1}_{U,B}\big) \circ \alpha^{-1}_{A,B,U}. \quad (3.17)$$

From the Deligne product, we obtain a natural isomorphism φ between $L, R : \mathcal{C} \boxtimes \mathcal{C} \boxtimes \mathcal{C} \to \mathcal{C}$ such that $\varphi_{A \boxtimes B \boxtimes U} = \tilde{\varphi}_{A,B,U}$. We will most often use φ in the form

$$\varphi_{X,U} := \varphi_{T(X)\boxtimes U} : T(X) \otimes_{\mathcal{C}} U \longrightarrow U \otimes_{\mathcal{C}} T(X); \quad X \in \mathcal{C} \boxtimes \mathcal{C}^{\mathrm{rev}}, Y \in \mathcal{C}. \quad (3.18)$$

There is an alternative way to define $\varphi_{X,U}$ by transporting the braiding from $\mathcal{C} \boxtimes \mathcal{C}^{\mathrm{rev}}$ to $\mathcal{C}$ with T. By the next lemma, these two possibilities give the same result.

Lemma 3.9 *For $X \in \mathcal{C} \boxtimes \mathcal{C}^{\mathrm{rev}}$ and $U \in \mathcal{C}$, the following diagram commutes.*

$$\begin{array}{ccccc} T(X) \otimes_{\mathcal{C}} U & \xrightarrow{\sim} & T(X) \otimes_{\mathcal{C}} T(U \boxtimes \mathbf{1}) & \xrightarrow{T_2} & T(X \otimes_{\mathcal{C}^2} (U \boxtimes \mathbf{1})) \\ \big\downarrow{\scriptstyle \varphi_{X,U}} & & & & \big\downarrow{\scriptstyle T(c_{X,U\boxtimes\mathbf{1}})} \\ U \otimes_{\mathcal{C}} T(X) & \xrightarrow{\sim} & T(U \boxtimes \mathbf{1}) \otimes_{\mathcal{C}} T(X) & \xrightarrow{T_2} & T((U \boxtimes \mathbf{1}) \otimes_{\mathcal{C}^2} X) \end{array} \quad (3.19)$$

Proof By Lemma 3.8 it is enough to verify commutativity of the diagram on product objects $X = A \boxtimes B$. Drawing the corresponding string diagrams using (3.12), (3.14) and (3.17) one finds the string diagram (3.16) for both paths. □

With the help of the above lemma, it is easy to use identities for the braiding on $\mathcal{C} \boxtimes \mathcal{C}^{\mathrm{rev}}$ to obtain identities for φ. We will need

$$\begin{aligned} \varphi_{X\otimes_{\mathcal{C}^2} Y,U} = \big[T(XY)U &\xrightarrow{T_2^{-1}\otimes id_U} (TXTY)U \\ &\xrightarrow{\sim} TX(TYU) \xrightarrow{id_{TX}\otimes\varphi_{Y,U}} TX(UTY) \\ &\xrightarrow{\sim} (TXU)TY \xrightarrow{\varphi_{X,U}\otimes id_{TY}} (UTX)TY \\ &\xrightarrow{\sim} U(TXTY) \xrightarrow{id_U\otimes T_2} UT(XY)\big], \end{aligned} \quad (3.20)$$

[13]Our convention for associators is $\alpha_{X,Y,Z} : X \otimes (Y \otimes Z) \to (X \otimes Y) \otimes Z$.

which follows form applying T to the hexagon identity $c_{X\otimes_{\mathcal{C}^2}Y,U\boxtimes\mathbf{1}} = (c_{X,U\boxtimes\mathbf{1}} \otimes_{\mathcal{C}^2} id_Y) \circ (id_X \otimes_{\mathcal{C}^2} c_{Y,U\boxtimes\mathbf{1}})$ (we have omitted the associators) and rearranging terms via Lemma 3.9.

Instead of $\varphi_{X,U}$, which takes one argument from $\mathcal{C} \boxtimes \mathcal{C}^{\text{rev}}$ and one from $\mathcal{C}$, we can use the diagram (3.19) to define $\hat{\varphi}_{X,Y}$, which takes both arguments from $\mathcal{C} \boxtimes \mathcal{C}^{\text{rev}}$ via

$$\begin{array}{ccc} T(X) \otimes_{\mathcal{C}} T(Y) & \xrightarrow{T_2} & T(X \otimes_{\mathcal{C}^2} Y) \\ {\scriptstyle \hat{\varphi}_{X,Y}} \Big\downarrow & & \Big\downarrow {\scriptstyle T(c_{X,Y})} \\ T(Y) \otimes_{\mathcal{C}} T(X) & \xrightarrow{T_2} & T(Y \otimes_{\mathcal{C}^2} X) \end{array} \tag{3.21}$$

The following observation will be important below.

Lemma 3.10 *For $X, Y \in \mathcal{C} \boxtimes \mathcal{C}^{\text{rev}}$ we have the identity*

$$\varphi_{X,T(Y)} = \hat{\varphi}_{X,Y} : T(X) \otimes_{\mathcal{C}} T(Y) \to T(Y) \otimes_{\mathcal{C}} T(X). \tag{3.22}$$

Proof From (3.19) and (3.21) we see that we have to establish commutativity of

$$\begin{array}{ccccccc} T(X \otimes_{\mathcal{C}^2} Y) & \xrightarrow{T_2^{-1}} & T(X) \otimes_{\mathcal{C}} T(Y) & \xrightarrow{\sim} & T(X) \otimes_{\mathcal{C}} T(T(Y) \boxtimes \mathbf{1}) & \xrightarrow{T_2} & T(X \otimes_{\mathcal{C}^2} (T(Y) \boxtimes \mathbf{1})) \\ \Big\downarrow {\scriptstyle T(c_{X,Y})} & & & & & & {\scriptstyle T(c_{X,T(Y)\boxtimes\mathbf{1}})} \Big\downarrow \\ T(Y \otimes_{\mathcal{C}^2} X) & \xrightarrow{T_2^{-1}} & T(Y) \otimes_{\mathcal{C}} T(X) & \xrightarrow{\sim} & T(T(Y) \boxtimes \mathbf{1}) \otimes_{\mathcal{C}} T(X) & \xrightarrow{T_2} & T((T(Y) \boxtimes \mathbf{1}) \otimes_{\mathcal{C}^2} X) \end{array} \tag{3.23}$$

By Lemma 3.8, it is enough to verify this for $X = A \boxtimes B$ and $Y = U \boxtimes V$ for all $A, B, U, V \in \mathcal{C}$. In this case, the above diagram reads (not writing $\otimes_{\mathcal{C}}$, brackets between objects, and associators)

$$\begin{array}{ccccccc} AUBV & \xrightarrow{id_A \otimes c_{U,B} \otimes id_V} & ABUV & \xrightarrow{\sim} & ABUV\mathbf{1} & \xrightarrow{id_A \otimes c^{-1}_{UV,B} \otimes id_{\mathbf{1}}} & AUVB\mathbf{1} \\ {\scriptstyle c_{A,U} \otimes c^{-1}_{V,B}} \Big\downarrow & & & & & & \Big\downarrow {\scriptstyle c_{A,UV} \otimes c^{-1}_{\mathbf{1},B}} \\ UAVB & \xrightarrow{id_U \otimes c_{A,V} \otimes id_B} & UVAB & \xrightarrow{\sim} & UV\mathbf{1}AB & \xrightarrow{id_{UV} \otimes c^{-1}_{\mathbf{1},A} \otimes id_B} & UVA\mathbf{1}B \end{array} \tag{3.24}$$

That this diagram commutes can be checked easily by drawing string diagrams. □

Remark 3.11 The functor T is *central* in the sense of Sect. 2 in [8]. Namely, it factors through the braided tensor functor G from $\mathcal{C} \boxtimes \mathcal{C}^{\text{rev}}$ to the monoidal centre of $\mathcal{C}$ as $T_{\mathcal{C}} = [\mathcal{C} \boxtimes \mathcal{C}^{\text{rev}} \xrightarrow{G} \mathcal{Z}(\mathcal{C}) \xrightarrow{\text{forget}} \mathcal{C}]$; we refer to Sect. 2 in [8] for details.

3.4 Algebras

We recall the definition of algebras in monoidal categories, and of commutative algebras in braided monoidal categories. In the category of vector spaces, these give the usual notions of algebras/commutative algebras.

Definition 3.12

(i) An *algebra* in a monoidal category $\mathcal{C}$ is an object $A \in \mathcal{C}$ together with a morphism $\mu : A \otimes A \to A$ which is associative in the sense that

$$\begin{array}{ccccc} A \otimes (A \otimes A) & \xrightarrow{id_A \otimes \mu} & A \otimes A & & \\ & & & \overset{\mu}{\searrow} & \\ \downarrow{\scriptstyle \alpha_{A,A,A}} & & & & A \\ & & & \underset{\mu}{\nearrow} & \\ (A \otimes A) \otimes A & \xrightarrow{\mu \otimes id_A} & A \otimes A & & \end{array} \tag{3.25}$$

commutes. A is called *unital* if it is equipped with a morphism $\iota : \mathbf{1} \to A$ such that

$$\begin{array}{ccccc} \mathbf{1} \otimes A & \xrightarrow{\iota \otimes id_A} & A \otimes A & \xleftarrow{id_A \otimes \iota} & A \otimes \mathbf{1} \\ & \searrow{\scriptstyle \lambda_A} & \downarrow{\scriptstyle \mu} & \swarrow{\scriptstyle \rho_A} & \\ & & A & & \end{array} \tag{3.26}$$

commutes. Here α is the associator of $\mathcal{C}$ and λ, ρ are the unit isomorphisms. An *algebra homomorphism* from (A, μ) to (A', μ') is a morphism $f : A \to A'$ such that $f \circ \mu = \mu' \circ (f \otimes f)$. If A and A' are unital, f is called *unital* if $f \circ \iota = \iota'$.

(ii) An algebra in a braided monoidal category is called *commutative* if $\mu \circ c_{A,A} = \mu$.

The tensor unit $\mathbf{1} \in \mathcal{C}$ with multiplication $\mu = \lambda_I = \rho_I$ and unit $\iota = id_{\mathbf{1}}$ is always a commutative unital algebra. A similar class of examples are objects $S \in \mathcal{C}$ such that $\mathcal{C}(S, S) = k \cdot id_S$ and $S \otimes_{\mathcal{C}} S \cong S$. Each isomorphism $S \otimes_{\mathcal{C}} S \to S$ is a commutative associative multiplication on S (not necessarily unital), and of course all these multiplications give isomorphic algebras, see Appendix B.2. In the $W_{2,3}$-model treated in Sect. 4, this will give three examples of algebras (namely the representations $\mathcal{W}$, $\mathcal{W}^*$ and $\mathcal{W}(0)$, see Sect. 4 for details).

Suppose $\mathcal{C}$ has property (C). By a *pairing* on an algebra A in $\mathcal{C}$ we mean a morphism $\pi : A \otimes A \to \mathbf{1}^*$. The pairing is called *invariant* if

$$\pi \circ (id_A \otimes \mu) = \pi \circ (\mu \otimes id_A) \circ \alpha_{A,A,A}. \tag{3.27}$$

Table 1 Relation between the algebraic notions of this section and the discussion of CFT in Sect. 2. These relations have been proved for non-logarithmic rational CFTs (see [13, 18, 28, 33]). In general the table should be understood as 'similarity in structure'. This table continues after some preparation with Table 2 below

Conformal field theory	Algebraic counterpart
Rep $\mathcal{V}$, for a vertex operator algebra $\mathcal{V}$ which is 'logarithmic-rational': the tensor product theory of [30] should apply and it should only have a finite number of irreducible sectors	A braided monoidal category $\mathcal{C}$ which is $\mathbb{C}$-linear, abelian, with right exact tensor product, and which satisfies the finiteness condition (PF) and has conjugates in the sense of condition (C)
(B, m, ω^*), a non-degenerate boundary theory as in Definitions 2.13 and 2.14	An algebra $B \in \mathcal{C}$ with associative product $m : B \otimes_{\mathcal{C}} B \to B$ and a map $\omega^* : B \to \mathbf{1}^*$ such that the pairing $\omega^* \circ m$ on B is non-degenerate
(F, M, Ω^*), a non-degenerate CFT on $\mathbb{C}$ as in Definitions 2.4 and 2.7	An algebra $F \in \mathcal{C} \boxtimes \mathcal{C}^{\text{rev}}$ with associative, commutative product $M : F \otimes_{\mathcal{C}^2} F \to F$ and a map $\Omega^* : F \to \mathbf{1}^* \boxtimes \mathbf{1}^*$ such that the pairing $\Omega^* \circ M$ is non-degenerate
(F, M), a CFT on $\mathbb{C}$ with background states as defined in Sect. 2.2	An algebra $F \in \mathcal{C} \boxtimes \mathcal{C}^{\text{rev}}$ with associative, commutative product $M : F \otimes_{\mathcal{C}^2} F \to F$

If A is unital, giving an invariant pairing is the same as giving a morphism $\tau : A \to \mathbf{1}^*$ via $\pi = \tau \circ \mu$. The notion of non-degeneracy of a paring on A is that of Definition 3.2.

A brief comparison between these algebraic notions and the discussion of CFT in Sect. 2 is given in Table 1.

It is not surprising that a monoidal functor between two monoidal categories transports algebras to algebras. However, also the weaker notion of a lax monoidal functor is sufficient for this purpose.

Definition 3.13 Let $\mathcal{A}$ and $\mathcal{B}$ be two monoidal categories and let $F : \mathcal{A} \to \mathcal{B}$ be a functor. Then F is called *lax monoidal* if it is equipped with morphisms $F_0 : \mathbf{1}_{\mathcal{B}} \to F(\mathbf{1}_{\mathcal{A}})$ and $F_{2;U,V} : F(U) \otimes_{\mathcal{B}} F(V) \to F(U \otimes_{\mathcal{A}} V)$, the latter natural in U, V, such that for all $U, V, W \in \mathcal{A}$,

$$\begin{array}{ccc}
F(U) \otimes_{\mathcal{B}} (F(V) \otimes_{\mathcal{B}} F(W)) & \xrightarrow{\alpha^{\mathcal{B}}_{FU,FV,FW}} & (F(U) \otimes_{\mathcal{B}} F(V)) \otimes_{\mathcal{B}} F(W) \\
\Big\downarrow {\scriptstyle id_{FU} \otimes F_{2;V,W}} & & \Big\downarrow {\scriptstyle F_{2;U,V} \otimes id_{FW}} \\
F(U) \otimes_{\mathcal{B}} F(V \otimes_{\mathcal{A}} W) & & F(U \otimes_{\mathcal{A}} V) \otimes_{\mathcal{B}} F(W) \\
\Big\downarrow {\scriptstyle F_{2;U,VW}} & & \Big\downarrow {\scriptstyle F_{2;UV,W}} \\
F(U \otimes_{\mathcal{A}} (V \otimes_{\mathcal{A}} W)) & \xrightarrow{F(\alpha^{\mathcal{A}}_{U,V,W})} & F((U \otimes_{\mathcal{A}} V) \otimes_{\mathcal{A}} W)
\end{array} \tag{3.28}$$

and

$$\begin{array}{ccc} \mathbf{1}_{\mathcal{B}} \otimes_{\mathcal{B}} F(U) & \xrightarrow{\lambda^{\mathcal{B}}_{FU}} & F(U) \\ {\scriptstyle F_0 \otimes id_{FU}} \downarrow & & \uparrow {\scriptstyle F(\lambda^{\mathcal{A}}_U)} \\ F(\mathbf{1}_{\mathcal{A}}) \otimes_{\mathcal{B}} F(U) & \xrightarrow{F_{2;\mathbf{1},U}} & F(\mathbf{1}_{\mathcal{A}} \otimes_{\mathcal{A}} U) \end{array} \quad , \quad \begin{array}{ccc} F(U) \otimes_{\mathcal{B}} \mathbf{1}_{\mathcal{B}} & \xrightarrow{\rho^{\mathcal{B}}_{FU}} & F(U) \\ {\scriptstyle id_{FU} \otimes F_0} \downarrow & & \uparrow {\scriptstyle F(\rho^{\mathcal{A}}_U)} \\ F(U) \otimes_{\mathcal{B}} F(\mathbf{1}_{\mathcal{A}}) & \xrightarrow{F_{2;U,\mathbf{1}}} & F(U \otimes_{\mathcal{A}} \mathbf{1}_{\mathcal{A}}) \end{array} \tag{3.29}$$

commute. If F_0 and F_2 are isomorphisms, F is called *strong monoidal* (or just *monoidal*).

Let $\mathcal{A}$, $\mathcal{B}$ be monoidal categories and let $F : \mathcal{A} \to \mathcal{B}$ be a lax monoidal functor. If (A, μ) is an algebra in $\mathcal{A}$, then the image object $F(A)$ also carries the structure of an algebra, with associative multiplication given by

$$\mu_{F(A)} = \big[F(A) \otimes_{\mathcal{B}} F(A) \xrightarrow{F_{2;A,A}} F(A \otimes_{\mathcal{A}} A) \xrightarrow{F(\mu)} F(A) \big]. \tag{3.30}$$

If A is unital with unit ι, then so is $F(A)$ with unit $F(\iota) \circ F_0$. If $f : A \to B$ is a homomorphism of algebras in $\mathcal{A}$, then $F(f) : F(A) \to F(B)$ is a homomorphism of algebras in $\mathcal{B}$. See Sect. 5 in [31] for details.

3.5 The Full Centre in $\mathcal{C} \boxtimes \mathcal{C}^{\mathrm{rev}}$

In this subsection, $\mathcal{C}$ is assumed to be a braided monoidal k-linear abelian category with conjugates as in (C), which satisfies the finiteness condition (PF), and which has a k-linear right exact tensor product functor. The assumptions (PF) and (C) will guarantee existence of the full centre of an algebra in $\mathcal{C}$, to be defined now (though much weaker conditions should be sufficient, too). Recall the definition of the functor $T : \mathcal{C} \boxtimes \mathcal{C}^{\mathrm{rev}} \to \mathcal{C}$ from Sects. 3.2 and 3.3, as well as the mixed braiding $\varphi_{X,A}$ from (3.18).

Definition 3.14 Let (A, μ_A) be an algebra in $\mathcal{C}$. The *full centre in* $\mathcal{C} \boxtimes \mathcal{C}^{\mathrm{rev}}$ is an object $Z(A) \in \mathcal{C} \boxtimes \mathcal{C}^{\mathrm{rev}}$ together with a morphism $z : T(Z(A)) \to A$ in $\mathcal{C}$ such that the following universal property holds: For all pairs (X, x) with $X \in \mathcal{C} \boxtimes \mathcal{C}^{\mathrm{rev}}$ and $x : T(X) \to A$ such that the diagram

$$\begin{array}{ccccc} T(X) \otimes_{\mathcal{C}} A & \xrightarrow{x \otimes_{\mathcal{C}} id_A} & A \otimes_{\mathcal{C}} A & & \\ & & & \searrow^{\mu_A} & \\ {\scriptstyle \varphi_{X,A}} \downarrow & & & & A \\ & & & \nearrow_{\mu_A} & \\ A \otimes_{\mathcal{C}} T(X) & \xrightarrow{id_A \otimes_{\mathcal{C}} x} & A \otimes_{\mathcal{C}} A & & \end{array} \tag{3.31}$$

Table 2 Continuation of Table 1

Conformal field theory	Algebraic counterpart
$\tilde{b}$, a 'candidate bulk-boundary map' from a 'candidate space of bulk fields' $\tilde{F}$ to the space of boundary fields B of a boundary theory, satisfying in particular the centrality condition from Sect. 2.5	An object $\tilde{F} \in \mathcal{C} \boxtimes \mathcal{C}^{\text{rev}}$ and an algebra $B \in \mathcal{C}$ together with a morphism $\tilde{b} : T(\tilde{F}) \to B$ such that $\tilde{b} \in \text{Cent}(\tilde{F}, B)$
$\mathcal{P}$, the category of pairs from Sect. 2.5 and the terminal object $(F(B), b(B))$ in it from (2.33), interpreted as the maximal bulk theory compatible with the given boundary theory B	The category $\mathcal{C}_{\text{full center}}(B)$ for a given algebra $B \in \mathcal{C}$ and the terminal object (Z, z) in it, where $Z \in \mathcal{C} \boxtimes \mathcal{C}^{\text{rev}}$ is the full centre of B and z the corresponding map $T(Z) \to B$
$(F, M, \Omega^*; B, m, \omega^*; b)$, a CFT on the upper half plane as in Definition 2.11, for which the CFT on $\mathbb{C}$ and the boundary theory are non-degenerate	A commutative algebra (F, M) in $\mathcal{C} \boxtimes \mathcal{C}^{\text{rev}}$ with non-degenerate pairing $\Omega^* \circ M$, a not necessarily commutative algebra (B, m) in $\mathcal{C}$ with non-degenerate pairing $\omega^* \circ m$, and an algebra map $b : T(F) \to B$, such that $b \in \text{Cent}(F, B)$

in $\mathcal{C}$ commutes, there exists a unique morphism $\zeta_{(X,x)} : X \to Z(A)$ such that

$$\begin{array}{ccc} T(X) & \xrightarrow{T(\zeta_{(X,x)})} & T(Z(A)) \\ & {\scriptstyle x}\searrow \quad \swarrow{\scriptstyle z} & \\ & A & \end{array} \tag{3.32}$$

commutes.

The existence of the full centre will be proved in Theorem 3.24 below.

For later use we give a name to the space of maps for which the diagram (3.31) commutes. For A an algebra in $\mathcal{C}$ and $X \in \mathcal{C} \boxtimes \mathcal{C}^{\text{rev}}$

$$\text{Cent}(X, A) := \big\{x : T(X) \to A \,|\, (3.31) \text{ commutes}\big\} \tag{3.33}$$

('Cent' for centrality condition, cf. Table 2).

Remark 3.15

(i) The above definition can be recast into describing the full centre in $\mathcal{C} \boxtimes \mathcal{C}^{\text{rev}}$ as a terminal object. Namely, consider the category $\mathcal{C}_{\text{full center}}(A)$ whose objects are pairs (X, x) with $X \in \mathcal{C} \boxtimes \mathcal{C}^{\text{rev}}$ and $x \in \text{Cent}(X, A)$, and whose morphisms are maps $f : X \to Y$ in $\mathcal{C} \boxtimes \mathcal{C}^{\text{rev}}$ such that

$$\begin{array}{ccc} T(X) & \xrightarrow{T(f)} & T(Y) \\ & {\scriptstyle x}\searrow \quad \swarrow{\scriptstyle y} & \\ & A & \end{array} \tag{3.34}$$

commutes. By definition, the full centre (Z, z) of an algebra A is a terminal object in $\mathcal{C}_{\text{full center}}(A)$.

(ii) The full centre was introduced in [46] and Definition 4.9 in [14] in the case that $\mathcal{C}$ is a modular category and is (in particular) an object in $\mathcal{C} \boxtimes \mathcal{C}^{\text{rev}}$. The notion was then generalised to arbitrary monoidal categories (not necessarily braided or abelian) in Sect. 4 in [7], where the full centre, if it exists, is an object in the monoidal centre $\mathcal{Z}(\mathcal{C})$ of $\mathcal{C}$. If $\mathcal{C}$ is modular, $\mathcal{Z}(\mathcal{C}) \cong \mathcal{C} \boxtimes \mathcal{C}^{\text{rev}}$ by Theorem 7.10 in [39], and the two definitions agree (cf. Sect. 8 in [7]). In general, $\mathcal{Z}(\mathcal{C})$ and $\mathcal{C} \boxtimes \mathcal{C}^{\text{rev}}$ may not be equivalent (but T factors through $\mathcal{Z}(\mathcal{C})$, cf. Remark 3.11). For this reason, we added the suffix 'in $\mathcal{C} \boxtimes \mathcal{C}^{\text{rev}}$' to the name 'full centre' in Definition 3.14. However, because we will only ever use the full centre in $\mathcal{C} \boxtimes \mathcal{C}^{\text{rev}}$ and never the full centre in $\mathcal{Z}(\mathcal{C})$, we will drop the suffix 'in $\mathcal{C} \boxtimes \mathcal{C}^{\text{rev}}$' from now on.

Let (Z, z) be the full centre of an algebra A in $\mathcal{C}$ as in Definition 3.14. Suppose we are given a morphism $\mu_Z : Z \otimes_{\mathcal{C}^2} Z \to Z$; this will later be an associative, commutative product, but let us not demand that yet. Equation (3.30) defines a product $\mu_{T(Z)}$ on $T(Z)$. Suppose further that z intertwines $\mu_{T(Z)}$ and μ_A, i.e.

$$\begin{array}{ccccc} T(Z) \otimes_{\mathcal{C}} T(Z) & & \xrightarrow{\quad \mu_{T(Z)} \quad} & & T(Z) \\ & \nwarrow^{T^{-1}_{2;Z,Z}} & & \nearrow^{T(\mu_Z)} & \\ {\scriptstyle z \otimes_{\mathcal{C}} z} \downarrow & & T(Z \otimes_{\mathcal{C}^2} Z) & & \downarrow {\scriptstyle z} \\ A \otimes_{\mathcal{C}} A & & \xrightarrow{\quad \mu_A \quad} & & A \end{array} \tag{3.35}$$

commutes (we included also the definition of $\mu_{T(Z)}$). This diagram can be read in a second way: Starting from $T(Z \otimes_{\mathcal{C}^2} Z)$ and following the two paths to A, we see that it is an instance of (3.32). If we can establish (3.31) for the left path, the universal property of (Z, z) provides us with a unique choice for μ_Z, which, as we will see, is automatically associative and commutative. This is done in the next statement, which is just Proposition 4.1 in [7] with $\mathcal{Z}(\mathcal{C})$ replaced by $\mathcal{C} \boxtimes \mathcal{C}^{\text{rev}}$. Even the proof works in the same way. Still, as the full centre is one of the main players in this paper we include parts of the proof in Appendix B.3.

Theorem 3.16 *Let (Z, z) be the full centre of an algebra $A \in \mathcal{C}$. There exists a unique product $\mu_Z : Z \otimes Z \to Z$ such that* (3.35) *commutes. This product is associative and commutative. If A has a unit ι_A, then there exists a unique map $\iota_Z : \mathbf{1} \to Z$ such that*

$$\begin{array}{ccc} T(\mathbf{1}) & \xrightarrow{T(\iota_Z)} & T(Z) \\ {\scriptstyle (F_0)^{-1}} \downarrow & & \downarrow {\scriptstyle z} \\ \mathbf{1} & \xrightarrow{\iota_A} & A \end{array} \tag{3.36}$$

commutes. This map is a unit for the product μ_Z.

In particular, $z : T(Z) \to A$ is an algebra map. It is unital if A is unital.

3.6 The Right Adjoint R of T

As in the previous subsection, $\mathcal{C}$ is assumed to be k-linear abelian and braided monoidal, to satisfy (PF) and (C), and to have a k-linear right exact tensor product. The aim of this section is to show the existence of the right adjoint $R : \mathcal{C} \to \mathcal{C} \boxtimes \mathcal{C}^{\text{rev}}$ of the functor $T : \mathcal{C} \boxtimes \mathcal{C}^{\text{rev}} \to \mathcal{C}$ and give an explicit expression for it. In the next subsection, the adjoint R will be used to give an explicit description of the full centre and thereby prove its existence.

Consider the functor $\mathcal{C}(T(-), \mathbf{1}^*)$ from $\mathcal{C} \boxtimes \mathcal{C}^{\text{rev}}$ to $\mathcal{V}ect$. Define $R_{\mathbf{1}^*}$ (if it exists) to be the representing object of this functor. That is, there is a natural (in X) isomorphism of functors $\mathcal{C} \boxtimes \mathcal{C}^{\text{rev}} \to \mathcal{V}ect$,

$$\chi_X : \mathcal{C}\big(T(X), \mathbf{1}^*\big) \xrightarrow{\sim} \mathcal{C}^2(X, R_{\mathbf{1}^*}). \tag{3.37}$$

Here and below, $\mathcal{C}^2(X, Y)$ denotes the space of morphisms from X to Y in $\mathcal{C} \boxtimes \mathcal{C}^{\text{rev}}$. We will now show that $R_{\mathbf{1}^*}$ may be written as the cokernel of a morphism between two projective objects in $\mathcal{C} \boxtimes \mathcal{C}^{\text{rev}}$; in particular, $R_{\mathbf{1}^*}$ exists.

Let P be a projective generator of $\mathcal{C}$ (which exists by (PF)). By Corollary 3.7, also $\mathcal{C} \boxtimes \mathcal{C}^{\text{rev}}$ satisfies property (PF) and $P \boxtimes P$ is a projective generator of $\mathcal{C} \boxtimes \mathcal{C}^{\text{rev}}$. Define the linear subspace $N \subset \mathcal{C}^2(P \boxtimes P, P \boxtimes P^*)$ to consist of all $f : P \boxtimes P \to P \boxtimes P^*$ such that

$$\big[P \otimes_{\mathcal{C}} P \xrightarrow{T(f)} P \otimes_{\mathcal{C}} P^* \xrightarrow{\beta_P} \mathbf{1}^*\big] = 0, \tag{3.38}$$

where β_P is the non-degenerate pairing defined in (3.1). Let $\{u_1, \dots, u_{|N|}\}$ be a basis of N (the space is finite-dimensional by (PF)). Define the map $n : (P \boxtimes P)^{\oplus |N|} \to P \boxtimes P^*$ as $n = \sum_{i=1}^{|N|} u_i \circ \pi_i$, with π_i the projection to the ith direct summand. Define R' to be the cokernel of n, so that we have the exact sequence

$$(P \boxtimes P)^{\oplus |N|} \xrightarrow{n} P \boxtimes P^* \xrightarrow{\text{cok}(n)} R' \longrightarrow 0. \tag{3.39}$$

Now consider the diagram

$$\begin{array}{ccccccc} T((P \boxtimes P)^{\oplus |N|}) & \xrightarrow{T(n)} & P \otimes P^* & \xrightarrow{T(\text{cok}(n))} & T(R') & \longrightarrow & 0 \\ & & \Big\downarrow{\scriptstyle \beta_P} & \swarrow{\scriptstyle \exists! r'} & & & \\ & & \mathbf{1}^* & & & & \end{array} \tag{3.40}$$

Since by its construction in (3.7), T is right exact, the top row of the diagram is exact, i.e. $T(\mathrm{cok}(n))$ is the cokernel of $T(n)$. Because $\beta_P \circ T(n) = \sum_{i=1}^{|N|} \beta_P \circ T(u_i) \circ T(\pi_i) = 0$ (by definition of the u_i), from the universal property of the cokernel we obtain the arrow $r' : T(R') \to \mathbf{1}^*$. The next theorem, whose proof can be found in Appendix B.4, states that R' is the object we are looking for.

Theorem 3.17 *The object R' just constructed represents the functor $\mathcal{C}(T(-), \mathbf{1}^*)$, i.e. one may take $R_{\mathbf{1}^*} = R'$.*

We will soon use the object $R_{\mathbf{1}^*}$ to construct the entire adjoint functor R, but first we would like to state one important property of $R_{\mathbf{1}^*}$. Given a natural transformation $(\nu_U : U \to U)_{U \in \mathcal{C}}$ of the identity functor on $\mathcal{C}$, set

$$\tilde{\nu}_U = \big[U \xrightarrow{\delta_U} U^{**} \xrightarrow{(\nu_{U^*})^*} U^{**} \xrightarrow{\delta_U^{-1}} U\big], \tag{3.41}$$

where δ is the natural isomorphism $Id \Rightarrow (-)^{**}$ from condition (C). Then $\tilde{\nu}$ is again natural in U. In particular, both $\boxtimes \circ (\nu \times id)$ and $\boxtimes \circ (id \times \tilde{\nu})$ are natural transformations of $\boxtimes : \mathcal{C} \times \mathcal{C}^{\mathrm{rev}} \to \mathcal{C} \boxtimes \mathcal{C}^{\mathrm{rev}}$. Via the defining equivalence (3.2), these give two natural transformations $\nu \boxtimes id$ and $id \boxtimes \tilde{\nu}$ of the identity functor on $\mathcal{C} \boxtimes \mathcal{C}^{\mathrm{rev}}$.

Theorem 3.18 *Let $R_{\mathbf{1}^*} \in \mathcal{C} \boxtimes \mathcal{C}^{\mathrm{rev}}$ be as above and let $\nu : Id_{\mathcal{C}} \Rightarrow Id_{\mathcal{C}}$ be a natural transformation. Then $(\nu \boxtimes id)_{R_{\mathbf{1}^*}} = (id \boxtimes \tilde{\nu})_{R_{\mathbf{1}^*}}$.*

The theorem is proved in Appendix B.4.

Remark 3.19 Let $\mathcal{V}$ be a vertex operator algebra such that $\mathcal{C} = \mathrm{Rep}(\mathcal{V})$ satisfies the conditions set out in the beginning of this subsection. Then $\exp(2\pi i L_0)$ acting on some representation $A \in \mathcal{C}$ is an example of a natural transformation of the identity functor (it commutes with all modes of all fields in the VOA, and it can be moved past all intertwiners $f : A \to B$, i.e. it is natural in A). Theorem 3.18 states in this case that $\exp(2\pi i \cdot L_0 \otimes_{\mathbb{C}} id)$ and $\exp(2\pi i \cdot id \otimes_{\mathbb{C}} L_0)$ act in the same way on the $\mathcal{V} \otimes_{\mathbb{C}} \mathcal{V}$-module $R_{\mathbf{1}^*}$. In other words,

$$\exp\big\{2\pi i (L_0 \otimes_{\mathbb{C}} id - id \otimes_{\mathbb{C}} L_0)\big\}\big|_{R_{\mathbf{1}^*}} = id_{R_{\mathbf{1}^*}}. \tag{3.42}$$

In CFT terms this means that in a situation where $R_{\mathbf{1}^*}$ is the space of bulk fields,[14] the partition function is invariant under the T-transformation $\tau \mapsto \tau + 1$.

We now turn to the right adjoint R. The involution $(-)^*$ on $\mathcal{C}$ induces an involution on $\mathcal{C} \boxtimes \mathcal{C}^{\mathrm{rev}}$, which we also denote by $(-)^*$, and which also satisfies condition

[14]For this to be possible, we must have $Z(\mathbf{1}^*) = R(\mathbf{1}^*)$ (we will see in (3.43) that $R(\mathbf{1}^*) = R_{\mathbf{1}^*}$). By Lemma 3.25 below this is true if $\mathbf{1}^* \cong \mathbf{1}$. We expect that $Z(\mathbf{1}^*) = R(\mathbf{1}^*)$ also holds in the $W_{2,3}$-model (where $\mathbf{1}^* \ncong \mathbf{1}$), see Sect. 4.

(C) (see Appendix B.5). We can use $R_{\mathbf{1}^*}$ and the involution $(-)^*$ to define a functor $R : \mathcal{C} \to \mathcal{C} \boxtimes \mathcal{C}^{\text{rev}}$ as

$$R(U) = \big((U^* \boxtimes \mathbf{1}) \otimes_{\mathcal{C}^2} (R_{\mathbf{1}^*})^*\big)^*, \qquad R(f) = \big((f^* \boxtimes id_{\mathbf{1}}) \otimes_{\mathcal{C}^2} id_{(R_{\mathbf{1}^*})^*}\big)^*. \quad (3.43)$$

Note that $R(\mathbf{1}^*) \cong R_{\mathbf{1}^*}$.

Theorem 3.20 *The functor R is a right adjoint for T.*

The proof and the adjunction isomorphisms are given in Appendix B.5. On general grounds, the functor R, being adjoint to a monoidal functor, is lax monoidal (see, e.g., Lemma 2.7 in [34]). The structure maps R_0 and R_2 can equally be found in Appendix B.5. Thus, for an algebra $A \in \mathcal{C}$, $R(A)$ is an algebra in $\mathcal{C} \boxtimes \mathcal{C}^{\text{rev}}$ with multiplication (3.30).

Remark 3.21

(i) Since R is a right adjoint functor, it is left exact. This can also be seen explicitly from (3.43), namely

$$R = \Big[\mathcal{C}^{\text{op}} \xrightarrow{(-)^*} \mathcal{C} \xrightarrow{(-)\boxtimes \mathbf{1}} \mathcal{C} \boxtimes \mathcal{C}^{\text{rev}} \xrightarrow{(-)\otimes_{\mathcal{C}^2} (R_{\mathbf{1}^*})^*} \mathcal{C} \boxtimes \mathcal{C}^{\text{rev}} \xrightarrow{(-)^*} (\mathcal{C} \boxtimes \mathcal{C}^{\text{rev}})^{\text{op}}\Big], \quad (3.44)$$

where $(-)^*$ is exact, $(-) \boxtimes \mathbf{1}$ is exact (see the beginning of the proof of Theorem 3.6), and $(-) \otimes_{\mathcal{C}^2} (R_{\mathbf{1}^*})^*$ is right exact. Thus R is a right exact functor $\mathcal{C}^{\text{op}} \to (\mathcal{C} \boxtimes \mathcal{C}^{\text{rev}})^{\text{op}}$ which is the same as a left exact functor $\mathcal{C} \to \mathcal{C} \boxtimes \mathcal{C}^{\text{rev}}$.

(ii) Suppose there are isomorphisms $(U \otimes_{\mathcal{C}} V)^* \to V^* \otimes_{\mathcal{C}} U^*$, natural in U and V. Then, firstly, $\mathbf{1} \cong (\mathbf{1}^*)^* \cong (\mathbf{1} \otimes_{\mathcal{C}} \mathbf{1}^*)^* \cong \mathbf{1} \otimes_{\mathcal{C}} \mathbf{1}^* \cong \mathbf{1}^*$. Secondly, the above formula for R simplifies to $R(U) = R_{\mathbf{1}} \otimes_{\mathcal{C}^2} (U \boxtimes \mathbf{1})$, and analogously for $R(f)$. In this formulation, R is clearly right exact, so that together with (i) we see that R is exact. Similarly, the natural isomorphism $(-)^* \circ T \circ (-)^* \cong T$ shows that T is exact.

3.7 Left Centre and Full Centre

In this subsection, we will express the full centre of an algebra A as the 'left centre'—to be defined momentarily—of the adjoint functor R applied to A. As before, $\mathcal{C}$ is assumed to be k-linear abelian and braided monoidal, to satisfy (PF) and (C), and to have a k-linear right exact tensor product.

In the braided setting, one distinguishes three different notions of the centre of an algebra: the left centre, the right centre and the full centre. From these, the left and right centre are subobjects of the algebra itself, while—as we have seen in Definition 3.14—the full centre lives in a different category. The left and right centres were introduced in [50] and appeared in various incarnations in [7, 17, 40]. The following definition is taken from Sect. 5 in [7].

Definition 3.22 Let B be an algebra in a braided monoidal category $\mathcal{B}$. The *left centre* of B is terminal among all pairs $(U \in \mathcal{B}, u : U \to B)$, such that

$$\begin{array}{ccccc} U \otimes_{\mathcal{B}} B & \xrightarrow{u \otimes_{\mathcal{B}} id_B} & B \otimes_{\mathcal{B}} B & & \\ & & & \overset{\mu_B}{\searrow} & \\ {\scriptstyle c_{U,B}} \downarrow & & & & B \\ & & & \underset{\mu_B}{\nearrow} & \\ B \otimes_{\mathcal{B}} U & \xrightarrow{id_B \otimes_{\mathcal{B}} u} & B \otimes_{\mathcal{B}} B & & \end{array} \tag{3.45}$$

commutes. Equivalently, the left centre is an object $C_l(B)$ in $\mathcal{B}$ together with a morphism $e : C_l(B) \to B$ such that the pair $(C_l(B), e)$ satisfies (3.45), and such that for every other pair (U, u) satisfying (3.45) there exists a unique arrow $\tilde{u} : U \to C_l(B)$ such that $e \circ \tilde{u} = u$.

Remark 3.23

(i) There is an analogous definition for the right centre, see [40, 50].
(ii) If the category $\mathcal{B}$ is in addition abelian and has conjugates as in (C), the left centre of an algebra B can be expressed as a kernel. This shows at the same time that the left centre exists and that $e : C_l(B) \to B$ is injective, i.e. $C_l(B)$ is a subobject of B. See Appendix B.6 for details.
(iii) $C_l(B)$ carries a unique algebra structure such that $e : C_l(B) \to B$ is an algebra map. This algebra structure on $C_l(B)$ is commutative. If A is unital, so is $C_l(A)$. See [7, 40, 50] for details.

In the previous subsection we gave the direct definition of the full centre as first formulated in Sect. 4 in [7]. The original definition in Definition 4.9 in [14] proceeds in two steps: first, one applies the adjoint R of T to the algebra A and second, one finds the left centre of $R(A)$. The same works in the present setting, as we now show. The proof is the same as in Theorem 5.4 in [7], we reproduce an adapted version in Appendix B.6. Denote the adjunction natural transformation $TR \Rightarrow Id$ by ε, cf. (B.46).

Theorem 3.24 *Let A be an algebra in $\mathcal{C}$. The pair (Z, z) with*

$$Z = C_l\big(R(A)\big), \qquad z = \big[T\big(C_l\big(R(A)\big)\big) \xrightarrow{T(e)} T\big(R(A)\big) \xrightarrow{\varepsilon_A} A\big] \tag{3.46}$$

is the full centre of A.

In particular, since R exists by Theorem 3.20 and the left centre exists by Remark 3.23(ii), the full centre of an algebra exists under the assumptions set out in the beginning of this subsection.

Even if A is a commutative algebra, $R(A)$ need not be commutative and one still needs to take the left centre to arrive at $Z(A)$. However, the next lemma gives a simple condition in addition to commutativity which guarantees $Z(A) = R(A)$; this will be useful in Sect. 4. An object $S \in \mathcal{C}$ is called *transparent* if $c_{U,S} \circ c_{S,U} = id_{S \otimes U}$ for all $U \in \mathcal{C}$.

Lemma 3.25 *If (S, μ_S) is a commutative algebra in $\mathcal{C}$ and S is transparent in $\mathcal{C}$, then we can take $(Z(S), z) = (R(S), \varepsilon_S)$. In particular, $Z(\mathbf{1}) = R(\mathbf{1})$.*

Proof From (3.17) one checks that for transparent S we have $\tilde{\varphi}_{A,B,S} = c_{A \otimes B,S}$. Thus also $\varphi_{X,S} = c_{T(X),S}$. Condition (3.31) is then true for all $x : T(X) \to S$ as by commutativity of S we have $\mu_S \circ c_{S,S} = \mu_S$. But then the universal property of the full centre reduces to that in Lemma B.2(ii) with $U = S$, $R' = Z(S)$ and $r' = z$. By part (i) of that lemma, $R' = R(S)$ and $r' = \xi^{-1}_{R(S),S}(id_{R(S)}) = \varepsilon_S$, see (B.44) and (B.46). □

Remark 3.26 A different approach to finding algebraic counterparts to logarithmic CFTs on $\mathbb{C}$ is taken in [19]. There, the category $\mathcal{C}$ is chosen to be H-Mod for a certain Hopf algebra H (in more detail, finite dimensional representations of a finite-dimensional factorisable ribbon Hopf algebra) and $\mathcal{C} \boxtimes \mathcal{C}^{\text{rev}} \cong H$-Bimod (see Appendix A.3 in [19]). In H-Bimod the coregular bimodule H^* is studied and shown to be a commutative Frobenius algebra (Propositions 2.10 and 3.1 in [19]). In addition, H^* satisfies certain modular invariance properties (Theorem 5.6 in [19]). In Appendix B in [19], the bimodule H^* is proposed to be a candidate bulk theory for a logarithmic CFT in case H-Mod $\cong$ Rep $\mathcal{V}$ for the vertex operator algebra $\mathcal{V}$ encoding the chiral symmetry. In the setting of the present paper, H^* corresponds to $R(\mathbf{1})$.

4 The $W_{2,3}$-Model with $c = 0$

In this section we look more closely at one particular class of examples, namely conformal field theories built from representations of the $W_{2,3}$ vertex operator algebra. This symmetry algebra was chosen because it demonstrates that the level of generality assumed in Sect. 3 is indeed needed in the treatment of interesting examples.

We start in Sect. 4.1 with a brief collection of what is known or expected about the representation theory of the $W_{2,3}$ vertex operator algebra. In Sect. 4.2 it is shown how the formalism of finding the maximal bulk theory for a given boundary theory can produce the (trivial) $c = 0$ Virasoro minimal model. A non-trivial bulk theory is discussed in Sects. 4.3–4.5; this bulk theory is logarithmic and can be understood as a ‘refinement’ of the $c = 0$ minimal model.

4.1 The W-Algebra and Some of Its Representations

Let Ver($h=0, c=0$) be the Virasoro Verma module generated by the state Ω with $L_0\Omega = C\Omega = 0$. It has a maximal proper submodule which is generated by the two vectors

$$n_1 = L_{-1}\Omega, \qquad n_2 = \left(L_{-2} - \frac{3}{2}L_{-1}L_{-1}\right)\Omega. \tag{4.1}$$

Since words in L_{-1} and L_{-2} acting on Ω span Ver$(0,0)$, the quotient Ver$(0,0)/\langle n_1, n_2\rangle$ is just $\mathbb{C}\Omega$ with trivial Vir-action. This describes the vacuum representation of the Virasoro minimal model with $c=0$, which is trivial in the sense that it is a two-dimensional topological field theory for the commutative algebra $\mathbb{C}$, cf. Remark 2.6. The module $\mathcal{V} \equiv$ Ver$(0,0)/\langle n_1\rangle$ is infinite dimensional and carries the structure of a vertex operator algebra with Virasoro element $T = L_{-2}\Omega \neq 0$ (note that Ver$(0,0)$ is not itself a vertex operator algebra because the vacuum Ω is not annihilated by the translation operator L_{-1}). The VOA $\mathcal{V}$ has an infinite number of distinct irreducible representations (see Theorem 4.4 in [15] and Sect. 2.3 in [38]). To be able to apply the discussion in Sect. 3, we can pass to a larger VOA $\mathcal{W} \supset \mathcal{V}$, which is the chiral symmetry algebra for the $W_{2,3}$-model [1, 12], and is obtained as an extension of $\mathcal{V}$ by two fields of weight 15. Its character reads

$$\chi_{\mathcal{W}}(q) = 1 + q^2 + q^3 + 2q^4 + 2q^5 + 4q^6 + 4q^7 + 7q^8 + 8q^9 + 12q^{10} \\ + 14q^{11} + 21q^{12} + 24q^{13} + 34q^{14} + 44q^{15} + 58q^{16} + \cdots. \tag{4.2}$$

It turns out that $\chi_{\mathcal{W}}(q)$ differs from the character of $\mathcal{V}$ only starting from q^{15}, namely $\chi_{\mathcal{V}}(q) = \cdots + 41q^{15} + 55q^{16} + \cdots$ (so e.g. there are three new fields at weight 15). The VOA $\mathcal{W}$ is C_2-cofinite [1] and has 13 irreducible representations [1, 12], which we label by their lowest L_0-weight:

	$s=1$	$s=2$	$s=3$
$r=1$	$0, 2, 7$	$0, 1, 5$	$\frac{1}{3}, \frac{10}{3}$
$r=2$	$\frac{5}{8}, \frac{33}{8}$	$\frac{1}{8}, \frac{21}{8}$	$\frac{-1}{24}, \frac{35}{24}$

(4.3)

Here, the two entries '0' refer to the same irreducible representation $\mathcal{W}(0) \equiv \mathbb{C}\Omega$. We write $\mathcal{W}(h)$ for the irreducible $\mathcal{W}$-representation of lowest L_0-weight h. Their characters (from [12]) are listed in our notation in Appendix A.1 in [25].

At this point we note the first oddity of the $W_{2,3}$-model: the vertex operator algebra $\mathcal{W}$ is not one of the irreducible representations: it is indecomposable but not irreducible. Indeed, Ω is a cyclic vector (hence indecomposability) and the stress tensor $T = L_{-2}\Omega$ generates a $\mathcal{W}$-subrepresentation (on the level of Virasoro modes, this follows since $L_nT = 0$ for $n > 0$, where $L_2T = 0$ is a special feature of $c = 0$). Specifically, $\mathcal{W}$ is the middle term in a non-split exact sequence

$$0 \longrightarrow \mathcal{W}(2) \longrightarrow \mathcal{W} \longrightarrow \mathcal{W}(0) \longrightarrow 0. \tag{4.4}$$

This brings us to the second oddity. Denote by R^* the contragredient representation of a representation R (see, e.g. Definition I:2.35 in [30]). Then $\mathcal{W}^* \ncong \mathcal{W}$, as can be seen for example from their socle filtration[15]

$$\mathcal{W}: \begin{matrix} 0 \\ | \\ 2 \end{matrix}, \qquad \mathcal{W}^*: \begin{matrix} 2 \\ | \\ 0 \end{matrix}. \tag{4.5}$$

The above diagrams show the semi-simple quotients of successive submodules in the socle filtration. For example, the largest semi-simple subrepresentation of $\mathcal{W}$ is $\mathcal{W}(2)$. Quotienting by $\mathcal{W}(2)$, one obtains a representation whose largest semi-simple subrepresentation is $\mathcal{W}(0)$, and this accounts for all of $\mathcal{W}$ (this is just the statement of the sequence (4.4) and the fact that it is non-split). For $\mathcal{W}^*$, the largest semi-simple subrepresentation is $\mathcal{W}(0)$ and the quotient is isomorphic to $\mathcal{W}(2)$.

We have now pretty much reached the frontier of established mathematical truth regarding the $W_{2,3}$-model. Hence it is time for the following

Disclaimer: The statements concerning the structure of the $W_{2,3}$-model in the remainder of Sect. 4 should be treated as conjectures, even if we refrain from writing 'conjecturally' in every sentence.

The first statement under the umbrella of the above disclaimer is: The tensor product theory of [30] turns $\mathcal{C} \equiv \mathrm{Rep}(\mathcal{W})$ into a braided monoidal category

+ which has property (PF) from Sect. 3.2,
+ whose tensor product functor is right exact in each argument,
+ whose contragredient functor $(-)^*$ has property (C) from Sect. 3.1.

The category $\mathrm{Rep}(\mathcal{W} \otimes_{\mathbb{C}} \mathcal{W})$ (with inverse convention for the braiding in the second factor) is just the Deligne product $\mathcal{C} \boxtimes \mathcal{C}^{\mathrm{rev}}$. Every irreducible $\mathcal{W}(h)$ has a projective cover, which we denote by $\mathcal{P}(h)$. The fusion rules of the representations generated from the 13 irreducibles and from $\mathcal{W}^*$ (and from two representations $\mathcal{Q}$, $\mathcal{Q}^*$ which have the socle filtration (4.5) with 2 replaced by 1) are listed in Appendix 4 in [25] (see also [10, 12, 43]); some of the fusion rules of the projective cover $\mathcal{P}(0)$ are given in Appendix B.1 in [26]. The fusion-tensor product of $\mathcal{C}$ will be denoted by $\otimes_f$.

The properties marked '+' above allow one to apply the formalism in Sect. 3. However, there are many other convenient properties which $\mathcal{C}$ does not have:

– $\mathcal{C}$ is not semi-simple (e.g. the sequence (4.4) is not split).

---

[15]The *socle* $\mathrm{soc}(M)$ of a module M is the largest semi-simple submodule contained in M. The *socle filtration* of M is the unique filtration $\{0\} = M_0 \subset M_1 \subset M_2 \subset \cdots \subset M_n = M$ where M_1 is the socle of M, M_2/M_1 is the socle of M/M_1 and in general M_{i+1}/M_i is semi-simple and equals the socle of M/M_i. The socle filtration is a unique version of the composition series. In the latter, one iteratively picks a simple submodule and quotients by it. The composition series hence involves choices.

- The tensor unit $\mathbf{1} \equiv \mathcal{W}$ in $\mathcal{C}$ is not simple (cf. (4.4)).
- $\mathbf{1} \equiv \mathcal{W}$ is not isomorphic to its conjugate $\mathbf{1}^* \equiv \mathcal{W}^*$ (cf. (4.5)).
- The involution $(-)^*$ is not monoidal, e.g. $(\mathcal{W} \otimes_f \mathcal{W}(0))^* = \mathcal{W}(0)$ and $\mathcal{W}^* \otimes_f \mathcal{W}(0)^* = 0$.
- The tensor product of $\mathcal{C}$ is not exact. For example, the functor $\mathcal{W}(0) \otimes_f (-)$ transports the exact sequence $0 \to \mathcal{W}(0) \to \mathcal{W}^* \to \mathcal{W}(2) \to 0$ to $0 \to \mathcal{W}(0) \to 0 \to 0 \to 0$, which is not exact.
- $\mathcal{C}$ is not rigid, i.e. not every object has a dual (in the categorical sense—not to be confused with the contragredient representation, which always exists). Examples are the irreducibles $\mathcal{W}(0)$, $\mathcal{W}(1)$, $\mathcal{W}(2)$, $\mathcal{W}(5)$, $\mathcal{W}(7)$, the contragredient $\mathcal{W}^*$ of the VOA, and the projective cover $\mathcal{P}(0)$; we refer to Sect. 1.1.1 in [25] and Appendix B.1 in [26] for details.
- Even if $U \in \mathcal{C}$ has a dual $U^\vee$, it may happen that $U^\vee \ncong U^*$, e.g. $\mathcal{W}^\vee = \mathcal{W}$ (the tensor unit is self-dual in any monoidal category), but $\mathcal{W}^* \ncong \mathcal{W}$.

For the rest of this subsection we take a look at the most intricate[16] of the $\mathcal{W}$-representations, the projective cover $\mathcal{P}(0)$ of $\mathcal{W}(0)$. It has the socle filtration (as argued for in Appendix B.2 in [26]):

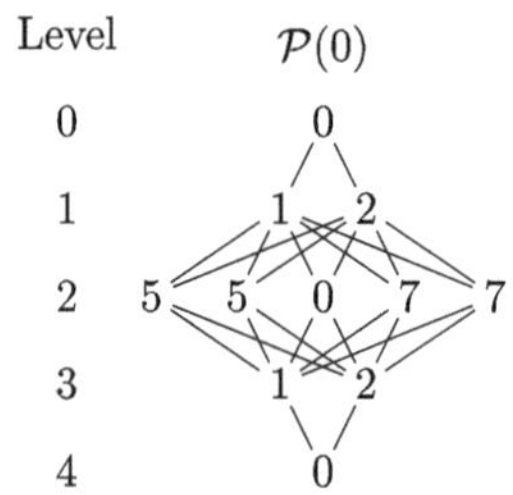

(4.6)

As before, the numbers in each row give the simple summands in the quotient of two consecutive layers of the socle filtration. The lines indicate the action of the W-modes; for $\mathcal{P}(0)$ they merely state that a vector at a given level can be transported into any of the lower lying submodules (this is not so for $\mathcal{P}(h)$ with $h = 1, 2, 5, 7$, see Appendix A.1 in [26]).

There are three quasi-primary states of generalised L_0-weight 0, no such states at weight 1, two states at weight 2, and infinitely more at higher weights.[17] It seems natural to us that the quasi-primary states up to weight 2 organise themselves under the Virasoro action as in the following diagram (the action is given up to constants,

---

[16] Also the $\mathcal{P}(h)$ with $h \in \{1, 2, 5, 7\}$ have a socle filtration with 5 levels (see Appendix A.1 in [26]). $\mathcal{P}(0)$ is 'most intricate' in the sense that it does not occur in the representations generated by fusion from the 13 irreducibles and its structure has only been found by indirect reasoning.

[17] The character of $\mathcal{P}(0)$ starts as $3 + 2q + 4q^2 + \cdots$. The two states at weight 1 are L_{-1}-descendants, as are two of the four states at weight 2.

see below for the full expressions)

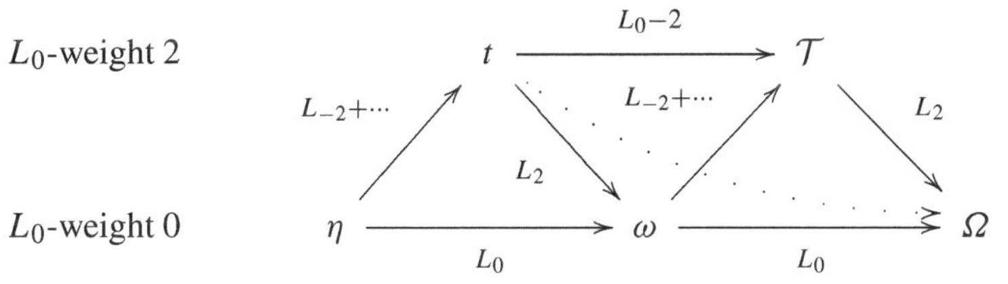

(4.7)

where $L_{-2}+\cdots$ stands for the operators defined by

$$t := \left(L_{-2} - \frac{3}{2}L_{-1}L_{-1} + \frac{9}{5}\left(L_{-2} + \frac{1}{6}L_{-1}L_{-1}\right)L_0\right)\eta,$$
$$\mathcal{T} := \left(L_{-2} - \frac{3}{2}L_{-1}L_{-1}\right)\omega. \tag{4.8}$$

In more detail, let us assume that we are given a Virasoro representation with the following properties:

1. It allows for a non-degenerate symmetric pairing such that $\langle v, L_m w\rangle = \langle L_{-m}v, w\rangle$.
2. It has a cyclic vector η which is primary (i.e. the Vir-action on η generates the entire representation and $L_m\eta = 0$ for all $m > 0$).
3. η generates a rank three Jordan cell for L_0 of generalised eigenvalue 0; we set

$$\omega := L_0\eta, \qquad \Omega := L_0\omega. \tag{4.9}$$

4. $\mathbb{C}\Omega$ is the trivial Virasoro representation: $L_m\Omega = 0$ for all $m \in \mathbb{Z}$.

Let us draw some conclusions from these assumptions. Firstly, $\langle\omega, \Omega\rangle = \langle L_0\eta, (L_0)^2\eta\rangle = \langle\eta, (L_0)^3\eta\rangle = 0$ and similarly $\langle\Omega, \Omega\rangle = 0$, so that we must have $\langle\eta, \Omega\rangle \neq 0$ by non-degeneracy. Hence also $\langle\omega, \omega\rangle = \langle L_0\eta, L_0\eta\rangle = \langle\eta, \Omega\rangle \neq 0$. Suppose that $\langle\eta, \omega\rangle \neq 0$. Then we can replace $\eta \rightsquigarrow \eta' := \eta + (\text{const})\omega$ such that $\langle\eta', L_0\eta'\rangle = 0$. Next, if $\langle\eta', \eta'\rangle \neq 0$ we replace $\eta' \rightsquigarrow \eta'' := \eta' + (\text{const})\Omega$ such that $\langle\eta'', \eta''\rangle = 0$. We will henceforth assume that both has been done. Altogether, the pairing on the states of generalised L_0-eigenvalue 0 is, for some normalisation constant $N \neq 0$:

$\langle\,,\,\rangle$	η	ω	Ω
η	0	0	N
ω	0	N	0
Ω	N	0	0

(4.10)

Secondly, using points 3. and 4. above, one verifies with a little patience that for t and $\mathcal{T}$ as defined in (4.8)

$$\begin{aligned} L_1 t &= 0, & L_1 \mathcal{T} &= 0, \\ L_2 t &= -5\omega + 9\Omega, & L_2 \mathcal{T} &= -5\Omega, \\ (L_0 - 2)t &= \mathcal{T}, & (L_0 - 2)\mathcal{T} &= 0. \end{aligned} \tag{4.11}$$

It is then easy to compute the pairing on the weight 2 states (the pairing of a quasi-primary with an L_{-1}-descendant vanishes; we give the pairing restricted to t, $\mathcal{T}$). For example, using invariance of the pairing and the relations (4.11) gives

$$\langle \mathcal{T}, t \rangle = \left\langle \omega, \left(L_2 - \frac{3}{2} L_1 L_1 \right) t \right\rangle = -5 \langle \omega, \omega \rangle. \tag{4.12}$$

Altogether, the pairing takes the form:

$\langle , \rangle$	t	$\mathcal{T}$
t	0	$-5N$
$\mathcal{T}$	$-5N$	0

(4.13)

The fact that $\langle t, t \rangle = 0$ is the motivation for the complicated choice of t in (4.8).

In summary, if the Vir-submodule of $\mathcal{P}(0)$ generated by a state η representing the top 0 in the socle filtration (4.6) indeed has properties 1.–4., then we have quasi-primary states ω, Ω, t, $\mathcal{T}$ defined as in (4.8) and (4.9) with the properties (4.10)–(4.13). We will return to this in the discussion of OPEs.

4.2 Computation of $R(\mathcal{W}(0))$

An instance where we can compute the value of the adjoint functor R directly is the one-dimensional $\mathcal{W}$-module $\mathcal{W}(0)$. Namely, as we will explain in the second half of this short subsection,

$$R\big(\mathcal{W}(0)\big) = \mathcal{W}(0) \boxtimes \mathcal{W}(0). \tag{4.14}$$

The object $\mathcal{W}(0)$ is transparent because it is a quotient of $\mathcal{W}$ and the tensor unit is always transparent (recall that $\otimes_f$ is right exact and hence preserves surjections). Furthermore, $\mathcal{W}(0) \otimes_f \mathcal{W}(0) \cong \mathcal{W}(0)$ so that Lemma B.1 implies that $\mathcal{W}(0)$ is a commutative associative algebra. Lemma 3.25 now tells us that the full centre is

$$Z\big(\mathcal{W}(0)\big) = \mathcal{W}(0) \boxtimes \mathcal{W}(0). \tag{4.15}$$

This result has an evident CFT interpretation. The algebra $\mathcal{W}(0)$ is a non-degenerate boundary theory in the sense of Definition 2.14. In fact, $\mathcal{W}(0)$ is nothing but the chiral symmetry algebra of the $c = 0$ Virasoro minimal model. According to the discussion in Sects. 2.5 and 3.5, $Z(\mathcal{W}(0))$ is the largest bulk theory that can be consistently and non-degenerately joined to the boundary theory $\mathcal{W}(0)$. It is then not surprising that this bulk theory is the $c = 0$ Virasoro minimal model, i.e. the trivial theory with one-dimensional state space.

The derivation of (4.14) is as follows. We first remark that the functor $\mathcal{W}(0) \otimes_f (-)$ from $\mathcal{C}$ to $\mathcal{C}$ is monoidal (combine $\mathcal{W}(0) \otimes_f \mathcal{W}(0) \cong \mathcal{W}(0)$ with Lemma B.1). The image of $\mathcal{W}(0) \otimes_f (-)$ lies in the full subcategory of $\mathcal{C}_0 \subset \mathcal{C}$ of objects isomorphic to direct sums of $\mathcal{W}(0)$ (thus $\mathcal{C}_0$ is a tensor-ideal). But $\mathcal{C}_0 \cong \mathcal{V}ect$ as monoidal categories via $N \mapsto \mathcal{C}(\mathcal{W}(0), N)$. We shall need $N \mapsto \mathcal{C}(N, \mathcal{W}(0))$ instead, which is a monoidal equivalence $\mathcal{C}_0^{\text{opp}} \cong \mathcal{V}ect$. Now

$$\begin{aligned} \mathcal{C}\big(\mathcal{W}(0) \otimes_f U, \mathcal{W}(0)\big) &\cong \mathcal{C}\big(\mathcal{W}(0) \otimes_f U \otimes_f \mathcal{W}(0), \mathcal{W}^*\big) \\ &\cong \mathcal{C}\big(U, \mathcal{W}(0)\big), \end{aligned} \tag{4.16}$$

so that the composition $\mathcal{C} \xrightarrow{\mathcal{W}(0) \otimes_f (-)} \mathcal{C}_0 \xrightarrow{\sim} \mathcal{V}ect$ is just $\mathcal{C}(-, \mathcal{W}(0))$. Since both functors are monoidal, we conclude that $\mathcal{C}(-, \mathcal{W}(0))$ is a monoidal functor $\mathcal{C}^{\text{opp}} \to \mathcal{V}ect$. Finally, for all $U, V \in \mathcal{C}$ we have

$$\begin{aligned} \mathcal{C}\big(T(U \boxtimes V), \mathcal{W}(0)\big) &= \mathcal{C}\big(U \otimes_f V, \mathcal{W}(0)\big) \\ &\overset{(1)}{\cong} \mathcal{C}\big(U, \mathcal{W}(0)\big) \otimes_{\mathbb{C}} \mathcal{C}\big(V, \mathcal{W}(0)\big) \\ &\overset{(2)}{\cong} \mathcal{C}^2\big(U \boxtimes V, \mathcal{W}(0) \boxtimes \mathcal{W}(0)\big), \end{aligned} \tag{4.17}$$

where (1) is monoidality of $\mathcal{C}(-, \mathcal{W}(0))$ and (2) is (3.4). By Lemma 3.8, this shows that for all $X \in \mathcal{C} \boxtimes \mathcal{C}^{\text{rev}}$ we have $\mathcal{C}(T(X), \mathcal{W}(0)) \cong \mathcal{C}^2(X, \mathcal{W}(0) \boxtimes \mathcal{W}(0))$. Thus by definition of the adjoint, $R(\mathcal{W}(0)) = \mathcal{W}(0) \boxtimes \mathcal{W}(0)$.

4.3 Computation of $R(\mathcal{W}^*)$

We now want to implement the general construction of Sect. 2.5 for more interesting boundary theories than the $\mathcal{W}(0)$-example treated in the previous subsection. That is, we should fix an associative algebra $A \neq \mathcal{W}(0)$ in $\mathcal{C}$ which has a non-degenerate pairing and compute its full centre $Z(A) \in \mathcal{C} \boxtimes \mathcal{C}^{\text{rev}}$ according to Definition 3.14.

A point to stress is that neither $\mathcal{W}$ nor $\mathcal{W}^*$ are non-degenerate boundary theories as in Definition 2.14. They are both associative (and commutative) algebras (cf. Lemma B.1), but neither allows for a non-degenerate pairing. This is evident from the socle filtration (4.5), as a necessary condition for a non-degenerate pairing on an algebra A is that $A^* \cong A$ (see Definition 3.2).

According to Theorem 3.10 in [25], one way to produce such an algebra $A \in \mathcal{C}$ is to take an object $U \in \mathcal{C}$ for which U^* is the categorical dual and set $A = U \otimes_f U^*$. There are (recall the above disclaimer) many such objects to choose from. The original idea was to choose A small in order to simplify the analysis, and one convenient choice which produces a particularly small A is $U = \mathcal{W}(\frac{5}{8})$ (which is self-contragredient). From Appendixes A.3 and A.4 in [25] we read off that $A = \mathcal{W}(\frac{5}{8}) \otimes_f \mathcal{W}(\frac{5}{8})$ has socle filtration

$$A: \quad \begin{array}{ccccc} & & 2 & & \\ & \diagup & | & \diagdown & \\ 7 & & 0 & & 7 \\ & \diagdown & | & \diagup & \\ & & 2 & & \end{array} \tag{4.18}$$

The next step would be to use expression (3.43) and Theorem 3.24 to obtain $Z(A)$ as the subobject $C_l(R(A))$ of $R(A) = ((A^* \boxtimes \mathcal{W}) \otimes_f R(\mathcal{W}^*)^*)^*$. Unfortunately, we do not control the tensor product and braiding on $\mathcal{C}$ well enough to carry out this computation.

Instead, let us have a closer look at $\mathcal{W}^*$. As we already remarked, $\mathcal{W}^*$ is not a non-degenerate boundary theory, but it is still a (non-unital) associative algebra and hence a boundary theory with background states as alluded to in Remark 2.15(ii). The full centre $Z(\mathcal{W}^*)$ is a commutative associative algebra and provides a bulk theory with background states as defined in Sect. 2.2. Indeed, it is by construction the maximal such theory that can be non-degenerately joined to the boundary theory $\mathcal{W}^*$ (this follows from (2.33), Table 2 and Remark 3.15(i)). However, since $Z(\mathcal{W}^*)$ is obtained from a 'non-standard' boundary theory, it is maybe not surprising that it will show some 'non-standard' features itself; this will be discussed in Sect. 4.5 below.

The first step towards $Z(\mathcal{W}^*)$ is to determine $R(\mathcal{W}^*)$. The method for this given in Sect. 3.6 has been carried out (recall the above disclaimer) in Sect. 2.2 in [26]. The result is as follows. As a $\mathcal{W} \otimes_{\mathbb{C}} \mathcal{W}$-representation, $R(\mathcal{W}^*)$ splits into 5 indecomposable summands,

$$R(\mathcal{W}^*) = \mathcal{H}_0 \oplus \mathcal{H}_{1/8} \oplus \mathcal{H}_{5/8} \oplus \mathcal{H}_{1/3} \oplus \mathcal{H}_{-1/24} \oplus \mathcal{H}_{35/24}, \tag{4.19}$$

where we have labelled the individual blocks $\mathcal{H}_h$ by the conformal weight of the lowest state. The blocks $\mathcal{H}_{-1/24}$ and $\mathcal{H}_{35/24}$ are irreducible and given by

$$\begin{aligned} \mathcal{H}_{-1/24} &= \mathcal{W}\left(\frac{-1}{24}\right) \otimes_{\mathbb{C}} \mathcal{W}\left(\frac{-1}{24}\right), \\ \mathcal{H}_{35/24} &= \mathcal{W}\left(\frac{35}{24}\right) \otimes_{\mathbb{C}} \mathcal{W}\left(\frac{35}{24}\right). \end{aligned} \tag{4.20}$$

The remaining blocks are not irreducible. The socle filtration of $\mathcal{H}_{1/8}$ reads

$$\begin{array}{c} \mathcal{W}\Big(\frac{1}{8}\Big) \otimes_{\mathbb{C}} \mathcal{W}\Big(\frac{1}{8}\Big) \oplus \mathcal{W}\Big(\frac{33}{8}\Big) \otimes_{\mathbb{C}} \mathcal{W}\Big(\frac{33}{8}\Big) \\ \downarrow \\ \mathcal{H}_{1/8} : 2 \cdot \mathcal{W}\Big(\frac{1}{8}\Big) \otimes_{\mathbb{C}} \mathcal{W}\Big(\frac{33}{8}\Big) \oplus 2 \cdot \mathcal{W}\Big(\frac{33}{8}\Big) \otimes_{\mathbb{C}} \mathcal{W}\Big(\frac{1}{8}\Big) \\ \downarrow \\ \mathcal{W}\Big(\frac{1}{8}\Big) \otimes_{\mathbb{C}} \mathcal{W}\Big(\frac{1}{8}\Big) \oplus \mathcal{W}\Big(\frac{33}{8}\Big) \otimes_{\mathbb{C}} \mathcal{W}\Big(\frac{33}{8}\Big). \end{array} \tag{4.21}$$

This can be organised in a more transparent fashion if we replace each direct sum by a little table where we indicate the multiplicity of each term as in

	$\frac{1}{8}$	$\frac{33}{8}$
$\frac{1}{8}$	1	0
$\frac{33}{8}$	0	1

$\longrightarrow$

	$\frac{1}{8}$	$\frac{33}{8}$
$\frac{1}{8}$	0	2
$\frac{33}{8}$	2	0

$\longrightarrow$

	$\frac{1}{8}$	$\frac{33}{8}$
$\frac{1}{8}$	1	0
$\frac{33}{8}$	0	1

(4.22)

The socle filtrations of $\mathcal{H}_{5/8}$ and $\mathcal{H}_{1/3}$ are the same, but with $\{\frac{1}{8}, \frac{33}{8}\}$ replaced by $\{\frac{5}{8}, \frac{21}{8}\}$ and $\{\frac{1}{3}, \frac{10}{3}\}$, respectively. The sector $\mathcal{H}_0$ is the most interesting, its socle filtration is (all empty entries are equal to '0')

Level 0

	0	1	2	5	7
0	1				
1		1			
2			1		
5				1	
7					1

$\longrightarrow$

Level 1

	0	1	2	5	7
0		1	1		
1	1			2	2
2	1			2	2
5		2	2		
7		2	2		

$\longrightarrow$

Level 2

	0	1	2	5	7
0	1			2	2
1		2	4		
2		4	2		
5	2			2	4
7	2			4	2

$\longrightarrow$

Level 3

	0	1	2	5	7
0		1	1		
1	1			2	2
2	1			2	2
5		2	2		
7		2	2		

$\longrightarrow$

Level 4

	0	1	2	5	7
0	1				
1		1			
2			1		
5				1	
7					1

.

(4.23)

Via (3.30), $R(\mathcal{W}^*)$ inherits the structure of an associative algebra from $\mathcal{W}^*$. To find the full centre we should compute $Z(\mathcal{W}^*) = C_l(R(\mathcal{W}^*)) \subset R(\mathcal{W}^*)$. Again, the lack of detailed knowledge of the braiding means we currently cannot do this. However, we know from Remark 3.19 that $\exp(2\pi i(L_0 - \overline{L}_0))$ acts as the identity on $R(\mathcal{W}^*)$. This implies that $m_R \circ c_{R,R} \circ c_{R,R} = m_R$ (abbreviating $R \equiv R(\mathcal{W}^*)$), i.e. taking one field all the way around another does not produce a monodromy. Our guess is that in fact $m_R \circ c_{R,R} = m_R$, i.e. $Z(\mathcal{W}^*) = R(\mathcal{W}^*)$, but as we already said, we cannot check this.

Finally, recall that the functor $R(-)$ is lax monoidal (cf. Sect. 3.6) and so the algebra map $\mathcal{W}^* \to \mathcal{W}(0)$ gives an algebra map $\pi : R(\mathcal{W}^*) \to R(\mathcal{W}(0))$, which we expect to be non-zero. As a consequence, there is an OPE-preserving surjection from the tentative bulk theory $Z(\mathcal{W}^*) = R(\mathcal{W}^*)$ to the $c = 0$ minimal model $\mathcal{W}(0) \boxtimes \mathcal{W}(0)$. In this sense, $R(\mathcal{W}^*)$ is a 'refinement' of the minimal model.

4.4 Modular Invariance

A second interesting feature of this construction is that it leads to a modular invariant partition function. It is a straightforward exercise to write down the vector space of modular invariant bilinear combinations of the 13 characters of irreducible $\mathcal{W}$-representations. Namely, make the general ansatz

$$\xi(M,\tau) := \sum_{a,b} M_{ab}\,\chi_{\mathcal{W}(a)}(q)\,\chi_{\mathcal{W}(b)}(\bar{q}); \quad q = e^{2\pi i\tau}, \tag{4.24}$$

where M is a 13×13-matrix and a, b run over the lowest L_0-weights of the 13 irreducibles. Then the condition $\xi(M,\tau+1) = \xi(M,\tau)$ already forces most entries of M to be zero. The known modular properties of the characters (see [12], or Appendix A.2 in [26] for the notation used here) turn $\xi(M,-1/\tau) = \xi(M,\tau)$ into a linear equation for M. In this way one finds that $\xi(M,\tau+1) = \xi(M,\tau) = \xi(M,-1/\tau)$ has a two-dimensional space of solutions given by (zeros are not written)

$M =$

	$\mathcal{W}(0)$	$\mathcal{W}(1)$	$\mathcal{W}(2)$	$\mathcal{W}(5)$	$\mathcal{W}(7)$	$\mathcal{W}(\frac{1}{3})$	$\mathcal{W}(\frac{10}{3})$	$\mathcal{W}(\frac{5}{8})$	$\mathcal{W}(\frac{21}{8})$	$\mathcal{W}(\frac{1}{8})$	$\mathcal{W}(\frac{33}{8})$	$\mathcal{W}(\frac{-1}{24})$	$\mathcal{W}(\frac{35}{24})$
$\mathcal{W}(0)$	α	2β	2β	2β	2β								
$\mathcal{W}(1)$	2β	4β	4β	4β	4β								
$\mathcal{W}(2)$	2β	4β	4β	4β	4β								
$\mathcal{W}(5)$	2β	4β	4β	4β	4β								
$\mathcal{W}(7)$	2β	4β	4β	4β	4β								
$\mathcal{W}(\frac{1}{3})$						2β	2β						
$\mathcal{W}(\frac{10}{3})$						2β	2β						
$\mathcal{W}(\frac{5}{8})$								2β	2β				
$\mathcal{W}(\frac{21}{8})$								2β	2β				
$\mathcal{W}(\frac{1}{8})$										2β	2β		
$\mathcal{W}(\frac{33}{8})$										2β	2β		
$\mathcal{W}(\frac{-1}{24})$												β	
$\mathcal{W}(\frac{35}{24})$													β

(4.25)

with $\alpha, \beta \in \mathbb{C}$. Summing up the entries of the tables in (4.22) and (4.23) level by level, one quickly checks that the above space of solutions is spanned by the characters of $R(\mathcal{W}(0))$ and $R(\mathcal{W}^*)$. In particular, we see that the character $\chi_{R(\mathcal{W}^*)}(q, \bar{q})$ is modular invariant.

To relate the character $\chi_{R(\mathcal{W}^*)}(q, \bar{q})$ to the partition function of $R(\mathcal{W}^*)$ we appeal to Remark 2.8(iv) and Sect. 2.3: The composition series (4.22) and (4.23) suggest that $R(\mathcal{W}^*) \cong R(\mathcal{W}^*)^*$, i.e. that $R(\mathcal{W}^*)$ is self-conjugate. Therefore, assuming inversion invariance of $R(\mathcal{W}^*)$, the construction in Remark 2.8(iv) provides us with non-degenerate two-point correlators on the Riemann sphere. According to Sect. 2.3 this allows one to express the torus amplitude as a trace over the space of states.

The partition function of $R(\mathcal{W}^*)$ follows a pattern also observed in supergroup WZW models and the $W_{1,p}$-models [24, 42], as well as in the study of modular properties of Hopf algebra modules [19] (cf. Remark 3.26). Namely, despite the complicated submodule structure of $R(\mathcal{W}^*)$ as given in (4.23), in terms of characters we simply have

$$\chi_{R(\mathcal{W}^*)}(q, \bar{q}) = \sum_h \chi_{\mathcal{W}(h)}(q) \chi_{\mathcal{P}(h)}(\bar{q}), \tag{4.26}$$

where the sum is over the weights of the 13 irreducibles.

In [41] it has been argued that this bilinear combination of characters is modular invariant for all $W_{p,q}$-models. Furthermore, it turns out that the function $\chi_{R(\mathcal{W}^*)}(q, \bar{q})$ can—up to a constant—be written as a linear combination of modular invariant free boson partition functions at $c = 1$ [41]. In this form, $\chi_{R(\mathcal{W}^*)}(q, \bar{q})$ has already appeared in the context of a model for dilute polymers [47].[18]

4.5 Correlators and OPEs in $R(\mathcal{W}^*)$

Finally, we want to explain the non-standard features of the putative bulk theory $R(\mathcal{W}^*)$ in more detail. In particular, we want to show that it does not have an identity field, nor a stress energy tensor. (However, the correlation functions are still invariant under infinitesimal conformal transformations.)

In order to understand these features let us study the OPEs of the low-lying fields. It follows from the socle filtration in (4.23) that there are three states of generalised conformal dimension $(0, 0)$. These are mapped into one another under the action of the zero modes. Denoting the relevant states again by η, ω and Ω, one would expect (as is also assumed in (4.7)) that the relevant zero mode can be taken to be L_0 or $\overline{L}_0$. Since locality requires that $L_0 - \overline{L}_0$ must be diagonalisable (cf. Remark 3.19), we then conclude that

$$L_0 \eta = \overline{L}_0 \eta = \omega, \qquad L_0 \omega = \overline{L}_0 \omega = \Omega, \qquad L_0 \Omega = \overline{L}_0 \Omega = 0. \tag{4.27}$$

[18] More precisely, $\chi_{R(\mathcal{W}^*)}(q, \bar{q}) = Z_c[\frac{3}{2}, 1] + 3$, where for $Z_c[\frac{3}{2}, 1]$ we refer to Eq. (38) in [47] and for the relation to polymers to Sect. 4.1.2 in [47]. We thank Hubert Saleur for a discussion on this point.

We can again define quasiprimary states t and $\mathcal{T}$ by (4.8), and likewise for $\bar{t}$ and $\overline{\mathcal{T}}$. It follows from (4.23) that $\mathcal{T}$ is a holomorphic field since there is no primary field of generalised dimension (2, 1) in the third or fourth level of the socle filtration and hence $\overline{L}_{-1}\mathcal{T} = 0$. By the same argument we also see that Ω is annihilated by all L_n and $\overline{L}_n$ modes. On the other hand, we cannot conclude that t is holomorphic, since there is a (2, 1) state in level 2 of the socle filtration (4.23); this is indeed expected since the diagram (4.7) still applies, and hence t is the 'logarithmic partner' of $\mathcal{T}$ (and thus should depend on both z and $\bar{z}$).

4.5.1 Some OPEs

The derivation of the OPEs and correlators presented below can be found in Appendix C, here we merely list the results. The simplest set of OPEs are those involving Ω:

$$\Omega(z)\phi(w) = \pi(\phi) \cdot \Omega(w), \qquad \text{for all } \phi \in F, \tag{4.28}$$

and the OPE does not contain subleading terms. Here π is the intertwiner $R(\mathcal{W}^*) \to R(\mathcal{W}(0)) \equiv \mathbb{C}$ introduced in the previous subsection. The map π is an algebra homomorphism, i.e. it is compatible with the OPE, and it is non-vanishing on the level 0 state η. We can normalise η such that $\pi(\eta) = 1$. In particular,

$$\Omega(z)\Omega(w) = \Omega(z)\omega(w) = 0, \qquad \Omega(z)\eta(w) = \Omega(w). \tag{4.29}$$

Since Ω is the only $sl(2,\mathbb{C})$-invariant field in $R(\mathcal{W}^*)$, this shows that $R(\mathcal{W}^*)$ has no identity field. Next we list some OPEs involving $\mathcal{T}$:

$$\begin{aligned} &\mathcal{T}(z)\omega(w) = \mathcal{O}\big((z-w)^0\big), \qquad \mathcal{T}(z)\mathcal{T}(0) = \mathcal{O}\big((z-w)^0\big),\\ &\mathcal{T}(z)\eta(w) = A \cdot \left(\frac{\Omega(w)}{(z-w)^2} + \frac{(\partial/\partial w)\omega(w)}{z-w}\right) + \mathcal{O}\big((z-w)^0\big),\\ &t(z)\mathcal{T}(w) = (A+1) \cdot \left(\frac{-5\Omega(w)}{(z-w)^4} + \frac{2\mathcal{T}(w)}{(z-w)^2} + \frac{(\partial/\partial w)\mathcal{T}(w)}{z-w}\right)\\ &\qquad + \mathcal{O}\big((z-w)^0\big), \end{aligned} \tag{4.30}$$

where $A \in \mathbb{C}$ is a so far undetermined constant. From this we see that $\mathcal{T}$—the only holomorphic field of weight $(2,0)$ in the space of fields F—does not behave as the stress tensor. For example, it has regular OPE with itself. However, a glance at (4.11) shows that the OPE of the field $\hat{t} = \frac{1}{A+1}t$ with $\mathcal{T}$ can be written as

$$\hat{t}(z)\mathcal{T}(w) = \sum_{n=-1}^{2} \frac{(L_n\mathcal{T})(w)}{(z-w)^{-n-2}} + \mathcal{O}\big((z-w)^0\big). \tag{4.31}$$

So in this OPE, $\hat{t}$ behaves as the stress tensor (but it is not the stress tensor as it is not holomorphic).

Finally, we give two more OPEs for fields of generalised weight $(0,0)$:

$$\begin{aligned}\omega(z)\omega(w) &= B\cdot\Omega(w)+\cdots\,,\\ \omega(z)\eta(w) &= \big(2(A-B)\ln|z-w|^2+C\big)\cdot\Omega(w)\\ &\quad+(1-B+2A)\cdot\omega(w)+\cdots\,,\end{aligned}\tag{4.32}$$

where $B,C\in\mathbb{C}$ are new constants which remain to be determined. The dots stand for terms which vanish for $|z-w|\to 0$ and which have no component of generalised weight $(0,0)$.

4.5.2 Some Correlators

Recall the intertwiner $\pi: R(\mathcal{W}^*)\to R(\mathcal{W}(0))\equiv\mathbb{C}$ from above. By our normalisation $\pi(\eta)=1$ and by (2.22) we have

$$\big\langle\eta(z_1)\cdots\eta(z_n)\big\rangle=1.\tag{4.33}$$

These are the correlators of the $c=0$ minimal model. If a state from the kernel of π is inserted, the correlator vanishes.

To obtain non-trivial correlators we have to allow background states as in Sect. 2.2. For example, the normalisation condition ${}^{\eta}\langle\Omega(0)\rangle=1$ and the OPE (4.28) imply the correlators

$${}^{\eta}\big\langle\phi_1(z_1)\cdots\phi_n(z_n)\Omega(w)\big\rangle=\pi(\phi_1)\cdot\pi(\phi_2)\cdots\pi(\phi_n)\tag{4.34}$$

for all $\phi_i\in F$, independent of the insertion points z_i and w. Another example is[19]

$${}^{t}\big\langle\mathcal{T}(0)\big\rangle={}^{\mathcal{T}}\big\langle t(0)\big\rangle=-5,\tag{4.35}$$

which follows immediately from (4.13) together with ${}^{\eta}\langle\Omega(0)\rangle=1$ which fixes $N=1$. Finally, from the OPEs (4.32) we can directly read off the two-point correlators

$$\begin{aligned}{}^{\omega}\big\langle\omega(z)\omega(w)\big\rangle&=0, & {}^{\omega}\big\langle\eta(z)\omega(w)\big\rangle&=1-B+2A,\\ {}^{\eta}\big\langle\omega(z)\omega(w)\big\rangle&=B, & {}^{\eta}\big\langle\eta(z)\omega(w)\big\rangle&=2(A-B)\ln|z-w|^2+C.\end{aligned}\tag{4.36}$$

In summary, we have seen that $R(\mathcal{W}^*)$ does not have an identity field or a stress tensor. Consequently, $R(\mathcal{W}^*)$ does not allow for an OPE-preserving embedding $\mathcal{W}\otimes_{\mathbb{C}}\mathcal{W}\to R(\mathcal{W}^*)$ as one might have expected from a $\mathcal{W}\otimes_{\mathbb{C}}\mathcal{W}$-symmetric theory. Nonetheless, by definition the n-point correlators are Vir $\oplus$ Vir-coinvariants

[19] The constant -5 found here is reminiscent of the b-value in the correlator of the stress tensor and its logarithmic partner (but recall that $\mathcal{T}$ is not a stress tensor). The value $b=-5$ has recently been observed in certain logarithmic bulk theories with $c=0$ [51].

(and also $\mathcal{W} \otimes_{\mathbb{C}} \mathcal{W}$-coinvariants). The above problems are closely related to the fact that the boundary theory $\mathcal{W}^*$ from which this construction starts only defines a boundary theory with background states, see Remark 2.15(ii). If one were to consider instead a usual non-degenerate boundary theory A with identity field as in (4.18), one would expect that the corresponding full centre $Z(A)$ is better behaved. In particular, the unit condition in Theorem 3.16 gives then a non-zero OPE-preserving map $\mathcal{W} \otimes_{\mathbb{C}} \mathcal{W} \to Z(A)$ which we expect to be an embedding, so that $Z(A)$ would have an identity field and a stress tensor.

Acknowledgements This paper is based on talks given by IR at the conferences *Conformal field theories and tensor categories* (13–17 June 2011, Beijing) and *Logarithmic CFT and representation theory* (3–7 October 2011, Paris). IR thanks the Beijing International Center for Mathematical Research at Peking University and the Centre Emile Borel at the Institut Henri Poincare in Paris for hospitality during one month stays around these conferences. MRG and SW thank the Centre Emile Borel for hospitality during the second conference. The authors thank Alexei Davydov, Jürgen Fuchs, Azat Gainutdinov, Jérôme Germoni, Antun Milas, Victor Ostrik, Hubert Saleur, Christoph Schweigert, Alexei Semikhatov and Carl Stigner for helpful discussions. The research of MRG is supported in part by the Swiss National Science Foundation. IR is partially supported by the Collaborative Research Centre 676 'Particles, Strings and the Early Universe—the Structure of Matter and Space-Time'. SW is supported by the SNSF scholarship for prospective researchers and by the World Premier International Research Center Initiative (WPI Initiative), MEXT, Japan.

Appendix A: Conditions B1–B5

In this appendix we write out conditions (B1)–(B5) referred to in Definition 2.11. Let $(F, M, \Omega^*; B, m, \omega^*; b)$ be as in that definition.

Below, we always take $(x_1, \dots, x_m) \in \mathbb{R}^m \backslash \text{diag}$, $(z_1, \dots, z_n) \in \mathbb{H}^n \backslash \text{diag}$ and $\psi_i \in B$, $\phi_j \in F$. The integers m, n are to be chosen such that all $U_{m,n}$ in the statement are defined (there has to be at least one field insertion; this field insertion can be a boundary field or a bulk field, i.e. $m, n \in \mathbb{Z}_{\geq 0}$, $m + n > 0$).

(B1) $U_{m,n}$ is smooth in each argument from $\mathbb{R}$ and $\mathbb{H}$, and linear in each argument from B and F.

(B2) $U_{m,n}$ is invariant under joint permutation[20] of $\mathbb{R}^m$ and B^m and $\mathbb{H}^n$ and F^n. Namely, for all $\sigma \in S_m$ and $\tau \in S_n$,

$$\begin{aligned} &U_{m,n}(x_1, \dots, x_m, z_1, \dots, z_n, \psi_1, \dots, \psi_m, \phi_1, \dots, \phi_n) \\ &= U_{m,n}(x_{\sigma(1)}, \dots, x_{\sigma(m)}, z_{\tau(1)}, \dots, z_{\tau(n)}, \\ &\quad \psi_{\sigma(1)}, \dots, \psi_{\sigma(m)}, \phi_{\tau(1)}, \dots, \phi_{\tau(n)}). \end{aligned} \tag{A.1}$$

[20] Since there is a natural ordering of points on the boundary line $\mathbb{R}$, instead of imposing invariance under joint permutations one could require the sequence of points $x_1, \dots, x_m$ to be strictly increasing. We prefer the present formulation since it avoids awkwardness in other places. For example in (B3c) below the parameter x would otherwise have to be placed according to the ordering. We also note that the permutation symmetry has nothing to do with commutativity of boundary (or bulk) fields. Indeed, there is no commutativity requirement on boundary fields, and the commutativity of bulk fields is instead a consequence of single valuedness of the functions C_n in Sect. 2.1.

Because there are three maps describing a short distance expansion in the defining data, namely M, m, b, there are three ways to link the $U_{k,l}$ for different k, l. These are listed in the next three conditions. We denote the canonical projection $\overline{F} \to \bigoplus_{d \le \Delta} F^{(d)}$ by P_Δ and the canonical projection $\overline{B} \to \bigoplus_{d \le h} B^{(d)}$ by P_h.

(B3a) (*Bulk OPE*) Suppose that $n \ge 2$ and that $|z_1 - z_2| < |z_i - z_2|$ for all $i > 2$ and $|z_1 - z_2| < |x_j - z_2|$ for all j. Then we can take the OPE of $\phi_1(z_1)$ and $\phi_2(z_2)$, reducing the number of bulk fields by one:

$$U_{m,n}(\dots, z_1, z_2, \dots, \phi_1, \phi_2, \dots) = \lim_{\Delta\to\infty} U_{m,n-1}\big(\dots, z_2, \dots, P_\Delta \circ M_{z_1-z_2}(\phi_1 \otimes \phi_2), \dots\big) \qquad \text{(A.2)}$$

(B3b) (*Boundary OPE*) Suppose that $m \ge 2$ and that $x_1 > x_2$, and $|x_1 - x_2| < |x_i - x_2|$ for all $i > 2$ and $|x_1 - x_2| < |z_j - x_2|$ for all j. Then we can take the OPE of $\psi_1(x_1)$ and $\psi_2(x_2)$, reducing the number of boundary fields by one:

$$U_{m,n}(x_1, x_2, \dots, \psi_1, \psi_2, \dots) = \lim_{h\to\infty} U_{m-1,n}\big(x_2, \dots, P_h \circ m_{x_1-x_2}(\psi_1 \otimes \psi_2), \dots\big) \qquad \text{(A.3)}$$

(B3c) (*Bulk-boundary map*) Suppose that $n \ge 1$. Write $z_1 = x + iy$. Suppose further that $|x_i - x| > y$ for all i and $|z_j - x| > y$ for all $j > 0$. Then we can expand $\phi_1(z_1)$ in terms of boundary fields at x, exchanging one bulk field for one boundary field:

$$\begin{aligned} &U_{m,n}(x_1, \dots, x_m, z_1, \dots, z_n, \psi_1, \dots, \psi_m, \phi_1, \dots, \phi_n) \\ &\quad= \lim_{h\to\infty} U_{m+1,n-1}\big(x, x_1, \dots, x_m, z_2, \dots, z_n, \\ &\qquad P_h \circ b_y(\phi_1), \psi_1, \dots, \psi_m, \phi_2, \dots, \phi_n\big) \end{aligned} \qquad \text{(A.4)}$$

The relation between derivatives and L_{-1} is as before,

(B4) The $U_{m,n}$ satisfy

$$\begin{aligned} \frac{d}{dz_1} U_{m,n}(\dots, z_1, \dots, \phi_1, \dots) &= U_{m,n}(\dots, z_1, \dots, L_{-1}\phi_1, \dots), \\ \frac{d}{d\bar{z}_1} U_{m,n}(\dots, z_1, \dots, \phi_1, \dots) &= U_{m,n}(\dots, z_1, \dots, \overline{L}_{-1}\phi_1, \dots), \\ \frac{d}{dx_1} U_{m,n}(x_1, \dots, \psi_1, \dots) &= U_{m,n}(x_1, \dots, L_{-1}\psi_1, \dots), \end{aligned} \qquad \text{(A.5)}$$

where $\frac{d}{dz_1}$ and $\frac{d}{d\bar{z}_1}$ are complex derivatives, and $\frac{d}{dx_1}$ is a real derivative.

Let f be a rational function on $\mathbb{C} \cup \{\infty\}$ which has poles at most in the set $\{x_1, \dots, x_m, z_1, \dots, z_n, \bar{z}_1, \dots, \bar{z}_n\}$ and ∞, and which satisfies the growth condition

$\lim_{\zeta\to\infty} \zeta^{-3} f(\zeta) = 0$. The expansion around each of these points is

$$f(\zeta) = \sum_{p=-\infty}^{\infty} f_p^k \cdot (\zeta - x_k)^{p+1} = \sum_{p=-\infty}^{\infty} g_p^{+,k} \cdot (\zeta - z_l)^{p+1}$$
$$= \sum_{p=-\infty}^{\infty} g_p^{-,k} \cdot (\zeta - \bar{z}_l)^{p+1}. \tag{A.6}$$

(B5) For all f as above,

$$\sum_{p=-\infty}^{\infty} \Bigg\{ \sum_{k=1}^{m} f_p^k U_{m,n}(\ldots, \psi_1, \ldots, L_p\psi_k, \ldots, \psi_m, \phi_1, \ldots, \phi_n)$$
$$+ \sum_{l=1}^{n} U_{m,n}\big(\ldots, \psi_1, \ldots, \psi_m, \phi_1, \ldots, (g_p^{+,l} L_p + g_p^{-,l} \overline{L}_p)\phi_l, \ldots, \phi_n\big) \Bigg\}$$
$$= 0. \tag{A.7}$$

As in (C5), only a finite number of summands in the sum over p are non-zero. There is a corresponding condition with L_p and $\overline{L}_p$ exchanged in the sum over bulk insertions.

The complicated looking set of conditions (B5) is obtained following the original argument in [4]: The fact that the boundary condition preserves conformal symmetry means that the correlator on the UHP satisfies the same conditions as the 'holomorphic part' of a bulk correlator with insertions at $\{x_1, \ldots, x_m, z_1, \ldots, z_n, \bar{z}_1, \ldots, \bar{z}_n\}$. In other words, an insertion at z in the upper half plane is duplicated to an insertion at z and $\bar{z}$. This prescription arises again from contour integration, as noted in the following remark.

Remark A.1 As in Remark 2.2, one can replace (B5) by the stronger requirement that there should exist a stress tensor, that is, a field $T^{\mathrm{bnd}} \in B^{(2)}$ such that $m_x(T \otimes \psi) = \sum_{m=-\infty}^{\infty} x^{-m-2} L_m \psi$. The CFT on the complex plane (F, M, Ω^*) is then equally required to be equipped with a stress tensor $T, \overline{T} \in F^{(2)}$. The statement 'the boundary condition respects conformal symmetry' means that the two components of the stress tensor in the bulk agree with the stress tensor on the boundary in the sense that

$$\lim_{y\to 0} \big\langle T(x+iy) \cdots \big\rangle = \big\langle T^{\mathrm{bnd}}(x) \cdots \big\rangle = \lim_{y\to 0} \big\langle \overline{T}(x+iy) \cdots \big\rangle \tag{A.8}$$

holds in all correlators. Define the meromorphic function $u(\zeta)$ on the complex plane as follows:

$$u(\zeta) = \begin{cases} \langle T(\zeta)\psi_1(x_1)\cdots\phi_1(z_1)\cdots\rangle; & \mathrm{Im}(\zeta) \geq 0 \\ \langle \overline{T}(\bar{\zeta})\psi_1(x_1)\cdots\phi_1(z_1)\cdots\rangle; & \mathrm{Im}(\zeta) < 0 \end{cases} \tag{A.9}$$

The conditions (A.7) arise from the contour integral $\frac{1}{2\pi i}\oint f(\zeta)u(\zeta)d\zeta = 0$, where the contour is a big circle enclosing $\{x_1,\dots,x_m,z_1,\dots,z_n,\bar{z}_1,\dots,\bar{z}_n\}$. Then the contour is deformed to a union of small circles around each of the x_i, z_i, $\bar{z}_i$ and the OPEs of the stress tensor are substituted.

Appendix B: Further Details on the Algebraic Reformulation

B.1 Proof of Theorem 3.5

Proof of Theorem 3.5 If $\mathcal{A} \cong \mathrm{Rep}_{f.d.}(A)$, condition (PF) follows by taking $P = A$, seen as a right module over itself. For the converse, pick a projective generator P. We can choose $A := \mathcal{A}(P, P)$ and define the functor $H : \mathcal{A} \to \mathrm{Rep}_{f.g.}(A)$ on objects and morphisms by

$$U \mapsto \mathcal{A}(P,U), \qquad [U \xrightarrow{f} V] \mapsto [P \xrightarrow{(-)} U \xrightarrow{f} V]. \tag{B.1}$$

The right action of $a \in A$ is given by $f \mapsto f \circ a$. Note that $H(P) = A$. The functor H is

- faithful: there exists a surjection $P^{\oplus m} \to U$ for some m, and so, given $g : U \to V$, if $[P \xrightarrow{s} U \xrightarrow{g} V] = 0$ for all s, then also $g = 0$.
- full: We need to show that every linear map $\varphi : \mathcal{A}(P,U) \to \mathcal{A}(P,V)$ such that $\varphi(f) \circ a = \varphi(f \circ a)$ for all $a \in A$, is of the form $\varphi(f) = \psi \circ f$ for some $\psi : U \to V$. Let $P^{\oplus k} \xrightarrow{K} P^{\oplus m} \xrightarrow{s} U$ the first two steps of a projective resolution (thus s is surjective and the image of K is the kernel of s). Let $s_1,\dots,s_m$ be the restriction of s to each summand. The pullback property along surjections,

$$\begin{array}{ccc} & & P \\ & \overset{\exists a}{\swarrow} & \downarrow f \\ P^{\oplus m} & \xrightarrow{s} & U \end{array} \tag{B.2}$$

shows that all f can be written as $\sum_i s_i \circ a_i$ for some $a_i \in A$. Thus $s_1,\dots,s_m$ generates $\mathcal{A}(P,U)$ as an A-module. Next, consider the cokernel diagram

$$\begin{array}{ccccc} P^{\oplus k} & \xrightarrow{K} & P^{\oplus m} & \xrightarrow{s} & U \\ & & {\scriptstyle\sum_i \varphi(s_i)} \downarrow & \underset{\exists ! \psi}{\swarrow} & \\ & & V & & \end{array} \tag{B.3}$$

The cokernel property can be applied because, denoting by $K_{ij} : P \to P$ the components of K, $\sum_i \varphi(s_i) \circ K_{ij} = \sum_i \varphi(s_i \circ K_{ij}) = 0$ as $\sum_i s_i \circ K_{ij} = 0$ for all j. The diagram then shows that $\varphi(s_i) = \psi \circ s_i$ for some $\psi : U \to V$. Since the s_i generate $\mathcal{A}(P, U)$, this fixes φ uniquely.

- essentially surjective: Let M be a finite-dimensional A-module. Let $A^{\oplus k} \to A^{\oplus m} \to M$ be the first two steps of a projective (in fact: free) resolution. In other words, $\mathcal{A}(P, P^{\oplus k}) \xrightarrow{\varphi} \mathcal{A}(P, P^{\oplus m}) \to M \to 0$ is exact for some A-module map φ. By fullness, there is a $\psi : P^{\oplus k} \to P^{\oplus m}$ such that $\varphi = \psi \circ (-)$. Since P is projective, the functor $\mathcal{A}(P, -)$ is exact, and so the exact sequence $P^{\oplus k} \xrightarrow{\psi} P^{\oplus m} \xrightarrow{\mathrm{cok}(\psi)} U \to 0$ gets mapped to the exact sequence $\mathcal{A}(P, P^{\oplus k}) \xrightarrow{\psi\circ(-)} \mathcal{A}(P, P^{\oplus m}) \to \mathcal{A}(P, U) \to 0$. Thus $M \cong \mathcal{A}(P, U)$ for some U. □

B.2 Idempotent Absolutely Simple Objects Are Algebras

Lemma B.1 *Let $\mathcal{M}$ be a k-linear monoidal category and let $S \in \mathcal{M}$ be such that $S \otimes S \cong S$ and $\mathcal{C}(S, S) = k \cdot id_S$. Pick an isomorphism $m : S \otimes S \to S$.*

(i) *The associator $\alpha_{S,S,S} : S \otimes (S \otimes S) \to (S \otimes S) \otimes S$ is $\alpha_{S,S,S} = (m^{-1} \otimes id_S) \circ (id_S \otimes m)$.*
(ii) *If $\mathcal{M}$ is in addition braided, the braiding on S is $c_{S,S} = id_{S\otimes S}$.*
(iii) *The pair (S, m) is an associative algebra in $\mathcal{M}$. If $\mathcal{M}$ is braided, this algebra is commutative.*
(iv) *If $n : S \otimes S \to S$ is an isomorphism, then (S, m) and (S, n) are isomorphic as algebras.*

Proof We will omit '⊗' between objects for better readability.

(i) The space $\mathcal{M}(S(SS), (SS)S)$ is isomorphic to $\mathcal{M}(S, S)$ and hence one-dimensional. Therefore, there has to exist a $\lambda \in k^\times$ such that

$$\alpha_{S,S,S} = \lambda \cdot \big[S(SS) \xrightarrow{id_S \otimes m} SS \xrightarrow{m^{-1}\otimes id_S} (SS)S\big]. \tag{B.4}$$

Naturality of the associator implies

$$\begin{aligned} &\big[U(VW) \xrightarrow{f\otimes(g\otimes h)} S(SS) \xrightarrow{\alpha_{S,S,S}} (SS)S\big] \\ &= \big[U(VW) \xrightarrow{\alpha_{U,V,W}} (UV)W \xrightarrow{(f\otimes g)\otimes h} (SS)S\big] \end{aligned} \tag{B.5}$$

Applying this to $f = m$, $g = h = id_S$, etc., allows one to solve for α with one entry being SS. The result is

$$\begin{aligned} \alpha_{SS,S,S} &= \lambda \cdot \big(m^{-1} \otimes id_S \otimes m\big) = \alpha_{S,S,SS}, \\ \alpha_{S,SS,S} &= \lambda \cdot \big\{\big(id_S \otimes m^{-1}\big) \circ m^{-1}\big\} \otimes \big\{m \circ (m \otimes id_S)\big\}. \end{aligned} \tag{B.6}$$

The pentagon with all four objects set to S reads

$$\alpha_{SS,S,S} \circ \alpha_{S,S,SS} = (\alpha_{S,S,S} \otimes id_S) \circ \alpha_{S,SS,S} \circ (id_S \otimes \alpha_{S,S,S}). \tag{B.7}$$

Substituting the expressions in terms of λ and m one quickly checks that the above identity simplifies to $\lambda^2 \cdot u = \lambda^3 \cdot u$, with $u = \{(m^{-1} \otimes id_S) \circ m^{-1}\} \otimes \{m \circ (id_S \otimes m)\} \neq 0$. Thus, $\lambda = 1$.

(ii) By assumption $\mathcal{M}(SS, SS)$ is one-dimensional, and hence there has to be an $\omega \in k^\times$ such that $c_{S,S} = \omega \cdot id_{SS}$. By naturality,

$$[UV \xrightarrow{f\otimes g} SS \xrightarrow{c_{S,S}} SS] = [UV \xrightarrow{c_{U,V}} VU \xrightarrow{g\otimes f} SS], \tag{B.8}$$

and applying this to $f = m$ and $g = id_S$ we can solve for $c_{SS,S}$. The result is $c_{SS,S} = \omega \cdot (m^{-1} \otimes id_S) \circ (id_S \otimes m)$. One of the two hexagons with all objects set to S reads

$$\alpha_{S,S,S} \circ c_{SS,S} \circ \alpha_{S,S,S} = (c_{S,S} \otimes id_S) \circ \alpha_{S,S,S} \circ (id_S \otimes c_{S,S}) \tag{B.9}$$

Substituting the expressions for $\alpha_{S,S,S}$ from (i) and $c_{S,S}$, $c_{SS,S}$ as above, this reduces to $\omega \cdot v = \omega^2 \cdot v$ with $v = (m^{-1} \otimes id_S) \circ (id_S \otimes m)$. Thus $\omega = 1$.

(iii) Associativity is $m \circ (id_S \otimes m) = m \circ (m \otimes id_S) \circ \alpha_{S,S,S}$, which holds by (i), and commutativity is trivial as $c_{S,S} = id_{SS}$ by (ii).

(iv) Since $\mathcal{M}(SS, S)$ is one-dimensional, we have $n = \lambda m$ for some $\lambda \in k^\times$. Take $f = \lambda id_S$. Then $f \circ n = m \circ (f \otimes f)$. □

B.3 Proof of Theorem 3.16

Proof of Theorem 3.16 *Existence*: Consider the composition (the left path in (3.35))

$$w := \big[T(Z \otimes_{\mathcal{C}^2} Z) \xrightarrow{T^{-1}_{2;Z,Z}} T(Z) \otimes_{\mathcal{C}} T(Z) \xrightarrow{z \otimes_{\mathcal{C}} z} A \otimes_{\mathcal{C}} A \xrightarrow{\mu_A} A\big]. \tag{B.10}$$

We need to check that the pair $(Z \otimes_{\mathcal{C}^2} Z, w)$ satisfies condition (3.31), i.e. that it is an object in $\mathcal{C}_{\text{full center}}(A)$. This amounts to commutativity of (brackets, associators and '$\otimes_{\mathcal{C}}$' are not written)

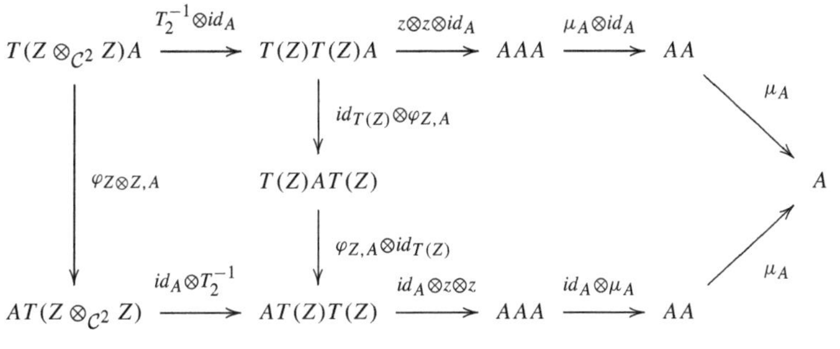

(B.11)

The left subdiagram is just (3.20), while the details for the right subdiagram are obtained by copying out the corresponding diagram in the proof of Proposition 4.1 in [7] in the present setting; we omit the details.

By the universal property of (Z, z), there exists a unique morphism $Z \otimes_{\mathcal{C}^2} Z \to Z$ such that (3.32) commutes. We define this morphism to be μ_Z.

Commutativity: We will show below that $c_{Z,Z}$ is an arrow from $(Z \otimes_{\mathcal{C}^2} Z, w)$ to itself in $\mathcal{C}_{\text{full center}}(A)$. This provides us with two arrows from $(Z \otimes_{\mathcal{C}^2} Z, w)$ to (Z, z) in $\mathcal{C}_{\text{full center}}(A)$, namely μ_Z and $\mu_Z \circ c_{Z,Z}$. By uniqueness, they have to be equal, establishing commutativity.

That $c_{Z,Z}$ is an endomorphism of $(Z \otimes_{\mathcal{C}^2} Z, w)$ amounts to commutativity of the diagram

$$\begin{array}{ccccccccc}
T(Z \otimes_{\mathcal{C}^2} Z) & \xrightarrow{T_2^{-1}} & T(Z) \otimes_{\mathcal{C}} T(Z) & \xrightarrow{id_{T(Z)} \otimes z} & T(Z) \otimes_{\mathcal{C}} A & \xrightarrow{z \otimes id_A} & A \otimes A & & \\
 & & & & & & & \searrow^{\mu_A} & \\
{\scriptstyle T(c_{Z,Z})}\downarrow & & {\scriptstyle \hat{\varphi}_{Z,Z} = \varphi_{Z,T(Z)}}\downarrow & & {\scriptstyle \varphi_{Z,A}}\downarrow & & & & A \\
 & & & & & & & \nearrow_{\mu_A} & \\
T(Z \otimes_{\mathcal{C}^2} Z) & \xrightarrow{T_2^{-1}} & T(Z) \otimes_{\mathcal{C}} T(Z) & \xrightarrow{z \otimes id_{T(Z)}} & A \otimes_{\mathcal{C}} T(Z) & \xrightarrow{id_A \otimes z} & A \otimes A & &
\end{array} \tag{B.12}$$

Starting from the left, the first square commutes by definition (3.21) of $\hat{\varphi}_{Z,Z}$. By Lemma 3.10, this is equal to $\varphi_{Z,T(Z)}$. The second square is then just naturality of $\varphi_{Z,T(Z)}$. The third square is property (3.31) for z.

Associativity: In the proof of associativity, we will not write out tensor product symbols and brackets between objects, and we omit all associators. We will show the equality of the two maps $a = \mu_Z \circ (\mu_Z \otimes id_Z)$ and $b = \mu_Z \circ (id_Z \otimes \mu_Z)$ from ZZZ to Z via the terminal object property. Define the map

$$y := \left[T(ZZZ) \xrightarrow{\sim} T(Z)T(Z)T(Z) \xrightarrow{z \otimes z \otimes z} AAA \xrightarrow{\text{mult.}} A \right], \tag{B.13}$$

where the first isomorphism is constructed from T_2 and associators, and 'mult.' stands for any order of multiplying the three factors via μ_A. That $y \in \text{Cent}(ZZZ, A)$ is checked by an analogous argument as that giving commutativity of (B.11). We now need to verify that a and b are maps from (ZZZ, y) to (Z, z). This will imply $a = b$ and hence associativity of μ_Z. That $T(a) : T(ZZZ) \to T(Z)$ makes (3.31) commute amounts to commutativity of

$$\begin{array}{ccccc}
T(ZZZ) & \xrightarrow{T(\mu_Z \otimes id_Z)} & T(ZZ) & \xrightarrow{T(\mu_Z)} & T(Z) \\
\downarrow{\scriptstyle \sim} & & \downarrow{\scriptstyle \sim} & & \| \\
T(Z)T(Z)T(Z) & \xrightarrow{\mu_{TZ} \otimes id_{TZ}} & T(Z)T(Z) & \xrightarrow{\mu_{TZ}} & T(Z) \\
\downarrow{\scriptstyle z \otimes z \otimes z} & & \downarrow{\scriptstyle z \otimes z} & & \downarrow{\scriptstyle z} \\
AAA & \xrightarrow{\mu_A \otimes id_A} & AA & \xrightarrow{\mu_A} & A
\end{array} \tag{B.14}$$

The top two squares commute by definition of μ_{TZ} in (3.30), the bottom two squares commute because z is an algebra map (since it satisfies (3.35)). The argument for $T(b)$ is similar.

Unitality: The construction of the unit for Z rests on the observation that

$$\varphi_{\mathbf{1},U} = \big[T(\mathbf{1})U \xrightarrow{T_0^{-1}\otimes id_U} \mathbf{1}U \xrightarrow{\lambda_U} U \xrightarrow{\rho_U^{-1}} U\mathbf{1} \xrightarrow{id_U\otimes T_0} UT(\mathbf{1})\big], \tag{B.15}$$

which can be checked directly from (3.17). Define the map

$$u := \big[T(\mathbf{1}) \xrightarrow{T_0^{-1}} \mathbf{1} \xrightarrow{\iota_A} A\big]. \tag{B.16}$$

To see that $u \in \mathrm{Cent}(\mathbf{1}, A)$, we need to establish commutativity of

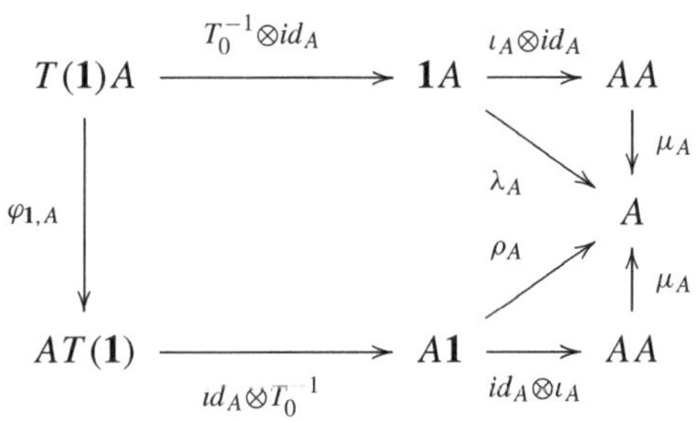

(B.17)

The pentagon is (B.15) and the remaining triangles amount to the unit property of ι_A. Thus there exists a unique $\iota_Z : \mathbf{1} \to Z$ such that $u = z \circ T(\iota_Z)$. The unit property of ι_Z follows by verifying that $\mu \circ (id_Z \otimes \iota_Z) \circ \rho_Z^{-1}$, $\mu \circ (\iota_Z \otimes id_Z) \circ \lambda_Z^{-1}$ and id_Z are morphisms $Z \to Z$ in the category $\mathcal{C}_{\text{full center}}(A)$ and hence are all equal. We refer to Proposition 4.1 in [7] for details. □

B.4 Proofs for Theorems 3.17 and 3.18

The proof of Theorem 3.17 requires three lemmas. The first one gives an alternative characterisation of a representing object.

Lemma B.2 *Let $U \in \mathcal{C}$, $R' \in \mathcal{C}\boxtimes\mathcal{C}^{\text{rev}}$ and $r' : T(R') \to U$. The following are equivalent*:

(i) *The object R' represents the functor $\mathcal{C}(T(-), U)$ such that the natural isomorphism $\mathcal{C}(T(-), U) \to \mathcal{C}^2(-, R')$ maps r' to $id_{R'}$.*
(ii) *The pair (R', r') satisfies the following universal property*: *For all pairs (X, x) with $X \in \mathcal{C}\boxtimes\mathcal{C}^{\text{rev}}$ and $x : T(X) \to U$, there exists a unique morphism $\tilde{x} : X \to$*

R' such that the diagram

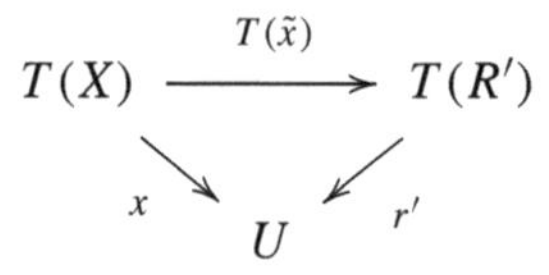

(B.18)

commutes.

Proof (i) $\Rightarrow$ (ii): Denote the natural isomorphism by $\chi_- : \mathcal{C}(T(-), U) \to \mathcal{C}^2(-, R')$. Naturality amounts to the following two equivalent identities, for all $f : X \to Y$, $y : T(Y) \to U$, and for $b = \chi_Y(y)$,

$$\chi_X\big(y \circ T(f)\big) = \chi_Y(y) \circ f, \qquad \chi_Y^{-1}(b) \circ T(f) = \chi_X^{-1}(b \circ f). \tag{B.19}$$

Suppose we are given (X, x). We need to show existence and uniqueness of $\tilde{x}$.

Existence: Choose $\tilde{x} = \chi_X(x)$. Commutativity of (B.18) follows since $r' \circ T(\tilde{x}) = \chi_{R'}^{-1}(id_{R'}) \circ T(\chi_X(x)) = \chi_X^{-1}(id_{R'} \circ \chi_X(x)) = x$.

Uniqueness: Suppose (B.18) holds for some $a : X \to R'$ in place of $\tilde{x}$, i.e. $r' \circ T(a) = x$. By naturality, $r' \circ T(a) = \chi_{R'}^{-1}(id_{R'}) \circ T(a) = \chi_X^{-1}(a)$. Thus $\chi_X^{-1}(a) = x$, which is equivalent to $a = \chi_X(x)$.

(ii) $\Rightarrow$ (i): Given $x : T(X) \to U$, we define the map $\chi_X : \mathcal{C}(T(X), U) \to \mathcal{C}^2(X, R')$ to be $\chi_X(x) = \tilde{x}$. By uniqueness of $\tilde{x}$, this is well-defined. Since for $(X, x) = (R', r')$ we can choose $\tilde{x} = id_{R'}$, the collection of maps χ_- satisfies $\chi_{R'}(r') = id_{R'}$, as required. It remains to see that χ_X is a bijection for each X and that it is natural in X.

Naturality: We will check the first identity in (B.19). By uniqueness of $\tilde{x}$ in (B.18) it is enough to check that also $\chi_Y(y) \circ f$ provides an arrow from $(X, y \circ T(f))$ to (R', r'), i.e. that the diagram

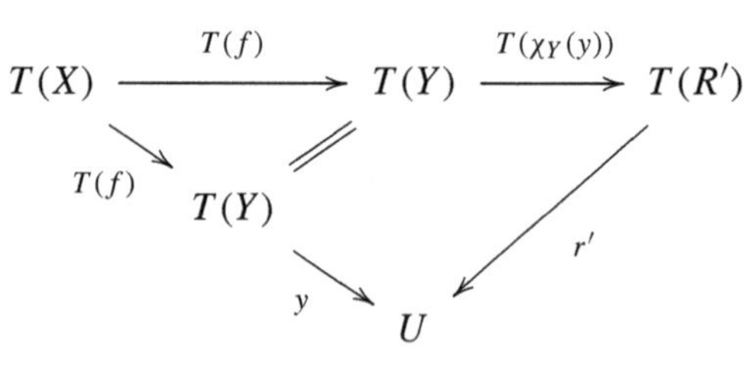

(B.20)

commutes, which it does by definition of χ_Y.

Surjectivity: Given $a : X \to R'$, by naturality and $\chi_{R'}(r') = id_{R'}$ one has $\chi_X(r' \circ T(a)) = \chi_X(r') \circ a = a$.

Injectivity: Suppose $\chi_X(x) = 0$. Then by definition also $x = r' \circ T(0) = 0$. □

The second lemma allows one to rewrite any pairing in terms of the canonical non-degenerate pairings defined in (3.1).

Lemma B.3 *Let* $p : U \otimes V \to \mathbf{1}^*$.

(i) *There exist unique maps* $f : U \to V^*$ *and* $g : V \to U^*$ *such that* $p = \beta_U \circ (id_U \otimes g)$ *and* $p = \beta_{V^*} \circ (f \otimes \delta_V)$.
(ii) *For all* $h : U \to V$ *we have* $\beta_V \circ (h \otimes id_{V^*}) = \beta_U \circ (id_U \otimes h^*)$.

Proof Both parts follow from naturality of δ and π in condition (C). The latter amounts to the statement that for all $a : X \to U$, $b : Y \to V$ and $q : U \to V^*$,

$$\begin{aligned} \pi_{X,V}(q \circ a) &= \pi_{U,V}(q) \circ (a \otimes id_V), \\ \pi_{U,Y}\big(b^* \circ q\big) &= \pi_{U,V}(q) \circ (id_U \otimes b). \end{aligned} \tag{B.21}$$

For part (ii) we compute $\pi_{V,V^*}(\delta_V) \circ (h \otimes id_{V^*}) = \pi_{U,V^*}(\delta_V \circ h) = \pi_{U,V^*}(h^{**} \circ \delta_U) = \pi_{U,U^*}(\delta_U) \circ (id_U \otimes h^*)$. For part (i) set $f = \pi_{U,V}^{-1}(p)$ and $g = f^* \circ \delta_V$. Then

$$\begin{aligned} \beta_U \circ (id_U \otimes g) &= \pi_{U,U^*}(\delta_U) \circ \big(id_U \otimes \big(f^* \circ \delta_V\big)\big) \overset{(1)}{=} \pi_{U,V}\big(\big(f^* \circ \delta_V\big)^* \circ \delta_U\big) \\ &= \pi_{U,V}\big((\delta_V)^* \circ f^{**} \circ \delta_U\big) \overset{(2)}{=} \pi_{U,V}\big((\delta_V)^* \circ \delta_{V^*} \circ f\big) \\ &\overset{(3)}{=} \pi_{U,V}(f) = p, \end{aligned} \tag{B.22}$$

where (1) is naturality of π, (2) is naturality of δ and (3) is $(\delta_V)^* = (\delta_{V^*})^{-1}$, which is required by condition (C). The identity $\beta_{V^*} \circ (f \otimes \delta_V) = p$ is checked along the same lines (use naturality to move f and δ inside π). Uniqueness of f and g is implied by non-degeneracy of β_U and β_{V^*}, see the text below Definition 3.2. □

For the third lemma, recall the space N, the basis $\{u_1, \ldots, u_{|N|}\}$ of N and the map $n = \sum_i u_i \circ \pi_i$ defined in Sect. 3.6.

Lemma B.4 *For any* $f : (P \boxtimes P)^{\oplus m} \to P \boxtimes P^*$ *such that* $\beta_P \circ T(f) = 0$, *there exists a (typically non-unique)* $\varphi : (P \boxtimes P)^{\oplus m} \to (P \boxtimes P)^{\oplus |N|}$ *such that*

$$\begin{array}{ccc} (P \boxtimes P)^{\oplus m} & \xrightarrow{\quad f \quad} & P \boxtimes P^* \\ & {\scriptstyle \exists\varphi} \searrow \quad \nearrow {\scriptstyle n} & \\ & (P \boxtimes P)^{\oplus |N|} & \end{array} \tag{B.23}$$

commutes.

Proof Denote by

$$(P \boxtimes P)^{\oplus m} \underset{e_i}{\overset{p_i}{\rightleftarrows}} P \boxtimes P, \qquad (P \boxtimes P)^{\oplus |N|} \underset{\iota_i}{\overset{\pi_i}{\rightleftarrows}} P \boxtimes P \tag{B.24}$$

the embedding and projection maps of the two direct sums. Let $f_j = f \circ e_j$. By assumption, $f_j \in N$. Thus we can write $f_j = \sum_{i=1}^{|N|} A_{ij} u_i$ for some $A_{ij} \in k$. Define

$$\varphi = \sum_{i=1}^{|N|} \sum_{j=1}^{m} A_{ij} \cdot \iota_i \circ p_j : (P \boxtimes P)^{\oplus m} \to (P \boxtimes P)^{\oplus |N|}. \tag{B.25}$$

Then indeed $n \circ \varphi = \sum_{i,j,k} A_{ij} u_k \circ \pi_k \circ \iota_i \circ p_j = \sum_{i,j} A_{ij} u_i \circ p_j = \sum_j f_j \circ p_j = f$. □

Proof of Theorem 3.17 We will show that R' satisfies condition (ii) in Lemma B.2 (with $U = \mathbf{1}^*$). Namely, suppose we are given a pair (X, x) with $X \in \mathcal{C} \boxtimes \mathcal{C}^{\mathrm{rev}}$ and $x : T(X) \to \mathbf{1}^*$. We need to show that there exists a unique $\tilde{x} : X \to R'$ such that $x = r' \circ T(\tilde{x})$.

- *Existence*: Let

$$(P \boxtimes P)^{\oplus k} \xrightarrow{K} (P \boxtimes P)^{\oplus m} \xrightarrow{\mathrm{cok}(K)} X \tag{B.26}$$

be the first two steps of a projective resolution of X. That is, we have a surjection $s : (P \boxtimes P)^{\oplus m} \to X$ whose kernel is the image of $K : (P \boxtimes P)^{\oplus k} \to (P \boxtimes P)^{\oplus m}$ (and hence $s = \mathrm{cok}(K)$). Define

$$p = \left[T\big((P \boxtimes P)^{\oplus m}\big) \xrightarrow{T(\mathrm{cok}(K))} TX \xrightarrow{x} \mathbf{1}^*\right]. \tag{B.27}$$

Let $\pi_i : (P \boxtimes P)^{\oplus m} \to P \boxtimes P$ be the projection to the ith summand. Then $T(\pi_i) : T((P \boxtimes P)^{\oplus m}) \to P \otimes_{\mathcal{C}} P$ and if we can define p_i via

$$p = \sum_i \left[T\big((P \boxtimes P)^{\oplus m}\big) \xrightarrow{T(\pi_i)} P \otimes_{\mathcal{C}} P \xrightarrow{p_i} \mathbf{1}^*\right]. \tag{B.28}$$

By Lemma B.3(i), there exists a $q_i : P \to P^*$ such that $p_i = [P \otimes_{\mathcal{C}} P \xrightarrow{id \otimes q_i} P \otimes_{\mathcal{C}} P^* \xrightarrow{\beta_P} \mathbf{1}^*]$. Define $\tilde{p} := \sum_i (id_P \boxtimes q_i) \circ \pi_i$. Then

$$\begin{array}{ccc} T((P \boxtimes P)^{\oplus m}) & \xrightarrow{T(\tilde{p})} & P \boxtimes P^* \\ {\scriptstyle T(\mathrm{cok}(K))}\downarrow & \searrow^{p} & \downarrow{\scriptstyle \beta_P} \\ TX & \xrightarrow{x} & \mathbf{1}^* \end{array} \tag{B.29}$$

commutes by construction. It follows that $\beta_P \circ T(\tilde{p} \circ K) = x \circ T(\mathrm{cok}(K) \circ K) = 0$. From Lemma B.4 we get a map u such that subdiagram (1) in the following diagram commutes:

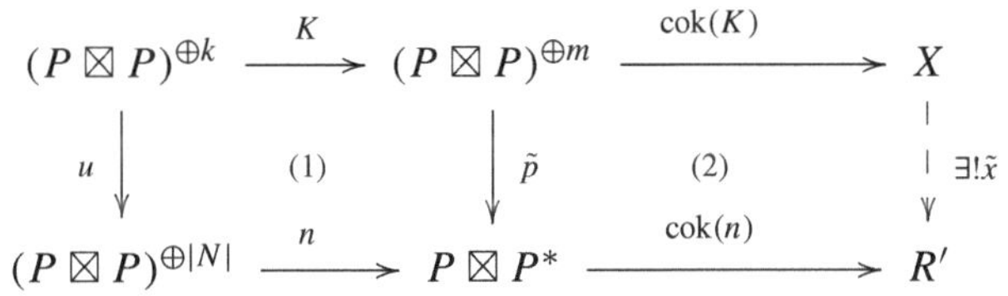

(B.30)

The existence of u implies that $\mathrm{cok}(n) \circ \tilde{p} \circ K = 0$, so that by the universal property of $\mathrm{cok}(K)$ there exists a unique $\tilde{x} : X \to R'$ such that subdiagram (2) commutes. This is the $\tilde{x}$ we are looking for. It remains to show that $x = r' \circ T(\tilde{x})$. Since $\mathrm{cok}(K)$ is a surjection and since T is right exact, also $T(\mathrm{cok}(K))$ is a surjection, and it is sufficient to verify $x \circ T(\mathrm{cok}(K)) = r' \circ T(\tilde{x}) \circ T(\mathrm{cok}(K))$, i.e. commutativity of

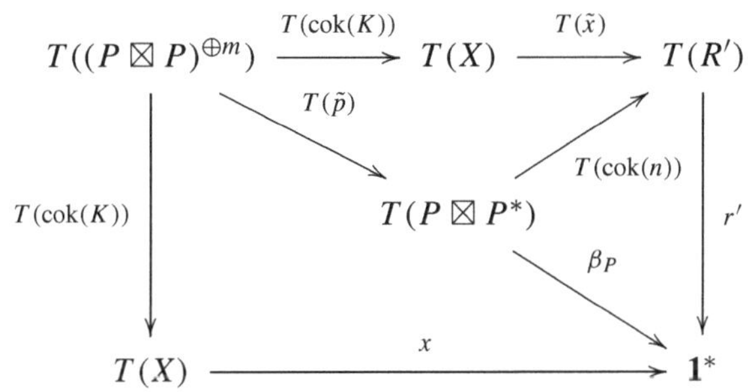

(B.31)

Commutativity of the top square is T applied to square (2) in (B.30); the right triangle is the definition of r' in (3.40); finally, the bottom left square is (B.29).

- *Uniqueness*: We will show that if a map $f : X \to R'$ satisfies $r' \circ T(f) = 0$, then $f = 0$. This implies that the $\tilde{x}$ constructed above is unique. Write $g = f \circ \mathrm{cok}(K)$. It is enough to show that $g = 0$. Consider the diagram

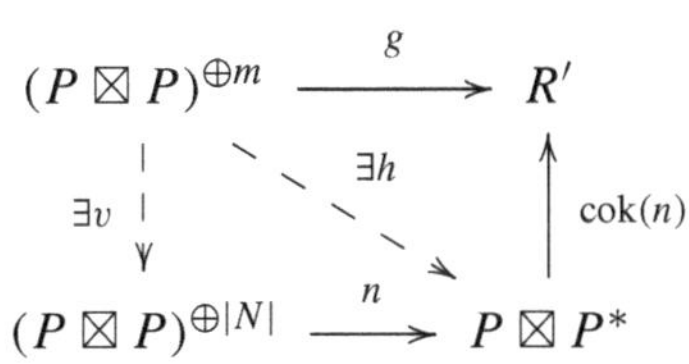

(B.32)

Since $(P \boxtimes P)^{\oplus m}$ is projective, we can pull back g along the surjection $\mathrm{cok}(n)$, giving us the existence of h. By (3.40) we have $\beta_P = r' \circ T(\mathrm{cok}(n))$, so that $\beta_P \circ T(h) = r' \circ T(\mathrm{cok}(n)) \circ T(h) = r' \circ T(g) = r' \circ T(f) \circ T(\mathrm{cok}(K)) = 0$ by assumption on f. Hence we can apply Lemma B.4 to obtain the map v in (B.32). Altogether, $g = \mathrm{cok}(n) \circ n \circ v = 0$. □

Remark B.5 Because of the finiteness assumption (PF), there is a finite number of isomorphism classes of simple objects in $\mathcal{C}$. Let $\{U_i | i \in \mathcal{I}\}$ be a choice of representatives. Furthermore, each U_i has a projective cover P_i. For the projective generator, we can choose $P = \bigoplus_{i\in\mathcal{I}} P_i$, so that R' arises as a quotient of $P \boxtimes P^*$. In fact, one can choose a 'smaller' starting point, namely

$$Q := \bigoplus_{i\in\mathcal{I}} P_i \boxtimes P_i^* \tag{B.33}$$

(but then the above proof would have involved more indices). To describe the map whose cokernel to take, define the subspace

$$M = \left\{ f : P \boxtimes P \to Q \,\middle|\, \sum_{i\in\mathcal{I}} \beta_{P_i} \circ f = 0 \right\} \subset \mathcal{C}^2(P \boxtimes P, Q). \tag{B.34}$$

Denote by $\iota_i : P_i \to P$ and $\pi_i : P \to P_i$ the embedding and restriction map of the direct sum. Pick a basis $\{v_j\}$ of M and define $m : (P \boxtimes P)^{\oplus|M|} \to Q$ as $m = \sum_{l=1}^{|M|} v_l \circ p_l$, with p_l the lth projection $(P \boxtimes P)^{\oplus|M|} \to P \boxtimes P$. Set $R'' = \mathrm{cok}(m)$. Then in fact

$$R' \cong R'', \tag{B.35}$$

with R' defined as in (3.39). To see this, define $\pi : P \boxtimes P^* \to Q$, $\pi = \bigoplus_{i\in\mathcal{I}} \pi_i \boxtimes \iota_i^*$ and $\iota : Q \to P \boxtimes P^*$, $\iota = \bigoplus_{i\in\mathcal{I}} \iota_i \boxtimes \pi_i^*$. These maps make the two diagrams contained in

$$\begin{array}{ccc} P \otimes_{\mathcal{C}} P^* & \xrightarrow{\beta_P} & \\ T(\pi)\big\downarrow \big\uparrow T(\iota) & & \mathbf{1}^* \\ \bigoplus_{i\in\mathcal{I}} P_i \otimes_{\mathcal{C}} P_i^* & \xrightarrow{\sum_{i\in\mathcal{I}} \beta_{P_i}} & \end{array} \tag{B.36}$$

commute. For example, $\beta_P = \sum_{i\in\mathcal{I}} \beta_P \circ (\iota_i \otimes_{\mathcal{C}} id_{P^*}) \circ (\pi_i \otimes_{\mathcal{C}} id_{P^*}) = \sum_{i\in\mathcal{I}} \beta_{P_i} \circ (\pi_i \otimes_{\mathcal{C}} \iota_i^*) = \sum_{i\in\mathcal{I}} \beta_{P_i} \circ T(\pi)$. We can now construct maps between the two cokernels using their universal properties. Consider the diagram

$$\begin{array}{ccccc} (P \boxtimes P)^{\oplus|N|} & \xrightarrow{n} & P \boxtimes P^* & \xrightarrow{\mathrm{cok}(n)} & R' \\ & & \pi \big\downarrow \big\uparrow \iota & & \exists! \,\big\downarrow\, \big\uparrow\, \exists! \\ (P \boxtimes P)^{\oplus|M|} & \xrightarrow{m} & Q & \xrightarrow{\mathrm{cok}(m)} & R'' \end{array} \tag{B.37}$$

The diagram (B.36) tells us that $(\sum_{i\in\mathcal{I}} \beta_{P_i}) \circ T(\pi \circ n) = \beta_P \circ T(n) = 0$. Thus the image of $\pi \circ n$ lies in the image of m (by an argument analogous to the one in Lemma B.4), so that $\mathrm{cok}(m) \circ \pi \circ n = 0$. The universal property gives a unique map $R' \to R''$. Similarly one checks that $\mathrm{cok}(n) \circ \iota \circ m = 0$, giving the map $R'' \to R'$. By uniqueness, these are inverse to each other.

The next lemma prepares the proof of Theorem 3.18.

Lemma B.6 *For all* $u \in \mathcal{C}(P, P)$ *we have* $\mathrm{cok}(n) \circ (u \boxtimes id - id \boxtimes u^*) = 0$.

Proof Pick an $m \in \mathbb{Z}_{>0}$ such that there is a surjection $s : (P \boxtimes P)^{\oplus m} \to P \boxtimes P^*$. Let $f = (u \boxtimes id - id \boxtimes u^*) \circ s$. Then the statement $\mathrm{cok}(n) \circ (u \boxtimes id - id \boxtimes u^*) = 0$ is equivalent to $\mathrm{cok}(n) \circ f = 0$. We will show the latter. By Lemma B.3(ii), we have

$$\left[T\big((P \boxtimes P)^{\oplus m}\big) \xrightarrow{T(s)} P \otimes P^* \xrightarrow{u \otimes id - id \otimes u^*} P \otimes P^* \xrightarrow{\beta_P} \mathbf{1}^*\right] = 0. \qquad \text{(B.38)}$$

We can thus apply Lemma B.4 and obtain a map $\tilde{f} : (P \boxtimes P)^{\oplus m} \to (P \boxtimes P)^{\oplus |N|}$ such that $f = n \circ \tilde{f}$. Hence $\mathrm{cok}(n) \circ f = \mathrm{cok}(n) \circ n \circ \tilde{f} = 0$. □

Proof of Theorem 3.18 Since $\nu \boxtimes id - id \boxtimes \tilde{\nu}$ is a natural transformation of the identity functor on $\mathcal{C} \boxtimes \mathcal{C}^{\mathrm{rev}}$, the diagram

$$\begin{array}{ccc} P \boxtimes P^* & \xrightarrow{\mathrm{cok}(n)} & R' \\ {\scriptstyle (\nu\boxtimes id - id\boxtimes\tilde{\nu})_{P\boxtimes P^*}}\big\downarrow & & \big\downarrow{\scriptstyle (\nu\boxtimes id - id\boxtimes\tilde{\nu})_{R'}} \\ P \boxtimes P^* & \xrightarrow{\mathrm{cok}(n)} & R' \end{array} \qquad \text{(B.39)}$$

commutes. Now note that for all $U \in \mathcal{C}$,

$$\tilde{\nu}_{U^*} = (\delta_{U^*})^{-1} \circ (\nu_{U^{**}})^* \circ \delta_{U^*} = (\delta_U)^* \circ (\nu_{U^{**}})^* \circ \big(\delta_U^{-1}\big)^* = (\nu_U)^*. \qquad \text{(B.40)}$$

Thus $(\nu \boxtimes id - id \boxtimes \tilde{\nu})_{P \boxtimes P^*} = \nu_P \boxtimes id - id \boxtimes (\nu_P)^*$. By Lemma B.6, the lower path in the above diagram is zero. Since $\mathrm{cok}(n)$ is surjective, this implies $(\nu \boxtimes id - id \boxtimes \tilde{\nu})_{R'} = 0$. □

B.5 Adjoint to the Tensor Product

We first need to establish the compatibility of condition (C) and the Deligne product. We do this under the assumption that we are given two categories $\mathcal{C}$, $\mathcal{D}$ which satisfy condition (PF) from Sect. 3.2 (rather than (F) for we need to invoke Proposition 5.5 in [9]), which are monoidal with k-linear right exact tensor product, and both have conjugates according to condition (C).

Since $(-)^*$ is an equivalence, it is exact. Thus $\mathcal{C}^{\mathrm{opp}} \times \mathcal{D}^{\mathrm{opp}} \xrightarrow{(-)^* \times (-)^*} \mathcal{C} \times \mathcal{D} \xrightarrow{\boxtimes} \mathcal{C} \boxtimes \mathcal{D}$ factors through a functor $\mathcal{C}^{\mathrm{opp}} \boxtimes \mathcal{D}^{\mathrm{opp}} \to \mathcal{C} \boxtimes \mathcal{D}$. By Proposition 5.5 in [9], we may take $\mathcal{C}^{\mathrm{opp}} \boxtimes \mathcal{D}^{\mathrm{opp}} = (\mathcal{C} \boxtimes \mathcal{D})^{\mathrm{opp}}$. Altogether, we get a contragredient involutive functor on $\mathcal{C} \boxtimes \mathcal{D}$, which we also denote by $(-)^*$. By definition, on 'factorised objects' it satisfies

$$(C \boxtimes D)^* = C^* \boxtimes D^*. \qquad \text{(B.41)}$$

Lemma B.7 $(-)^*: \mathcal{C} \boxtimes \mathcal{D} \to \mathcal{C} \boxtimes \mathcal{D}$ *satisfies property* (C).

Proof The natural isomorphisms $\delta^{\mathcal{C}}$ and $\delta^{\mathcal{D}}$ between the exact functors $\boxtimes$ and $\boxtimes \circ \{(-)^{**} \times (-)^{**}\}$ from $\mathcal{C} \times \mathcal{D}$ to $\mathcal{C} \boxtimes \mathcal{D}$ provide a natural isomorphism δ from *Id* to $(-)^{**}$ on $\mathcal{C} \boxtimes \mathcal{D}$ with the required property $(\delta_X)^* = (\delta_{X^*})^{-1}$.

For the existence of π, we stress again (Proposition 5.5 in [9]): $\mathcal{C}^{\mathrm{opp}} \boxtimes \mathcal{D}^{\mathrm{opp}} = (\mathcal{C} \boxtimes \mathcal{D})^{\mathrm{opp}}$. Thus there is an equivalence of functor categories between k-linear right exact functors in each argument $\mathcal{C}^{\mathrm{opp}} \times \mathcal{D}^{\mathrm{opp}} \to \mathcal{E}^{\mathrm{opp}}$ and right exact functors $(\mathcal{C} \boxtimes \mathcal{D})^{\mathrm{opp}} \to \mathcal{E}^{\mathrm{opp}}$. But this is the same as saying that there is an equivalence of functor categories between k-linear functors $\mathcal{C} \times \mathcal{D} \to \mathcal{E}$, left exact in each argument, and left exact functors $\mathcal{C} \boxtimes \mathcal{D} \to \mathcal{E}$. Given our assumptions on $\mathcal{C}$ and $\mathcal{D}$, by Corollary 5.4 in [9], the functor $\boxtimes : \mathcal{C} \times \mathcal{D} \to \mathcal{C} \boxtimes \mathcal{D}$ itself is exact in each argument.

Recall that an abelian category $\mathcal{A}$, the Hom-functor from $\mathcal{A}^{\mathrm{opp}} \times \mathcal{A}$ to abelian groups given by $(A, B) \mapsto \mathcal{A}(A, B)$ is left exact in each argument. Thus the functor $\mathcal{C}^{\mathrm{opp}} \times \mathcal{C}^{\mathrm{opp}} \to \mathcal{V}ect$ given by $(U, V) \mapsto \mathcal{C}(U, V^*)$ is left exact in each argument. The maps $\pi_{U,V}$ provide a natural isomorphism from this functor to the functor $(U, V) \mapsto \mathcal{C}(U \otimes V, \mathbf{1}^*)$, which therefore is also left exact in both arguments (even though it involves the right exact tensor product). The same reasoning applies to $\mathcal{D}$. The combined functors

$$\begin{aligned} &\mathcal{C}^{\mathrm{opp}} \times \mathcal{D}^{\mathrm{opp}} \times \mathcal{C}^{\mathrm{opp}} \times \mathcal{D}^{\mathrm{opp}} \to \mathcal{V}ect, \\ &(U, A, V, B) \mapsto \mathcal{C}(U, V^*) \otimes_k \mathcal{D}(A, B^*) \quad \text{and} \\ &(U, A, V, B) \mapsto \mathcal{C}(U \otimes V, \mathbf{1}^*) \otimes_k \mathcal{D}(A \otimes B, \mathbf{1}^*), \end{aligned} \tag{B.42}$$

are equally left exact in each argument, and thus give two functors $(\mathcal{C} \boxtimes \mathcal{D})^{\mathrm{opp}} \times (\mathcal{C} \boxtimes \mathcal{D})^{\mathrm{opp}} \to (\mathcal{C} \boxtimes \mathcal{D} \boxtimes \mathcal{C} \boxtimes \mathcal{D})^{\mathrm{opp}} \to \mathcal{V}ect$ which are left exact in each argument. In view of (3.4), these functors are necessarily given by

$$(X, Y) \mapsto \mathcal{C} \boxtimes \mathcal{D}\big(X, Y^*\big) \quad \text{and} \quad (X, Y) \mapsto \mathcal{C} \boxtimes \mathcal{D}\big(X \otimes Y, \mathbf{1}^* \boxtimes \mathbf{1}^*\big). \tag{B.43}$$

The equivalence (3.2) of functor categories—which as we saw above also holds for the corresponding categories of left exact functors—now shows that the natural isomorphism $\pi^{\mathcal{C}}_{U,V} \otimes_k \pi^{\mathcal{D}}_{A,B}$ between the functors (B.42) provides a natural isomorphism $\pi_{X,Y}$ between the functors (B.43). □

Recall the definition of the functor R in terms of the conjugates on $\mathcal{C}$ and $\mathcal{C} \boxtimes \mathcal{C}^{\mathrm{rev}}$ given in (3.43).

Proof of Theorem 3.20 The natural isomorphisms

$$\xi_{X,U} : \mathcal{C}\big(T(X), U\big) \xrightarrow{\sim} \mathcal{C}^2\big(X, R(U)\big) \tag{B.44}$$

are provided by the composition

$$
\begin{aligned}
\mathcal{C}\big(T(X), W\big) &\xrightarrow[\pi \text{ and } \delta]{\sim} \mathcal{C}\big(T(X) \otimes_{\mathcal{C}} W^*, \mathbf{1}^*\big) \\
&\xrightarrow{\sim} \mathcal{C}\big(T(X) \otimes_{\mathcal{C}} T\big(W^* \boxtimes \mathbf{1}\big), \mathbf{1}^*\big) \\
&\xrightarrow[T_2]{\sim} \mathcal{C}\big(T\big(X \otimes_{\mathcal{C}^2} \big(W^* \boxtimes \mathbf{1}\big)\big), \mathbf{1}^*\big) \\
&\xrightarrow[\chi \text{ from (3.37)}]{\sim} \mathcal{C}^2\big(X \otimes_{\mathcal{C}^2} \big(W^* \boxtimes \mathbf{1}\big), R_{\mathbf{1}^*}\big) \\
&\xrightarrow[\text{Lemma B.7}]{\sim} \mathcal{C}^2\big(\big[X \otimes_{\mathcal{C}^2} \big(W^* \boxtimes \mathbf{1}\big)\big] \otimes_{\mathcal{C}^2} (R_{\mathbf{1}^*})^*, \mathbf{1}^* \boxtimes \mathbf{1}^*\big) \\
&\xrightarrow[\text{assoc}]{\sim} \mathcal{C}^2\big(X \otimes_{\mathcal{C}^2} \big[\big(W^* \boxtimes \mathbf{1}\big) \otimes_{\mathcal{C}^2} (R_{\mathbf{1}^*})^*\big], \mathbf{1}^* \boxtimes \mathbf{1}^*\big) \\
&\xrightarrow[\text{Lemma B.7}]{\sim} \mathcal{C}^2\big(X, \big[\big(W^* \boxtimes \mathbf{1}\big) \otimes_{\mathcal{C}^2} (R_{\mathbf{1}^*})^*\big]^*\big) \\
&\equiv \mathcal{C}^2\big(X, R(W)\big).
\end{aligned}
\tag{B.45}
$$

□

The adjunction natural transformations are, in terms of the isomorphism (B.44),

$$
\begin{aligned}
\eta_X &:= \xi_{X,T(X)}(id_{T(X)}) : X \to R\big(T(X)\big), \\
\varepsilon_U &:= \xi^{-1}_{R(U),U}(id_{R(U)}) : T\big(R(U)\big) \to U,
\end{aligned}
\tag{B.46}
$$

and the satisfy the adjunction properties, for $X \in \mathcal{C} \boxtimes \mathcal{C}^{\mathrm{rev}}$ and $U \in \mathcal{C}$,

$$
\begin{aligned}
&\big[R(U) \xrightarrow{\eta_{R(U)}} RTR(U) \xrightarrow{R(\varepsilon_U)} R(U)\big] = id_{R(U)}, \\
&\big[T(X) \xrightarrow{T(\eta_X)} TRT(X) \xrightarrow{\varepsilon_{T(X)}} T(X)\big] = id_{T(X)}.
\end{aligned}
\tag{B.47}
$$

The functor R is lax monoidal (and colax monoidal), with R_0 and R_2 given by

$$
\begin{aligned}
R_0 &= \big[\mathbf{1} \boxtimes \mathbf{1} \xrightarrow{\eta_{\mathbf{1}\boxtimes\mathbf{1}}} R\big(T(\mathbf{1} \boxtimes \mathbf{1})\big) \xrightarrow{R(T_0^{-1})} R(\mathbf{1})\big], \\
R_{2;U,V} &= \big[RU \otimes_{\mathcal{C}^2} RV \xrightarrow{\eta_{RU \otimes RV}} R\big(T(RU \otimes_{\mathcal{C}^2} RV)\big) \\
&\quad \xrightarrow{T^{-1}_{2;RU,RV}} R\big(TR(U) \otimes_{\mathcal{C}} TR(V)\big) \xrightarrow{R(\varepsilon_U \otimes \varepsilon_V)} R(U \otimes_{\mathcal{C}} V)\big],
\end{aligned}
\tag{B.48}
$$

see, e.g., Definition 2.1 and Lemma 2.7 in [34].

B.6 More Details on Centres

This appendix contains an auxiliary result which implies the existence of the left centre in abelian monoidal categories which have conjugates as in (C), and it contains the proof of Theorem 3.24.

Let $\mathcal{A}$ be an abelian monoidal category with conjugates according to condition (C). Let $m : A \otimes B \to C$ be a morphism in $\mathcal{A}$. Consider the category $\mathcal{Q}$ whose objects are pairs (U, u) where $u : U \to A$ is such that $m \circ (u \otimes id_B) = 0$. Morphisms $f : (U, u) \to (V, v)$ in $\mathcal{Q}$ are maps $f : U \to V$ in $\mathcal{A}$ such that $v \circ f = u$.

Using conjugates, from m we obtain a morphism $\tilde{m} : A \to (B \otimes C^*)^*$ via

$$\begin{aligned} \mathcal{A}(AB, C) &\xrightarrow{\sim} \mathcal{A}(AB, C^{**}) \xrightarrow{\sim} \mathcal{A}((AB)C^*, \mathbf{1}^*) \\ &\xrightarrow{\sim} \mathcal{A}(A(BC^*), \mathbf{1}^*) \xrightarrow{\sim} \mathcal{A}(A, (BC^*)^*). \end{aligned} \tag{B.49}$$

When applied to m, this chain of natural isomorphisms yields

$$\tilde{m} = \pi^{-1}_{A,BC^*}\big[\pi_{AB,C^*}(\delta_C \circ m) \circ \alpha_{A,B,C^*}\big]. \tag{B.50}$$

Naturality in A (cf. (B.21)) implies that for any $u : U \to A$ we have

$$\tilde{m} \circ u = \pi^{-1}_{U,BC^*}\big[\pi_{UB,C^*}\big(\delta_C \circ m \circ (u \otimes id_B)\big) \circ \alpha_{U,B,C^*}\big]. \tag{B.51}$$

Lemma B.8 *The following are equivalent.*

(i) (K, k) *is terminal in* $\mathcal{Q}$.
(ii) $k : K \to A$ *is the kernel of* $\tilde{m} : A \to (B \otimes C^*)^*$.

Proof (i) $\Rightarrow$ (ii): Let $u : U \to A$ be a morphism such that $\tilde{m} \circ u = 0$. By (B.51), then also $m \circ (u \otimes id_B) = 0$. Thus (U, u) is an object in $\mathcal{Q}$. By terminality there is a unique arrow $f : U \to K$ such that $k \circ f = u$. This is the thought-for f in the universal property of the kernel.

(ii) $\Rightarrow$ (i): Let (U, u) be an object in $\mathcal{Q}$. Then $m \circ (u \otimes id_B) = 0$ and as above we see that $\tilde{m} \circ u = 0$. By the universal property of the kernel, there is a unique map $f : U \to K$ such that $k \circ f = u$. Thus there is a unique morphism $f : (U, u) \to (K, k)$. □

This lemma implies the existence of the left centre of an algebra B in the category $\mathcal{A}$. Indeed, the universal property of the left centre from Definition 3.22 amounts to the terminal object condition in the category $\mathcal{Q}$ from above with the choice

$$m = \mu_B \circ (id_{B\otimes B} - c_{B,B}) : B \otimes B \to B. \tag{B.52}$$

As the kernel of $\tilde{m}$ exists in $\mathcal{A}$, so does the terminal object in $\mathcal{Q}$ and hence the left centre.

Next, we turn to the proof of Theorem 3.24.

Proof of Theorem 3.24 We need to check that $(Z, z) \equiv (C_l(R(A)), \varepsilon_A \circ T(e))$ satisfies the universal property in Definition 3.14. Let thus (X, x) be a pair such that (3.31) commutes.

By Lemma B.2 with $U = A$, $R' = R(A)$ and $r' = \varepsilon_A$, there is a unique map $\tilde{x} : X \to R(A)$ such that (B.18) commutes, i.e. such that $\varepsilon_A \circ T(\tilde{x}) = x$. This map is given by (use (B.46) and naturality of ξ)

$$\tilde{x} = \big[X \xrightarrow{\xi_{X,A}(x)} R(A)\big] = \big[X \xrightarrow{\eta_X} R\big(T(X)\big) \xrightarrow{R(x)} R(A)\big]. \tag{B.53}$$

We will see below that the map $\tilde{x}$ satisfies the condition for the universal property of the left centre, that is, diagram (3.45) commutes for the pair $(X, \tilde{x})$. Thus, the map $\tilde{x}$ factors as

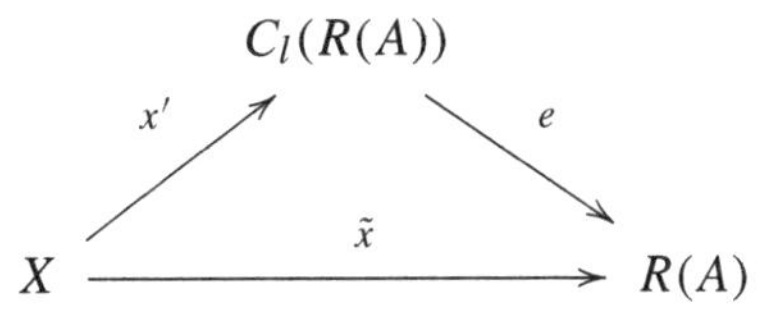

(B.54)

Since $\tilde{x}$ is unique and e is mono, also x' is unique. That x' makes (3.32) commute follows from $x = \varepsilon_A \circ T(\tilde{x}) = \varepsilon_A \circ T(e) \circ T(x') = z \circ T(x')$. This shows that (Z, z) is the full centre of A.

It remains to check that $\tilde{x}$ satisfies (3.45). For convenience, we reproduce (minimally adapted to our setting) the proof given in Theorem 5.4 in [7]. Substituting the expression (B.53) for $\tilde{x}$, we need to show commutativity of

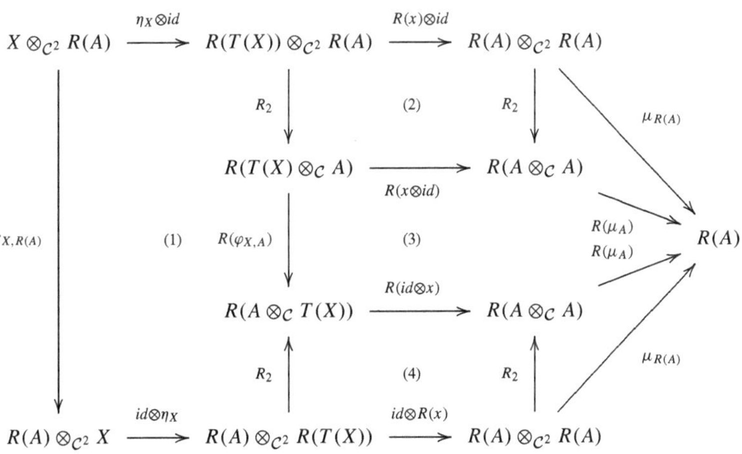

(B.55)

The rightmost triangles are the Definition (3.30) of $\mu_{R(A)}$. Squares 2 and 4 are naturality of R_2. Subdiagram 3 is R applied to the defining property (3.31) of x. Subdiagram 1 is somewhat tedious and is further analysed in Fig. 4. In explaining the commutativity of the various cells, let us start with the key step: the two ways of writing the arrow between cells 6 and 7, which amounts to R applied to Lemma 3.10. Using $R(\hat{\varphi}_{X,RA})$, cell 6 is R applied to the definition of $\hat{\varphi}$ in (3.21), and using

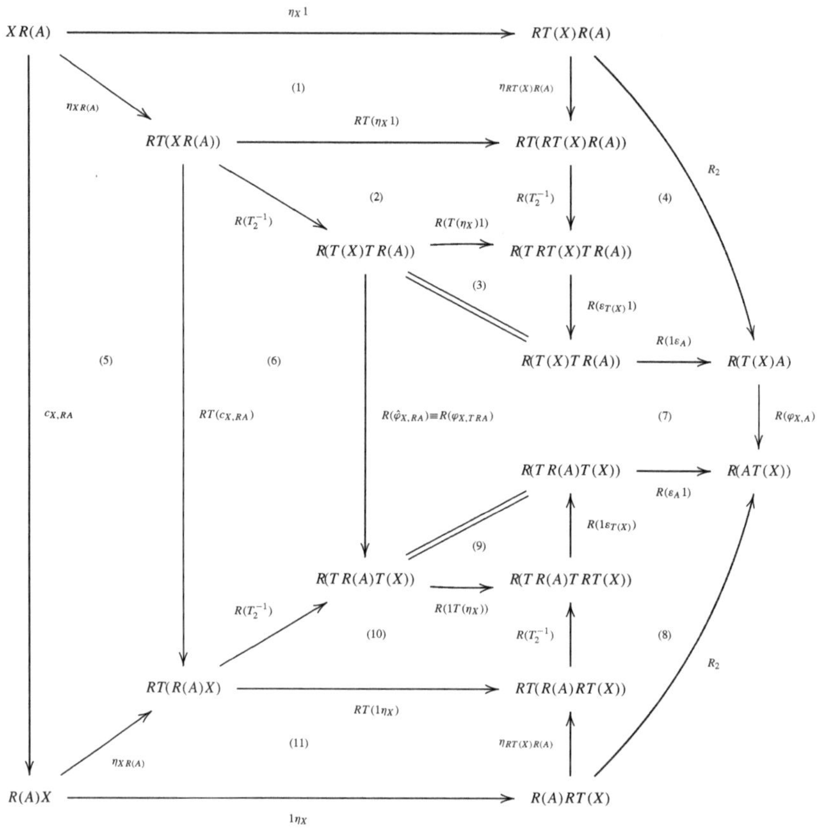

Fig. 4 Commutativity of subdiagram (1) in (B.55). All $\otimes$ have been omitted, instead of *id* the shorthand 1 in used, and only a minimum of brackets is given. The commutativity of the individual cells is explained in the main text

$R(\varphi_{X,TRA})$, cell 7 is R applied to naturality of $\varphi_{X,-}$. The remaining cells are as follows: cells 1, 5, 11 are naturality of η, cells 2, 10 are naturality of T_2, cells 3, 9 are the adjunction property (B.47), cells 4, 8 are the definition (B.48) of R_2. □

Appendix C: OPEs in $R(\mathcal{W}^*)$

C.1 OPEs Involving Ω

We will demonstrate that for all fields $\phi \in R(\mathcal{W}^*)$ one has

$$M_z(\Omega \otimes \phi) = M_z(\phi \otimes \Omega) = \pi(\phi) \cdot \Omega, \tag{C.1}$$

where $M_z(\phi \otimes \psi)$ is the OPE as introduced in (2.4). We will also use the conventional notation $\phi(z)\psi(0)$ for the OPE. To establish (C.1) we will first show that $C_2(u|z,0,\Omega,L_m\phi) = 0$ for all $\phi \in F$, $u \in F'$, and $m \in \mathbb{Z}$ (recall the notation from Sect. 2.2). Suppose the contrary and let $M \in \mathbb{Z}$ be the largest integer such that $C_2(u|z,0,\Omega,L_M\phi) \neq 0$. Choose $N > 0$ such that $L_n u = 0$ for all $n \geq N$. Apply property (C5′) for $f(\zeta) = (z-\zeta)^{-N-M}\zeta^{M+1}$. Since Ω is annihilated by all L_n, one checks that (2.17) becomes $0 = z^{-N-M} C_2(u|z,0,\Omega,L_M\phi)$, in contradiction to our assumption. Similarly one checks that $C_2(u|z,0,\Omega,\overline{L}_m\phi) = 0$ for all u, ϕ, m. Let Δ, N be such that $(L_0 + \overline{L}_0 - \Delta)^N \phi = 0$. Then

$$0 = C_2\big(u|z,0,\Omega,(L_0+\overline{L}_0-\Delta)^N\phi\big) = (-\Delta)^N C_2(u|z,0,\Omega,\phi), \tag{C.2}$$

and so the OPE $\Omega(z)\phi(0)$ can only be non-vanishing for $\phi \in F^{(0)}$. Since $\omega = L_0\eta$ and $\Omega = L_0^2\eta$, the OPE vanishes for $\phi = \omega, \Omega$. To confirm (C.1) it only remains to check $\Omega(z)\eta(0) = \Omega(0)$. Using once more that Ω is annihilated by all Virasoro modes, we have $L_m M_z(\Omega \otimes \eta) = M_z(\Omega \otimes L_m\eta)$, which is zero by the above argument, and analogously $\overline{L}_m M_z(\Omega \otimes \eta) = 0$. This applies in particular to $m = -1$, and the intersection of the kernels of L_{-1} and $\overline{L}_{-1}$ is $\mathbb{C}\Omega$. Thus $\Omega(z)\eta(0) = a \cdot \Omega(0)$ for some $a \in \mathbb{C}$. But then also $\eta(z)\Omega(0) = a \cdot \Omega(0)$ and $\eta(z)\eta(w)\Omega(0) = a^2 \cdot \Omega(0)$. On the other hand, from $\pi(\eta(z)\eta(0)) = 1$ we see that $\eta(z)\eta(0) = \eta(0) + (\text{other fields})$. Thus $a = 1$.

C.2 OPEs Involving the Holomorphic Field $\mathcal{T}$

The next-simplest set of OPEs are those of the form $\mathcal{T}(z)\phi(0)$. Since $\mathcal{T}$ is holomorphic, this OPE does not involve logarithmic singularities (or it would not be single-valued). For example, the most general ansatz for the OPE with ω is

$$\begin{aligned}\mathcal{T}(z)\omega(0) &= z^{-2}\big(P\cdot\eta(0) + Q\cdot\omega(0) + R\cdot\Omega(0)\big)\\ &\quad + z^{-1}\big(S\cdot L_{-1}\eta(0) + U\cdot L_{-1}\omega(0)\big) + \mathcal{O}\big(z^0\big)\end{aligned} \tag{C.3}$$

for some constants $P, Q, R, S, U \in \mathbb{C}$. These constants are further constrained by the identity

$$\begin{aligned}L_m M_z(\mathcal{T}\otimes\phi) &= \sum_{k=0}^{3}\binom{m+1}{k} z^{m+1-k} M_z(L_{k-1}\mathcal{T}\otimes\phi) + M_z(\mathcal{T}\otimes L_m\phi)\\ &= z^m\left(2(m+1) + z\frac{\partial}{\partial z}\right) M_z(\mathcal{T}\otimes\phi) - \frac{5}{6}\big(m^3-m\big)z^{m-2}\pi(\phi)\cdot\Omega\\ &\quad + M_z(\mathcal{T}\otimes L_m\phi),\end{aligned} \tag{C.4}$$

where $m \in \mathbb{Z}$ and $\phi \in F$ are arbitrary. The first equality follows from (C5′) and the second uses (4.11) and (C.1). If one applies this identity for $m = 0$ and $m = 1$ to

(C.3), one quickly finds that $P = Q = S = 0$ and $R = U$. Thus

$$\mathcal{T}(z)\omega(0) = R \cdot \big(z^{-2}\Omega(0) + z^{-1}L_{-1}\omega(0)\big) + \mathcal{O}\big(z^0\big). \tag{C.5}$$

This also provides the two-point correlator

$$^{\eta}\big\langle \mathcal{T}(z)\omega(0)\big\rangle = R \cdot z^{-2}. \tag{C.6}$$

Actually, R is necessarily zero, though it will take us a little while to get there. Since the states of generalised weight $(1,0)$, $(2,0)$ and $(3,0)$ are Virasoro descendants of η, the same method allows one to determine the z^0 and z^1 coefficient in this OPE. The calculations become more lengthy, but the answer is simply

$$\begin{aligned}\mathcal{T}(z)\omega(0) = R \cdot \big(z^{-2}\Omega(0) + z^{-1}L_{-1}\omega(0) + L_{-2}\omega(0) + zL_{-3}\omega(0)\big)\\ + \mathcal{O}\big(z^2\big).\end{aligned} \tag{C.7}$$

Next we compute $\mathcal{T}(z)\mathcal{T}(0)$ by using (C.4) to move all Virasoro modes in $M_z(\mathcal{T} \otimes (L_{-2} - \frac{3}{2}L_{-1}L_{-1})\omega)$ to the left and by then inserting the OPE (C.7). A short calculation yields

$$\mathcal{T}(z)\mathcal{T}(0) = R \cdot \big\{-5z^{-4}\Omega(0) + 2z^{-2}\mathcal{T}(0) + z^{-1}(L_{-1}\mathcal{T})(0)\big\} + \mathcal{O}\big(z^0\big). \tag{C.8}$$

The OPEs (C.7) and (C.8) allow one to determine the three-point function $^{\eta}\langle \mathcal{T}(z)\mathcal{T}(w)\omega(0)\rangle$ by singularity subtraction. Thinking of the three-point function as a function of z, this function vanishes at infinity and has poles only at w and 0. Subtracting these poles we hence find a holomorphic function on $\mathbb{C}$ vanishing at infinity, i.e. a function which is identically zero:

$$\begin{aligned}0 = {}^{\eta}\big\langle \mathcal{T}(z)\mathcal{T}(w)\omega(0)\big\rangle - R \cdot \bigg(\frac{2}{(z-w)^2}{}^{\eta}\big\langle \mathcal{T}(w)\omega(0)\big\rangle\\ + \frac{1}{(z-w)}\frac{\partial}{\partial w}{}^{\eta}\big\langle \mathcal{T}(w)\omega(0)\big\rangle + \frac{1}{z}\Big(-\frac{\partial}{\partial w}\Big){}^{\eta}\big\langle \mathcal{T}(w)\omega(0)\big\rangle\bigg).\end{aligned} \tag{C.9}$$

Substituting (C.6), the result is

$$^{\eta}\big\langle \mathcal{T}(z)\mathcal{T}(w)\omega(0)\big\rangle = \frac{2 \cdot R^2}{zw(z-w)^2}. \tag{C.10}$$

Note that this function is invariant under the exchange of z and w as it has to be. Repeating the above steps to constrain the OPE $\mathcal{T}(z)\eta(0)$ leads to

$$\begin{aligned}\mathcal{T}(z)\eta(0) = z^{-2} \cdot \big\{R \cdot \omega(0) + A \cdot \Omega(0)\big\} + z^{-1} \cdot \big\{R \cdot (L_{-1}\eta)(0) + A \cdot (L_{-1}\omega)(0)\big\}\\ + \bigg\{R \cdot (L_{-2}\eta)(0) + (A+1) \cdot (L_{-2}\omega)(0) - \frac{3}{2}(L_{-1}L_{-1}\omega)(0)\bigg\}\end{aligned}$$

$$+z\cdot\Big\{R\cdot(L_{-3}\eta)(0)+(A+1)\cdot(L_{-3}\omega)(0)+(L_{-2}L_{-1}\omega)(0)$$
$$-\frac{3}{2}(L_{-1}L_{-1}L_{-1}\omega)(0)\Big\}+\mathcal{O}(z^2), \tag{C.11}$$

where $A\in\mathbb{C}$ is a new constant. The corresponding two-point correlator is

$$^{\eta}\big\langle\mathcal{T}(z)\eta(0)\big\rangle=A\cdot z^{-2}. \tag{C.12}$$

As before, the above OPE can be used to determine the OPE of $\mathcal{T}$ and t with the result

$$\begin{aligned}\mathcal{T}(z)t(0)=z^{-4}&\cdot\big\{-5R\cdot\omega(0)+\big(9R-5(A+1)\big)\cdot\Omega(0)\big\}\\&+z^{-2}\cdot\big\{2R\cdot t(0)+\big(R+2(A+1)\big)\cdot\mathcal{T}(0)\big\}\\&+z^{-1}\cdot\big\{R\cdot(L_{-1}t)(0)+(A+1)\cdot(L_{-1}\mathcal{T})(0)\big\}+\mathcal{O}(z^0).\end{aligned} \tag{C.13}$$

In a slightly tedious exercise the OPEs determine the three-point correlator of $\langle\mathcal{T}\mathcal{T}\eta\rangle$ by singularity subtraction to be

$$^{\eta}\big\langle\mathcal{T}(z)\mathcal{T}(w)\eta(0)\big\rangle=\frac{-5R}{(z-w)^4}+\frac{R^2}{z^2w^2}+\frac{2RA}{zw(z-w)^2}+\frac{2RA}{zw^3}. \tag{C.14}$$

Because of the last summand, this expression is only invariant under $z\leftrightarrow w$ if $R=0$ or $A=0$. To see that actually $R=0$ is required, one can compute (in another such tedious exercise)

$$\begin{aligned}&^{\eta}\big\langle\mathcal{T}(z)\mathcal{T}(w)t(0)\big\rangle-{}^{\eta}\big\langle\mathcal{T}(w)\mathcal{T}(z)t(0)\big\rangle\\&\quad=20R(A+1)\cdot\frac{(z-w)(z^3+(3/2)z^2w+(3/2)zw^2+w^3)}{z^5w^5}.\end{aligned} \tag{C.15}$$

With $R=0$ the correlators $^{\eta}\langle\mathcal{T}(z)\mathcal{T}(w)\phi(0)\rangle$ are zero for ϕ any of ω, η, $\mathcal{T}$, t. The OPEs (C.7), (C.8), (C.11) and (C.13) reproduce the formulas in (4.30).

C.3 OPEs of Generalised Weight Zero Fields

Next we consider the non-holomorphic OPE $\omega(z)\omega(0)$. The identity corresponding to (C.4) reads in this case

$$\begin{aligned}L_mM_z(\omega\otimes\phi)=z^{m+1}&\frac{\partial}{\partial z}M_z(\omega\otimes\phi)+(m+1)z^m\pi(\phi)\cdot\Omega\\&+M_z(\omega\otimes L_m\phi),\end{aligned} \tag{C.16}$$

for all $\phi \in F$ and $m \in \mathbb{Z}$, and analogously for $\overline{L}_m$. For $m = 0$ we find in particular that

$$\left(L_0 - z\frac{\partial}{\partial z}\right)M_z(\omega \otimes \omega) = 0 = \left(\overline{L}_0 - \bar{z}\frac{\partial}{\partial \bar{z}}\right)M_z(\omega \otimes \omega). \tag{C.17}$$

The general solution to this first order differential equation reads

$$M_z(\omega \otimes \omega) = \exp\{\ln(z)L_0 + \ln(\bar{z})\overline{L}_0\}\Psi \quad \text{for some } \Psi \in \overline{F}. \tag{C.18}$$

This shows that the leading term in the OPE is (take the component of Ψ in $F^{(0)}$ to be $X \cdot \eta + Y \cdot \omega + B \cdot \Omega$)

$$\begin{aligned}\omega(z)\omega(0) &= X \cdot \eta(0) + \{Y + X\ln(|z|^2)\}\omega(0)\\ &\quad + \left\{B + Y\ln(|z|^2) + \frac{1}{2}X \cdot (\ln(|z|^2))^2\right\}\Omega(0) + \cdots \end{aligned} \tag{C.19}$$

This expression can be used to compute the leading term in the OPE $\omega(z)\mathcal{T}(0)$, which we already know from (4.30) to be of order $\mathcal{O}(z^0)$. Using (C.16) to move the L_m modes past $\omega(z)$ one quickly finds the requirement that $X = Y = 0$. This reproduces (4.32). For $\eta(z)\omega(0)$ we use (C.4) in the form

$$\begin{aligned}L_m M_z(\eta \otimes \omega) &= z^{m+1}\frac{\partial}{\partial z}M_z(\eta \otimes \omega) + (m+1)z^m M_z(\omega \otimes \omega)\\ &\quad + M_z(\eta \otimes L_m\omega),\end{aligned} \tag{C.20}$$

and analogously for $\overline{L}_m$. We can again make a general ansatz for the leading term in the OPE $\eta(z)\omega(0)$ and use the knowledge of $\omega(z)\omega(0)$ and $\eta(z)\mathcal{T}(0)$ (from (4.30)) to constrain the coefficients. The result is as stated in (4.32), we skip the details.

References

1. Adamović, D., Milas, A.: On W-algebras associated to $(2, p)$ minimal models and their representations. Int. Math. Res. Not., **2010**, 3896–3934 (2010). arXiv:0908.4053
2. Belavin, A.A., Polyakov, A.M., Zamolodchikov, A.B.: Infinite conformal symmetry in two-dimensional quantum field theory. Nucl. Phys. B **241**, 333–380 (1984)
3. Cappelli, A., Itzykson, C., Zuber, J.B.: The ADE classification of minimal and A1(1) conformal invariant theories. Commun. Math. Phys. **113**, 1–26 (1987)
4. Cardy, J.L.: Conformal invariance and surface critical behavior. Nucl. Phys. B **240**, 514–532 (1984)
5. Cardy, J.L.: Operator content of two-dimensional conformal invariant theories. Nucl. Phys. B **270**, 186–204 (1986)
6. Cardy, J.L., Lewellen, D.C.: Bulk and boundary operators in conformal field theory. Phys. Lett. B **259**, 274–278 (1991)
7. Davydov, A.: Centre of an algebra. Adv. Math. **225**, 319–348 (2010). arXiv:0908.1250
8. Davydov, A., Müger, M., Nikshych, D., Ostrik, V.: The Witt group of non-degenerate braided fusion categories. J. Reine Angew. Math. (2012). doi:10.1515/crelle.2012.014. arXiv:1009.2117

9. Deligne, P.: Catégories tannakiennes. In: Grothendieck Festschrift, vol. II, pp. 111–195. Birkhäuser, Boston (2007)
10. Eberle, H., Flohr, M.: Virasoro representations and fusion for general augmented minimal models. J. Phys. A **39**, 15245 (2006). arXiv:hep-th/0604097
11. Etingof, P.I., Ostrik, V.: Finite tensor categories. arXiv:math/0301027
12. Feigin, B.L., Gainutdinov, A.M., Semikhatov, A.M., Tipunin, I.Y.: Logarithmic extensions of minimal models: characters and modular transformations. Nucl. Phys. B **757**, 303–343 (2006). arXiv:hep-th/0606196
13. Fjelstad, J., Fuchs, J., Runkel, I., Schweigert, C.: TFT construction of RCFT correlators. V: proof of modular invariance and factorisation. Theory Appl. Categ. **16**, 342–433 (2006). arXiv:hep-th/0503194
14. Fjelstad, J., Fuchs, J., Runkel, I., Schweigert, C.: Uniqueness of open/closed rational CFT with given algebra of open states. Adv. Theor. Math. Phys. **12**, 1283–1375 (2008). arXiv:hep-th/0612306
15. Frenkel, I.B., Zhu, Y.: Vertex operator algebras associated to representations of affine and Virasoro algebras. Duke Math. J. **66**, 123–168 (1992)
16. Friedan, D., Shenker, S.H.: The analytic geometry of two-dimensional conformal field theory. Nucl. Phys. B **281**, 509–545 (1987)
17. Fröhlich, J., Fuchs, J., Runkel, I., Schweigert, C.: Correspondences of ribbon categories. Adv. Math. **199**, 192–329 (2006). arXiv:math/0309465
18. Fuchs, J., Runkel, I., Schweigert, C.: Conformal correlation functions, Frobenius algebras and triangulations. Nucl. Phys. B **624**, 452–468 (2002). arXiv:hep-th/0110133
19. Fuchs, J., Schweigert, C., Stigner, C.: Modular invariant Frobenius algebras from ribbon Hopf algebra automorphisms. J. Algebra **363**, 29–72 (2012). arXiv:1106.0210
20. Gaberdiel, M.R.: An introduction to conformal field theory. Rep. Prog. Phys. **63**, 607–667 (2000). arXiv:hep-th/9910156
21. Gaberdiel, M.R., Goddard, P.: Axiomatic conformal field theory. Commun. Math. Phys. **209**, 549–594 (2000). arXiv:hep-th/9810019
22. Gaberdiel, M.R., Kausch, H.G.: A local logarithmic conformal field theory. Nucl. Phys. B **538**, 631–658 (1999). arXiv:hep-th/9807091
23. Gaberdiel, M.R., Runkel, I.: The logarithmic triplet theory with boundary. J. Phys. A **39**, 14745 (2006). arXiv:hep-th/0608184
24. Gaberdiel, M.R., Runkel, I.: From boundary to bulk in logarithmic CFT. J. Phys. A **41**, 075402 (2008). arXiv:0707.0388
25. Gaberdiel, M.R., Runkel, I., Wood, S.: Fusion rules and boundary conditions in the $c = 0$ triplet model. J. Phys. A **42**, 325403 (2009). arXiv:0905.0916
26. Gaberdiel, M.R., Runkel, I., Wood, S.: A modular invariant bulk theory for the $c = 0$ triplet model. J. Phys. A **44**, 015204 (2011). arXiv:1008.0082
27. Hu, P., Kriz, I.: Conformal field theory and elliptic cohomology. Adv. Math. **189**, 325–412 (2004)
28. Huang, Y.-Z., Kong, L.: Full field algebras. Commun. Math. Phys. **272**, 345–396 (2007). arXiv:math/0511328
29. Huang, Y.-Z., Lepowsky, J.: Tensor products of modules for a vertex operator algebra and vertex tensor categories. In: Brylinski, R., Brylinski, J.-L., Guillemin, V., Kac, V. (eds.) Lie Theory and Geometry: In Honor of Bertram Kostant, pp. 349–383. Birkhäuser, Boston (1994). arXiv:hep-th/9401119
30. Huang, Y.-Z., Lepowsky, J., Zhang, L.: Logarithmic tensor category theory for generalized modules for a conformal vertex algebra, I–VIII. arXiv:1012.4193, arXiv:1012.4196, arXiv:1012.4197, arXiv:1012.4198, arXiv:1012.4199, arXiv:1012.4202, arXiv:1110.1929, arXiv:1110.1931
31. Joyal, A., Street, R.: Braided tensor categories. Adv. Math. **102**, 20–78 (1993)
32. Kapustin, A., Orlov, D.: Vertex algebras, mirror symmetry, and D-branes: the case of complex tori. Commun. Math. Phys. **233**, 79–136 (2003). arXiv:hep-th/0010293

33. Kong, L.: Open-closed field algebras. Commun. Math. Phys. **280**, 207–261 (2008). arXiv:math/0610293
34. Kong, L., Runkel, I.: Cardy algebras and sewing constraints, I. Commun. Math. Phys. **292**, 871–912 (2009). arXiv:0807.3356
35. Lewellen, D.C.: Sewing constraints for conformal field theories on surfaces with boundaries. Nucl. Phys. B **372**, 654–682 (1992)
36. Longo, R., Rehren, K.H.: Local fields in boundary conformal QFT. Rev. Math. Phys. **16**, 909–960 (2004). arXiv:math-ph/0405067
37. Mac Lane, S.: Categories for the Working Mathematician, 2nd edn. Springer, Berlin (1998)
38. Milas, A.: Weak modules and logarithmic intertwining operators for vertex operator algebras. Contemp. Math. **297**, 201–227 (2002). arXiv:math/0101167
39. Müger, M.: From subfactors to categories and topology II. The quantum double of tensor categories and subfactors. J. Pure Appl. Algebra **180**, 159–219 (2003). arXiv:math/0111205
40. Ostrik, V.: Module categories, weak Hopf algebras and modular invariants. Transform. Groups, **8**, 177–206 (2003). arXiv:math/0111139
41. Pearce, P.A., Rasmussen, J.: Coset graphs in bulk and boundary logarithmic minimal models. Nucl. Phys. B **846**, 616–649 (2011). arXiv:1010.5328
42. Quella, T., Schomerus, V.: Free fermion resolution of supergroup WZNW models. J. High Energy Phys. **0709**, 085 (2007). arXiv:0706.0744
43. Rasmussen, J., Pearce, P.A.: W-extended fusion algebra of critical percolation. J. Phys. A **41**, 295208 (2008). arXiv:0804.4335
44. Runkel, I.: Boundary structure constants for the A-series Virasoro minimal models. Nucl. Phys. B **549**, 563–578 (1999). arXiv:hep-th/9811178
45. Runkel, I.: Structure constants for the D-series Virasoro minimal models. Nucl. Phys. B **579**, 561–589 (2000). arXiv:hep-th/9908046
46. Runkel, I., Fjelstad, J., Fuchs, J., Schweigert, C.: Topological and conformal field theory as Frobenius algebras. Contemp. Math. **431**, 225–248 (2007). arXiv:math/0512076
47. Saleur, H.: Polymers and percolation in two-dimensions and twisted $N = 2$ supersymmetry. Nucl. Phys. B **382**, 486–531 (1992). arXiv:hep-th/9111007
48. Segal, G.: The definition of conformal field theory. In: Tillmann, U. (ed.) Topology, Geometry and Quantum Field Theory. London Math. Soc. Lect. Note Ser., vol. 308, pp. 421–577 (2002)
49. Vafa, C.: Conformal theories and punctured surfaces. Phys. Lett. B **199**, 195–202 (1987)
50. Van Oystaeyen, F., Zhang, Y.H.: The Brauer group of a braided monoidal category. J. Algebra **202**, 96–128 (1998)
51. Vasseur, R., Gainutdinov, A.M., Jacobsen, J.L., Saleur, H.: The puzzle of bulk conformal field theories at central charge $c = 0$. Phys. Rev. Lett. **108**, 161602 (2012). arXiv:1110.1327

Logarithmic Tensor Category Theory for Generalized Modules for a Conformal Vertex Algebra, I: Introduction and Strongly Graded Algebras and Their Generalized Modules

Yi-Zhi Huang, James Lepowsky, and Lin Zhang

Abstract This is the first part in a series of papers in which we introduce and develop a natural, general tensor category theory for suitable module categories for a vertex (operator) algebra. This theory generalizes the tensor category theory for modules for a vertex operator algebra previously developed in a series of papers by the first two authors to suitable module categories for a "conformal vertex algebra" or even more generally, for a "Möbius vertex algebra." We do not require the module categories to be semisimple, and we accommodate modules with generalized weight spaces. As in the earlier series of papers, our tensor product functors depend on a complex variable, but in the present generality, the logarithm of the complex variable is required; the general representation theory of vertex operator algebras requires logarithmic structure. This work includes the complete proofs in the present generality and can be read independently of the earlier series of papers. Since this is a new theory, we present it in detail, including the necessary new foundational material. In addition, with a view toward anticipated applications, we develop and present the various stages of the theory in the natural, general settings in which the proofs hold, settings that are sometimes more general than what we need for the main conclusions. In this paper (Part I), we give a detailed overview of our theory, state our main results and introduce the basic objects that we shall study in this work. We include a brief discussion of some of the recent applications of this theory, and also a discussion of some recent literature.

Y.-Z. Huang (✉) · J. Lepowsky · L. Zhang
Department of Mathematics, Rutgers University, Piscataway, NJ 08854, USA
e-mail: yzhuang@math.rutgers.edu

J. Lepowsky
e-mail: lepowsky@math.rutgers.edu

L. Zhang
e-mail: linzhang@math.rutgers.edu

Y.-Z. Huang
Beijing International Center for Mathematical Research, Peking University, Beijing, China

C. Bai et al. (eds.), *Conformal Field Theories and Tensor Categories*,
Mathematical Lectures from Peking University, DOI 10.1007/978-3-642-39383-9_5,

In this paper, Part I of a series of eight papers, we give a detailed overview of logarithmic tensor category theory, state our main results and introduce the basic objects that we shall study in this work. We include a brief discussion of some of the recent applications of this theory, and also a discussion of some recent literature. The sections, equations, theorems and so on are numbered globally in the series of papers rather than within each paper, so that for example equation (a.b) is the b-th labeled equation in Sect. a, which is contained in the paper indicated as follows: The present paper, Part I, contains Sects. 1 and 2. In Part II [81], which contains Sect. 3, we develop logarithmic formal calculus and study logarithmic intertwining operators. In Part III [82], which contains Sect. 4, we introduce and study intertwining maps and tensor product bifunctors. In Part IV [83], which contains Sects. 5 and 6, we give constructions of the $P(z)$- and $Q(z)$-tensor product bifunctors using what we call "compatibility conditions" and certain other conditions. In Part V [84], which contains Sects. 7 and 8, we study products and iterates of intertwining maps and of logarithmic intertwining operators and we begin the development of our analytic approach. In Part VI [85], which contains Sects. 9 and 10, we construct the appropriate natural associativity isomorphisms between triple tensor product functors. In Part VII [86], which contains Sect. 11, we give sufficient conditions for the existence of the associativity isomorphisms. In Part VIII [87], which contains Sect. 12, we construct braided tensor category structure.

1 Introduction

A Brief Description of the Present Work In the representation theory of many important algebraic structures, such as Lie algebras, groups (or group algebras), commutative associative algebras, Hopf algebras or quantum groups, tensor product operations among modules play a central role. They not only give new modules from known ones, but they of course also provide a powerful tool for studying modules. More significantly, suitable categories of modules for such algebras, equipped with tensor product operations and appropriate natural isomorphisms, and so on, become symmetric or braided tensor categories, and this tensor category structure is always used, even when it is not explicitly discussed.

Vertex operator algebras, and more generally, vertex algebras, are a fundamental class of algebraic structures whose extensive theory has been developed and used in recent years to provide the means to illuminate and to solve many problems in a wide variety of areas of mathematics and theoretical physics. In particular, the representation theory of vertex (operator) algebras plays deep roles in the construction and study of infinite-dimensional Lie algebra representations, of structures linking sporadic finite simple groups to string theory and to the theory of modular functions, and of knot invariants and 3-manifold invariants, in mathematics; and of conformal field theory and string theory, in physics.

The present work is devoted to introducing and developing a natural, general tensor category theory for suitable module categories for a vertex (operator) algebra.

This tensor category theory, and consequently, the representation theory, of vertex (operator) algebras, is much, much more elaborate and more difficult than that of Lie algebras, commutative associative algebras, Hopf algebras or quantum groups. In fact, the vertex-operator-algebraic analogues of even the most elementary parts of the tensor product theory of an algebra such as one of those are highly nontrivial, and the theory needs to be developed with completely new ideas and strategies (and with great care!). The present theory was what we needed to carry out in order to obtain the appropriate vertex-operator-algebraic analogue of the following routine triviality in the representation theory of (for example) Lie algebras: "Given a Lie algebra $\mathfrak{g}$, consider the symmetric tensor category of $\mathfrak{g}$-modules." A vertex operator algebra "wants to be" the space of primitive elements of a Hopf algebra (as is a Lie algebra, for example; this immediately yields the tensor category of modules), but a vertex operator algebra is *not* the space of primitive elements of any Hopf algebra, and this is the beginning of why the problem of constructing a tensor product theory and a tensor category theory of modules for a vertex operator algebra was (and is) hard. Yet it is at least as important to have a theory of tensor products and tensor categories of modules for a vertex operator algebra as it is in classical theories such as Lie algebra theory (where such tensor products and tensor categories of modules exist "automatically").

In Lie algebra theory (among other theories), many important module categories are semisimple, that is, every module is completely reducible, while on the other hand, many important module categories are not. Earlier, the first two authors developed a theory of braided tensor categories for the module category of a what we call a "finitely reductive" vertex operator algebra satisfying certain additional conditions; finite reductivity means that the module category is semisimple and that certain finiteness conditions hold. But it is just as natural and important to develop a theory for non-semisimple module categories in vertex operator algebra theory as it is in Lie algebra theory. Also, in any one of the classical theories such as Lie algebra theory, observing that there is a tensor category of modules is just as easy for not-necessarily-semisimple modules as it is for semisimple modules. For these and many other reasons, we considered it a natural problem to generalize the tensor category theory for vertex operator algebras from the finitely reductive case to the general case.

The present work accomplishes this goal, culminating in the construction of a braided tensor category structure on a suitable module category, not assumed semisimple, for a vertex (operator) algebra. It turns out the non-semisimplicity of modules is intimately linked to the presence of logarithms in the basic ingredients of the theory, beginning with intertwining operators among modules, and this is why we call the present theory "logarithmic tensor category theory." We must in fact consider "generalized modules"—structures for which a certain basic operator has generalized eigenvectors in addition to ordinary eigenvectors. This basic operator is contained in a natural copy of the three-dimensional simple Lie algebra, which plays the role of the Lie algebra of the group of Möbius symmetries; this Lie algebra is in turn a subalgebra of a natural copy of the Virasoro algebra, a central extension of a Lie algebra of conformal symmetries. In this work, we carry out our theory for suitable categories of generalized modules for a "conformal vertex algebra," which includes a copy of the Virasoro algebra, and even more generally, for a "Möbius vertex algebra," which has the Möbius symmetries but not all of the conformal symmetries. The present theory explicitly includes the earlier finitely reductive theory as a special case; however, the present theory is (necessarily) much more elaborate and subtle than the finitely reductive theory.

In both the finitely reductive and the logarithmic generality, even the construction of the tensor product (generalized) modules is nontrivial; the correct tensor product module of two modules (when it exists) is not at all based on the tensor product vector space of the two underlying vector spaces. Moreover, the construction of the necessary natural associativity isomorphisms among triples of modules is highly nontrivial. While in classical tensor product theories the natural associativity isomorphisms among triples of modules are given by the usual trivial maps, in the tensor product theory of modules for a vertex (operator) algebra, the corresponding statement is not at all true, and indeed, there are not even any candidates for easy associativity isomorphisms. These and many related issues require the present tensor product and tensor category theory to be elaborate.

A crucial discovery in the work of the first two authors in the finitely reductive case was the existence of natural tensor products of two or more elements in the algebraic completions of tensor product modules. All of the categorical structures and properties are formulated, constructed and/or proved using tensor products of elements. In the finitely reductive case, tensor products of elements were defined using intertwining operators (without logarithms). In order to develop the tensor category theory in the general setting of the present work, it is again crucial to establish the existence of tensor products of elements and to prove the fundamental properties of these tensor product elements, and to do this, we are inevitably led to the development of the theory of logarithmic intertwining operators.

The structures of tensor product module, natural associativity isomorphisms, and resulting braided tensor category structure incorporating these, constructed in the present work, are assumed to exist in a number of research works in mathematics and physics. The results in the present work allow one to remove assumptions of this type. We provide a mathematical foundation for such results and for ongoing and future research involving the representation theory of vertex (operator) algebras.

In fact, what we actually construct in this work is structure much stronger than braided tensor category structure: The natural associativity isomorphisms are constructed by means of a "logarithmic operator product expansion" theorem for logarithmic intertwining operators. This logarithmic operator product expansion is in fact the starting point of "logarithmic conformal field theory," which has been studied extensively by physicists as well as mathematicians. Here, this logarithmic operator product expansion is established as a mathematical theorem.

Moreover, our constructions and proofs in this work actually give what the first two authors have called "vertex-tensor-categorical structure," in which the tensor product bifunctors depend crucially on complex variables. This structure is necessary for producing the desired braided tensor category structure, through the use of the tensor product elements and logarithmic operator product expansion mentioned above, and our construction of braided tensor category structure involves a "limiting process" in which the complex-analytic information is "forgotten" and only the "topological" information associated with braided tensor category structure is retained. When we perform this specialization to the "limiting case" of braided tensor category structure, tensor products of three or more elements are no longer defined.

The word "algebra" appears in the phrases "vertex operator algebra" and "vertex algebra," but beginning at the stage of the theory where one must compose intertwining operators, or rather, intertwining maps, among (generalized) modules, one must use analysis as well as algebra, starting even from the definition of composition of intertwining maps. The kind of algebra on which the theory is largely based, and which is needed throughout, is called "formal calculus," which we must in fact extensively develop in the course of the work. We must also enhance formal calculus with a great deal of analytic reasoning, and the synthesized theory is no longer "pure algebra."

This work includes the complete proofs in the present generality and can be read independently of the first two authors' earlier series of papers carrying out the finitely reductive theory. Since this is a new theory, we present it in detail, including the necessary new foundational material. In addition, we develop and present the various stages of the theory in the natural, general settings in which the proofs hold, settings that are sometimes more general than what we need for the main conclusions. This will allow for the future use of the intermediate results in a variety of directions.

Later in the Introduction, we mention some of the recent applications of the present theory, and we include a discussion of some recent literature. We state the main results of the present work at the end of the Introduction.

The main results presented here have been announced in [80].

Introduction In a series of papers ([53, 68, 71–74]), the first two authors have developed a tensor product and tensor category theory for modules for a vertex operator algebra under suitable conditions. A structure called "vertex tensor category structure" (see [71]), which is much richer than tensor category structure, has thereby been established for many important categories of modules for classes of vertex operator algebras, since the conditions needed for invoking the general theory have been verified for these categories. The most important such families of examples of this theory are listed in Sect. 1.1 below. In the present work, which has been announced in [80], we generalize this tensor category theory to a larger family of module categories, for a "conformal vertex algebra," or even more generally, for a "Möbius vertex algebra," under suitably relaxed conditions. A conformal vertex algebra is just a vertex algebra in the sense of Borcherds [15] equipped with a conformal vector satisfying the usual axioms; a Möbius vertex algebra is a variant of a "quasi-vertex operator algebra" as in [38]. Central features of the present work are that we do not require the modules in our categories to be completely reducible and that we accommodate modules with generalized weight spaces.

As in the earlier series of papers, our tensor product functors depend on a complex variable, but in the present generality, the logarithm of the complex variable is required. The first part of this work is devoted to the study of logarithmic intertwining operators and their role in the construction of the tensor product functors. The remainder of this work is devoted to the construction of the appropriate natural associativity isomorphisms between triple tensor product functors, to the proof of their fundamental properties, and to the construction of the resulting braided tensor category structure. This leads to vertex tensor category structure for further important families of examples, or, in the Möbius case, to "Möbius vertex tensor category" structure.

We emphasize that we develop our representation theory (tensor category theory) in a very general setting; the vertex (operator) algebras that we consider are very general, and the "modules" that we consider are very general. We call them "generalized modules"; they are not assumed completely reducible. Many extremely important (and well-understood) vertex operator algebras have semisimple module categories, but in fact, now that the theory of vertex operator algebras and of their representations is as highly developed as it has come to be, it is in fact possible, and very fruitful, to work in the greater generality. Focusing mainly on the representation theory of those vertex operator algebras for which every module is completely reducible would be just as restrictive as focusing, classically, on the representation theory of semisimple Lie algebras as opposed to the representation theory of Lie algebras in general. In addition, once we consider suitably general vertex (operator) algebras, it is unnatural to focus on only those modules that are completely reducible. As we explain below, such a general representation theory of vertex (operator) algebras requires logarithmic structure.

A general representation theory of vertex operator algebras is crucial for a range of applications, and we expect that it will be a foundation for future developments. One example is that the original formulation of the uniqueness conjecture [37] for the moonshine module vertex operator algebra $V^{\natural}$ (again see [37]) requires (general) vertex operator algebras whose modules might not be completely reducible. Another example is that this general theory is playing a deep role in the (mathematical) construction of conformal field theories (cf. [61–64, 98]), which in turn correspond to the perturbative part of string theory. Just as the classical (general) representation theory of groups, or of Lie groups, or of Lie algebras, is not about any particular group or Lie group or Lie algebra (although one of its central goals is certainly to understand the representations of particular structures), the general representation theory of suitably general vertex operator algebras is "background independent," in the terminology of string theory. In addition, the general representation theory of vertex (operator) algebras can be thought of as a "symmetry" theory, where vertex (operator) algebras play a role analogous to that of groups or of Lie algebras in classical theories; deep and well-known analogies between the notion of vertex operator algebra and the classical notion of, for example, Lie algebra are discussed in several places, including [37, 38] and [99].

As we mentioned above, the present work includes the complete proofs in the present generality and can be read independently of the earlier series of papers of the first two authors constructing tensor categories. Our treatment is based on the theory of vertex operator algebras and their modules as developed in [15, 21, 37, 38] and [99]. Throughout the work, we must, and do, develop new algebraic and analytic methods, including a synthesis of the "formal calculus" of vertex operator algebra theory with analysis.

1.1 Tensor Category Theory for Finitely Reductive Vertex Operator Algebras

The main families for which the conditions needed for invoking the first two authors' general tensor category theory have been verified, thus yielding vertex tensor category structure [71] on these module categories, include the module categories for the following classes of vertex operator algebras (or, in the last case, vertex operator superalgebras):

1. The vertex operator algebras V_L associated with positive definite even lattices L; see [15, 37] for these vertex operator algebras and see [19, 21] for the conditions needed for invoking the general tensor category theory.
2. The vertex operator algebras $L(k, 0)$ associated with affine Lie algebras and positive integers k; see [35] for these vertex operator algebras and [35, 75] for the conditions.
3. The "minimal series" of vertex operator algebras associated with the Virasoro algebra; see [35] for these vertex operator algebras and [54, 139] for the conditions.
4. Frenkel, Lepowsky and Meurman's moonshine module $V^\natural$; see [15, 36, 37] for this vertex operator algebra and [20] for the conditions.
5. The fixed point vertex operator subalgebra of $V^\natural$ under the standard involution; see [36, 37] for this vertex operator algebra and [20, 55] for the conditions.
6. The "minimal series" of vertex operator superalgebras (suitably generalized vertex operator algebras) associated with the Neveu-Schwarz superalgebra and also the "unitary series" of vertex operator superalgebras associated with the $N = 2$ superconformal algebra; see [88] and [3] for the corresponding $N = 1$ and $N = 2$ vertex operator superalgebras, respectively, and [2, 4, 77, 78] for the conditions.

In addition, vertex tensor category structure has also been established for the module categories for certain vertex operator algebras built from the vertex operator algebras just mentioned, such as tensor products of such algebras; this is carried out in certain of the papers listed above.

For all of the six classes of vertex operator algebras (or superalgebras) listed above, each of the algebras is "rational" in the specific sense of Huang-Lepowsky's work on tensor category theory. This particular "rationality" property is easily proved to be a sufficient condition for insuring that the tensor product modules exist; see for instance [72]. Unfortunately, the phrase "rational vertex operator algebra" also has several other distinct meanings in the literature. Thus we find it convenient at this time to assign a new term, "finite reductivity," to our particular notion of "rationality": We say that a vertex operator algebra (or superalgebra) V is *finitely reductive* if:

1. Every V-module is completely reducible.
2. There are only finitely many irreducible V-modules (up to equivalence).
3. All the fusion rules (the dimensions of the spaces of intertwining operators among triples of modules) for V are finite.

We choose the term "finitely reductive" because we think of the term "reductive" as describing the complete reducibility—the first of the conditions (that is, the algebra "(completely) reduces" every module); the other two conditions are finiteness conditions.

The vertex-algebraic study of tensor category structure on module categories for certain vertex algebras was stimulated by the work of Moore and Seiberg [111, 112], in which, in the study of what they termed "rational" conformal field theory, they obtained a set of polynomial equations based on the assumption of the existence of a suitable operator product expansion for "chiral vertex operators" (which correspond to intertwining operators in vertex algebra theory) and observed an analogy between the theory of this set of polynomial equations and the theory of tensor categories. Earlier, in [14], Belavin, Polyakov, and Zamolodchikov had already formalized the relation between the (nonmeromorphic) operator product expansion, chiral correlation functions and representation theory, for the Virasoro algebra in particular, and Knizhnik and Zamolodchikov [96] had established fundamental relations between conformal field theory and the representation theory of affine Lie algebras. As we have discussed in the introductory material in [71, 72] and [75], such study of conformal field theory is deeply connected with the vertex-algebraic construction and study of tensor categories, and also with other mathematical approaches to the construction of tensor categories in the spirit of conformal field theory. Concerning the latter approaches, we would like to mention that the method used by Kazhdan and Lusztig, especially in their construction of the associativity isomorphisms, in their breakthrough work in [91–95], is related to the algebro-geometric formulation and study of conformal-field-theoretic structures in the influential works of Tsuchiya-Ueno-Yamada [135], Drinfeld [23] and Beilinson-Feigin-Mazur [13]. See also the important work of Deligne [18], Finkelberg ([28, 29]), Bakalov-Kirillov [12] and Nagatomo-Tsuchiya [113] on the construction of tensor categories in the spirit of this approach to conformal field theory, and also the discussions in Remark 1.8 and in Sect. 1.5 below.

1.2 Logarithmic Tensor Category Theory

The semisimplicity of the module categories mentioned in the examples above is related to another property of these modules, namely, that each module is a direct sum of its "weight spaces," which are the eigenspaces of a special operator $L(0)$ coming from the Virasoro algebra action on the module. But there are important situations in which module categories are not semisimple and in which modules are not direct sums of their weight spaces. Notably, for the vertex operator algebras $L(k, 0)$ associated with affine Lie algebras, when the sum of k and the dual Coxeter number of the corresponding Lie algebra is not a nonnegative rational number, the vertex operator algebra $L(k, 0)$ is not finitely reductive, and, working with Lie algebra theory rather than with vertex operator algebra theory, Kazhdan and Lusztig constructed a natural braided tensor category structure on a certain category of modules of level k for the affine Lie algebra ([91–95]). This work of Kazhdan-Lusztig in fact motivated the first two authors to develop an analogous theory for vertex operator algebras rather than for affine Lie algebras, as was explained in detail in the introductory material in [68, 71–73], and [75]. However, this general theory, in its original form, did not apply to Kazhdan-Lusztig's context, because the vertex-operator-algebra modules considered in [53, 68, 71–74] are assumed to be the direct sums of their weight spaces (with respect to $L(0)$), and the non-semisimple modules considered by Kazhdan-Lusztig fail in general to be the direct sums of their weight spaces. Although their setup, based on Lie theory, and ours, based on vertex operator algebra theory, are very different (as was discussed in the introductory material in our earlier papers), we expected to be able to recover (and further extend) their results through our vertex operator algebraic approach, which is very general, as we discussed above. This motivated us, in the present work, to generalize the work of the first two authors by considering modules with generalized weight spaces, and especially, intertwining operators associated with such generalized kinds of modules. As we discuss below, this required us to use logarithmic intertwining operators and logarithmic formal calculus, and we have been able to construct braided tensor category structure, and even vertex tensor category structure, on important module categories that are not semisimple. Using the present theory, the third author ([141, 142]) has indeed recovered the braided tensor category structure of Kazhdan-Lusztig, and has also extended it to vertex tensor category structure. While in our theory, logarithmic structure plays a fundamental role, in this Kazhdan-Lusztig work, logarithmic structure does not show up explicitly.

From the viewpoint of the general representation theory of vertex operator algebras, it would be unnatural to study only semisimple modules or only $L(0)$-semisimple modules; focusing only on such modules would be analogous to focusing only on semisimple modules for general (nonsemisimple) finite-dimensional Lie algebras. And as we have pointed out, working in this generality leads to logarithmic structure; the general representation theory of vertex operator algebras requires logarithmic structure.

Logarithmic structure in conformal field theory was in fact first introduced by physicists to describe Wess-Zumino-Witten models on supergroups ([132, 133]) and disorder phenomena [51]. A lot of progress has been made on this subject. We refer the interested reader to the review articles [31, 42, 118] and [39], and references therein. One particularly interesting class of logarithmic conformal field theories is the class associated to the triplet $\mathcal{W}$-algebras $\mathcal{W}(1, p)$ introduced by Kausch [89], of central charge $1 - 6\frac{(p-1)^2}{p}$, $p = 2, 3, \ldots$. We will discuss these algebras, and generalizations of them, including references, in Sect. 1.5 below. The paper [40] initiated a study of a possible generalization of the Verlinde conjecture for rational conformal field theories to these theories; see also [32, 33, 47] and [49]. The paper [39] assumed the existence of braided tensor category structures on the categories of modules for the vertex operator algebras considered; together with [65], the present work gives a construction of these structures. The paper [17] used the results in the present work as announced in [80].

Here is how such logarithmic structure also arises naturally in the representation theory of vertex operator algebras: In the construction of intertwining operator algebras, the first author proved (see [59]) that if modules for the vertex operator algebra satisfy a certain cofiniteness condition, then products of the usual intertwining operators satisfy certain systems of differential equations with regular singular points. In addition, it was proved in [59] that if the vertex operator algebra satisfies certain finite reductivity conditions, then the analytic extensions of products of the usual intertwining operators have no logarithmic terms. In the case when the vertex operator algebra satisfies only the cofiniteness condition but not the finite reductivity conditions, the products of intertwining operators still satisfy systems of differential equations with regular singular points. But in this case, the analytic extensions of such products of intertwining operators might have logarithmic terms. This means that if we want to generalize the results in [53, 68, 71–74] and [59] to the case in which the finite reductivity properties are not always satisfied, we have to consider intertwining operators involving logarithmic terms.

Logarithmic structure also appears naturally in modular invariance results for vertex operator algebras and in the genus-one parts of conformal field theories. For a vertex operator algebra V satisfying certain finiteness and reductivity conditions, Zhu proved in [144] a modular invariance result for q-traces of products of vertex operators associated to V-modules. Zhu's result was generalized to the case involving twisted vertex operators by Dong, Li and Mason in [22] and to the case of q-traces of products of one intertwining operator and arbitrarily many vertex operators by Miyamoto in [107]. In [106], Miyamoto generalized Zhu's modular invariance result to a modular invariance result involving the logarithm of q for vertex operator algebras not necessarily satisfying the reductivity condition. In [60], for vertex operator algebras satisfying certain finiteness and reductivity conditions, by overcoming the difficulties one encounters if one tries to generalize Zhu's method, the first author was able to prove the modular invariance for q-traces of products and iterates of more than one intertwining operator, using certain differential equations and duality properties for intertwining operators. If the vertex operator algebra satisfies only Zhu's cofiniteness condition but not the reductivity condition, the q-traces of products and iterates of intertwining operators still satisfy the same differential equations, but now they involve logarithms of all the variables. To generalize the general Verlinde conjecture proved in [63] and the modular tensor category structure on the category of V-modules obtained in [64], one will need such general logarithmic modular invariance. See [39, 40, 47] and [49] for research in this direction.

In [104], Milas introduced and studied what he called "logarithmic modules" and "logarithmic intertwining operators." See also [105]. Roughly speaking, logarithmic modules are weak modules for a vertex operator algebra that are direct sums of generalized eigenspaces for the operator $L(0)$. We will call such weak modules "generalized modules" in this work. Logarithmic intertwining operators are operators that depend not only on powers of a (formal or complex) variable x, but also on its logarithm $\log x$.

The special features of the logarithm function make the logarithmic theory very subtle and interesting. In order to develop our logarithmic tensor category theory, we were required to considerably develop:

1. Formal calculus, beyond what had been developed in [37, 38, 53, 72–74] and [99], in particular. (Formal calculus has been developed in a great many works.)
2. What we may call "logarithmic formal calculus," which involves arbitrary powers of formal variables and of their formal logarithms. This logarithmic formal calculus has been extended and exploited by Robinson [129–131].
3. Complex analysis involving series containing *arbitrary real* powers of the variables.
4. Complex analysis involving series containing nonnegative integral powers of the logarithms of the variables, in the presence of arbitrary real powers of the variables.
5. A blending of these themes in order to formulate and to prove many interchange-of-limit results necessary for the construction of the ingredients of the logarithmic tensor category theory and for the proofs of the fundamental properties.

Our methods intricately combine both algebra and analysis, and must do so, since the statements of the results themselves are both algebraic and analytic. See Remark 1.7 below for a discussion of these methods and their roles in this work.

As we mentioned above, one important application of our generalization is to the category $\mathcal{O}_\kappa$ of certain modules for an affine Lie algebra studied by Kazhdan and Lusztig in their series of papers [91–95]. It has been shown in [141] and [142] by the third author that, among other things, all the conditions needed to apply our theory to this module category are satisfied. As a result, it is proved in [141] and [142] that there is a natural vertex tensor category structure on this module category, giving in particular a new construction, in the context of general vertex operator algebra theory, of the braided tensor category structure on $\mathcal{O}_\kappa$. This construction does not use the Knizhnik-Zamolodchikov equations. The methods used in [91–95] were very different; the Knizhnik-Zamolodchikov equations play an essential role in their construction, while the present theory is very general.

The triplet $\mathcal{W}$-algebras belong to a different class of vertex operator algebras, satisfying certain finiteness, boundedness and reality conditions. In this case, it has been shown in [65] by the first author that all the conditions needed to apply the theory carried out in the present work to the category of grading-restricted modules for the vertex operator algebra are also satisfied. Thus, by the results obtained in this work, there is a natural vertex tensor category structure on this category.

In addition to these logarithmic issues, another way in which the present work generalizes the earlier tensor category theory for module categories for a vertex operator algebra is that we now allow the algebras to be somewhat more general than vertex operator algebras, in order, for example, to accommodate module categories for the vertex algebras V_L where L is a nondegenerate even lattice that is not necessarily positive definite (cf. [15, 21]); see [141].

What we accomplish in this work, then, is the following: We generalize essentially all the results in [72–74] and [53] from the category of modules for a vertex operator algebra to categories of suitably generalized modules for a conformal vertex algebra or a Möbius vertex algebra equipped with an additional suitable grading by an abelian group. The algebras that we consider include not only vertex operator algebras but also such vertex algebras as V_L where L is a nondegenerate even lattice, and the modules that we consider are not required to be the direct sums of their weight spaces but instead are required only to be the (direct) sums of their "generalized weight spaces," in a suitable sense. In particular, in this work we carry out, in the present greater generality, the construction theory for the "$P(z)$-tensor product" functor originally done in [72, 73] and [74] and the associativity theory for this functor—the construction of the natural associativity isomorphisms between suitable "triple tensor products" and the proof of their important properties, including the isomorphism property—originally done in [53]. This leads, as in [71, 76], to the proof of the coherence properties for vertex tensor categories, and in the Möbius case, the coherence properties for Möbius vertex tensor categories.

For simplicity, we present our theory only for a conformal vertex algebra or a Möbius vertex algebra and not for their superalgebraic analogues, but in fact our theory generalizes routinely to a conformal vertex superalgebra or a Möbius vertex superalgebra equipped with an additional suitable grading by an abelian group; here we are referring only to the usual sign changes associated with the "odd" subspace of a vertex superalgebra, and not to any superconformal structure.

The general structure of much of this work essentially follows that of [72–74] and [53]. However, the results here are much stronger and much more general than in these earlier works, and in addition, many of the results here have no counterparts in those works. Moreover, many ideas, formulations and proofs in this work are necessarily quite different from those in the earlier papers, and we have chosen to give some proofs that are new even in the finitely reductive case studied in the earlier papers.

Some of the new ingredients that we are introducing into the theory here are: an analysis of logarithmic intertwining operators, including "logarithmic formal calculus"; a notion of "$P(z_1, z_2)$-intertwining map" and a study of its properties; new "compatibility conditions"; considerable generalizations of virtually all of the technical results in [72–74] and [53]; and perhaps most significantly, the analytic ideas and methods that are sketched in Remark 1.7 below.

The contents of the sections of this work are as follows: In the rest of this Introduction we compare classical tensor product and tensor category theory for Lie algebra modules with tensor product and tensor category theory for vertex operator algebra modules. One crucial difference between the two theories is that in the vertex operator algebra setting, the theory depends on an "extra parameter" z, which must be understood as a (nonzero) complex variable rather than as a formal variable (although one needs, and indeed we very heavily use, an extensive "formal calculus," or "calculus of formal variables," in order to develop the theory). We also discuss recent applications of the present theory and some related literature and state the main results of the present work. In Sect. 2 we recall and extend some basic concepts in the theory of vertex (operator) algebras. We use the treatments of [21, 37, 38] and [99]; in particular, the formal-calculus approach developed in these works is needed for the present theory. Readers can consult these works for further detail. We also set up notation and terminology that will be used throughout the present work, and we describe the main categories of (generalized) modules that we will consider. In Sect. 3 we introduce the notion of logarithmic intertwining operator as in [104] and present a detailed study of the basic properties of these operators. At the beginning of this section we introduce and prove results about logarithmic formal calculus, including a general "formal Taylor theorem." In Sects. 4 and 5 we present the notions of $P(z)$- and $Q(z)$-intertwining maps, and based on this, the definitions and constructions of $P(z)$- and $Q(z)$-tensor products, generalizing considerations in [72, 73] and [74]. The constructions of the tensor product functors require certain "compatibility conditions" and "local grading restriction conditions." The proofs of some of the results in Sect. 5 are postponed to Sect. 6. In Sect. 7 the convergence condition for products and iterates of intertwining maps introduced in [53] is generalized to the present context. More importantly, in this section we start to develop the complex analysis approach that we will heavily use in later sections. The new notion of $P(z_1, z_2)$-intertwining map, generalizing the corresponding concept in [53], is introduced and developed in Sect. 8. This will play a crucial role in the construction of the natural associativity isomorphisms. In Sect. 9 we prove important conditions that are satisfied by vectors in the dual space of the vector-space tensor product of three modules that arise from products and from iterates of intertwining maps. This leads us to study elements in this dual space satisfying suitable compatibility and local grading restriction conditions. In this section we extensively use our complex analysis approach, including, in particular, for proving that the order of many iterated summations can be interchanged. By relating the subspaces considered in Sect. 9, we construct the associativity isomorphisms in Sect. 10. In Sect. 11, we generalize certain sufficient conditions for the existence of the associativity isomorphisms in [53], and we prove the relevant conditions using differential equations. In Sect. 12, we establish the coherence properties of our braided tensor category structure.

1.3 The Lie Algebra Case

In this section and the next, we compare classical tensor product and tensor category theory for Lie algebra modules with the present theory for vertex operator algebra modules, and in fact it is heuristically useful to start by considering tensor product theory for Lie algebra modules in a somewhat unusual way in order to motivate our approach for the case of vertex (operator) algebras.

In the theory of tensor products for modules for a Lie algebra, the tensor product of two modules is defined, or rather, constructed, as the vector-space tensor product of the two modules, equipped with a Lie algebra module action given by the familiar diagonal action of the Lie algebra. In the vertex algebra case, however, the vector-space tensor product of two modules for a vertex algebra is *not* the correct underlying vector space for the tensor product of the vertex-algebra modules. In this section we therefore consider another approach to the tensor category theory for modules for a Lie algebra—an approach, based on "intertwining maps," that will show how the theory proceeds in the vertex algebra case. Then, in the next section, we shall lay out the corresponding "road map" for the tensor category theory in the vertex algebra case, which we then carry out in the body of this work.

We first recall the following elementary but crucial background about tensor product vector spaces: Given vector spaces W_1 and W_2, the corresponding tensor product structure consists of a vector space $W_1 \otimes W_2$ equipped with a bilinear map

$$W_1 \times W_2 \longrightarrow W_1 \otimes W_2,$$

denoted

$$(w_{(1)}, w_{(2)}) \mapsto w_{(1)} \otimes w_{(2)}$$

for $w_{(1)} \in W_1$ and $w_{(2)} \in W_2$, such that for any vector space W_3 and any bilinear map

$$B : W_1 \times W_2 \longrightarrow W_3,$$

there is a unique linear map

$$L : W_1 \otimes W_2 \longrightarrow W_3$$

such that

$$B(w_{(1)}, w_{(2)}) = L(w_{(1)} \otimes w_{(2)})$$

for $w_{(i)} \in W_i$, $i = 1, 2$. This universal property characterizes the tensor product structure $W_1 \otimes W_2$, equipped with its bilinear map $\cdot \otimes \cdot$, up to unique isomorphism. In addition, the tensor product structure in fact exists.

As was illustrated in [71], and as is well known, the notion of tensor product of modules for a Lie algebra can be formulated in terms of what can be called "intertwining maps": Let W_1, W_2, W_3 be modules for a fixed Lie algebra V. (We are calling our Lie algebra V because we shall be calling our vertex algebra V, and we would like to emphasize the analogies between the two theories.) An *intertwining map of type* $\binom{W_3}{W_1\, W_2}$ is a linear map $I : W_1 \otimes W_2 \longrightarrow W_3$ (or equivalently, from what we have just recalled, a bilinear map $W_1 \times W_2 \longrightarrow W_3$) such that

$$\pi_3(v)I(w_{(1)} \otimes w_{(2)}) = I\big(\pi_1(v)w_{(1)} \otimes w_{(2)}\big) + I\big(w_{(1)} \otimes \pi_2(v)w_{(2)}\big) \tag{1.1}$$

for $v \in V$ and $w_{(i)} \in W_i$, $i = 1, 2$, where π_1, π_2, π_3 are the module actions of V on W_1, W_2 and W_3, respectively. (Clearly, such an intertwining map is the same as a module map from $W_1 \otimes W_2$, equipped with the tensor product module structure, to W_3, but we are now temporarily "forgetting" what the tensor product module is.)

A *tensor product of the V-modules W_1 and W_2* is then a pair (W_0, I_0), where W_0 is a V-module and I_0 is an intertwining map of type $\binom{W_0}{W_1\, W_2}$ (which, again, could be viewed as a suitable bilinear map $W_1 \times W_2 \longrightarrow W_0$), such that for any pair (W, I) with W a V-module and I an intertwining map of type $\binom{W}{W_1\,W_2}$, there is a unique module homomorphism $\eta : W_0 \longrightarrow W$ such that $I = \eta \circ I_0$. This universal property of course characterizes (W_0, I_0) up to canonical isomorphism. Moreover, it is obvious that the tensor product in fact exists, and may be constructed as the vector-space tensor product $W_1 \otimes W_2$ equipped with the diagonal action of the Lie algebra, together with the identity map from $W_1 \otimes W_2$ to itself (or equivalently, the canonical bilinear map $W_1 \times W_2 \longrightarrow W_1 \otimes W_2$). We shall denote the tensor product (W_0, I_0) of W_1 and W_2 by $(W_1 \boxtimes W_2, \boxtimes)$, where it is understood that the image of $w_{(1)} \otimes w_{(2)}$ under our canonical intertwining map $\boxtimes$ is $w_{(1)} \boxtimes w_{(2)}$. Thus $W_1 \boxtimes W_2 = W_1 \otimes W_2$, and under our identifications, $\boxtimes = 1_{W_1 \otimes W_2}$.

Remark 1.1 This classical explicit construction of course shows that the tensor product functor exists for the category of modules for a Lie algebra. For vertex algebras, it will be relatively straightforward to *define* the appropriate tensor product functor(s) (see [71–74]), but it will be a nontrivial matter to *construct* this functor (or more precisely, these functors) and thereby prove that the (appropriate) tensor product of modules for a (suitable) vertex algebra exists. The reason why we have formulated the notion of tensor product module for a Lie algebra in the way that we just did is that this formulation motivates the correct notion of tensor product functor(s) in the vertex algebra case.

Remark 1.2 Using this explicit construction of the tensor product functor and our notation $w_{(1)} \boxtimes w_{(2)}$ for the tensor product of elements, the standard natural associativity isomorphisms among tensor products of triples of Lie algebra modules are expressed as follows: Since $w_{(1)} \boxtimes w_{(2)} = w_{(1)} \otimes w_{(2)}$, we have

$$(w_{(1)} \boxtimes w_{(2)}) \boxtimes w_{(3)} = (w_{(1)} \otimes w_{(2)}) \otimes w_{(3)},$$
$$w_{(1)} \boxtimes (w_{(2)} \boxtimes w_{(3)}) = w_{(1)} \otimes (w_{(2)} \otimes w_{(3)})$$

for $w_{(i)} \in W_i$, $i = 1, 2, 3$, and so the canonical identification between $w_{(1)} \otimes (w_{(2)} \otimes w_{(3)})$ and $(w_{(1)} \otimes w_{(2)}) \otimes w_{(3)}$ gives the standard natural isomorphism

$$\begin{aligned} (W_1 \boxtimes W_2) \boxtimes W_3 &\to W_1 \boxtimes (W_2 \boxtimes W_3) \\ (w_{(1)} \boxtimes w_{(2)}) \boxtimes w_{(3)} &\mapsto w_{(1)} \boxtimes (w_{(2)} \boxtimes w_{(3)}). \end{aligned} \tag{1.2}$$

This collection of natural associativity isomorphisms of course satisfies the classical coherence conditions for associativity isomorphisms among multiple nested tensor product modules—the conditions that say that in nested tensor products involving any number of tensor factors, the placement of parentheses (as in (1.2), the case of three tensor factors) is immaterial; we shall discuss coherence conditions in detail later. Now, as was discovered in [53], it turns out that maps analogous to (1.2) can also be constructed in the vertex algebra case, giving natural associativity isomorphisms among triples of modules for a (suitable) vertex operator algebra. However, in the vertex algebra case, the elements "$w_{(1)} \boxtimes w_{(2)}$," which indeed exist (under suitable conditions) and are constructed in the theory, lie in a suitable "completion" of the tensor product module rather than in the module itself. Correspondingly, it is a nontrivial matter to construct the triple-tensor-product elements on the two sides of (1.2); in fact, one needs to prove certain convergence, under suitable additional conditions. Even after the triple-tensor-product elements are constructed (in suitable completions of the triple-tensor-product modules), it is a delicate matter to construct the appropriate natural associativity maps, analogous to (1.2), to prove that they are well defined, and to prove that they are isomorphisms. In the present work, we shall generalize these matters (in a self-contained way) from the context of [53] to a more general one. In the rest of this section, for triples of modules for a Lie algebra, we shall now describe a construction of the natural associativity isomorphisms that will seem roundabout and indirect, but this is the method of construction of these isomorphisms that will give us the correct "road map" for the corresponding construction (and theorems) in the vertex algebra case, as in [72–74] and [53].

A significant feature of the constructions in the earlier works (and in the present work) is that the tensor product of modules W_1 and W_2 for a vertex operator algebra V is the contragredient module of a certain V-module that is typically a *proper* subspace of $(W_1 \otimes W_2)^*$, the dual space of the vector-space tensor product of W_1 and W_2. In particular, our treatment, which follows, of the Lie algebra case will use contragredient modules, and we will therefore restrict our attention to *finite-dimensional* modules for our Lie algebra. It will be important that the double-contragredient module of a Lie algebra module is naturally isomorphic to the original module. We shall sometimes denote the contragredient module of a V-module W by W', so that $W'' = W$. (We recall that for a module W for a Lie algebra V, the corresponding contragredient module W' consists of the dual vector space W^* equipped with the action of V given by: $(v \cdot w^*)(w) = -w^*(v \cdot w)$ for $v \in V$, $w^* \in W^*$, $w \in W$.)

Let us, then, now restrict our attention to finite-dimensional modules for our Lie algebra V. The dual space $(W_1 \otimes W_2)^*$ carries the structure of the classical contragredient module of the tensor product module. Given any intertwining map of type $\binom{W_3}{W_1\, W_2}$, using the natural linear isomorphism

$$\mathrm{Hom}(W_1 \otimes W_2, W_3) \xrightarrow{\sim} \mathrm{Hom}\big(W_3^*, (W_1 \otimes W_2)^*\big) \tag{1.3}$$

we have a corresponding linear map in $\mathrm{Hom}(W_3^*, (W_1 \otimes W_2)^*)$, and this must be a map of V-modules. In the vertex algebra case, given V-modules W_1 and W_2, it turns out that with a suitable analogous setup, the union in the vector space $(W_1 \otimes W_2)^*$ of the ranges of all such V-module maps, as W_3 and the intertwining map vary (and with W_3^* replaced by the contragredient module W_3'), is the correct candidate for the contragredient module of the tensor product module $W_1 \boxtimes W_2$. Of course, in the Lie algebra situation, this union is $(W_1 \otimes W_2)^*$ itself (since we are allowed to take $W_3 = W_1 \otimes W_2$ and the intertwining map to be the canonical map), but in the vertex algebra case, this union is typically much smaller than $(W_1 \otimes W_2)^*$. In the vertex algebra case, we will use the notation $W_1 \boxbslash W_2$ to designate this union, and if the tensor product module $W_1 \boxtimes W_2$ in fact exists, then

$$W_1 \boxtimes W_2 = (W_1 \boxbslash W_2)', \tag{1.4}$$

$$W_1 \boxbslash W_2 = (W_1 \boxtimes W_2)'. \tag{1.5}$$

Thus in the Lie algebra case we will write

$$W_1 \boxbslash W_2 = (W_1 \otimes W_2)^*, \tag{1.6}$$

and (1.4) and (1.5) hold. (In the Lie algebra case we prefer to write $(W_1 \otimes W_2)^*$ rather than $(W_1 \otimes W_2)'$, because in the vertex algebra case, $W_1 \otimes W_2$ is typically not a V-module, and so we will not be allowed to write $(W_1 \otimes W_2)'$ in the vertex algebra case.)

The subspace $W_1 \boxbslash W_2$ of $(W_1 \otimes W_2)^*$ was in fact further described in the following terms in [72] and [74], in the vertex algebra case: For any map in $\mathrm{Hom}(W_3', (W_1 \otimes W_2)^*)$ corresponding to an intertwining map according to (1.3), the image of any $w'_{(3)} \in W_3'$ under this map satisfies certain subtle conditions, called the "compatibility condition" and the "local grading restriction condition"; these conditions are not "visible" in the Lie algebra case. These conditions in fact precisely describe the proper subspace $W_1 \boxbslash W_2$ of $(W_1 \otimes W_2)^*$. We will discuss such conditions further in Sect. 1.4 and in the body of this work. As we shall explain, this idea of describing elements in certain dual spaces was also used in constructing the natural associativity isomorphisms between triples of modules for a vertex operator algebra in [53].

In order to give the reader a guide to the vertex algebra case, we now describe the analogue for the Lie algebra case of this construction of the associativity isomorphisms. To construct the associativity isomorphism from $(W_1 \boxtimes W_2) \boxtimes W_3$ to $W_1 \boxtimes (W_2 \boxtimes W_3)$, it is equivalent (by duality) to give a suitable isomorphism from $W_1 \boxbslash (W_2 \boxtimes W_3)$ to $(W_1 \boxtimes W_2) \boxbslash W_3$ (recall (1.4), (1.5)).

Rather than directly constructing an isomorphism between these two V-modules, it turns out that we want to embed both of them, separately, into the single space $(W_1 \otimes W_2 \otimes W_3)^*$. Note that $(W_1 \otimes W_2 \otimes W_3)^*$ is naturally a V-module, via the contragredient of the diagonal action, that is,

$$\begin{aligned}\big(\pi(v)\lambda\big)(w_{(1)} \otimes w_{(2)} \otimes w_{(3)}) = &-\lambda\big(\pi_1(v)w_{(1)} \otimes w_{(2)} \otimes w_{(3)}\big)\\ &-\lambda\big(w_{(1)} \otimes \pi_2(v)w_{(2)} \otimes w_{(3)}\big)\\ &-\lambda\big(w_{(1)} \otimes w_{(2)} \otimes \pi_3(v)w_{(3)}\big),\end{aligned} \tag{1.7}$$

for $v \in V$ and $w_{(i)} \in W_i$, $i = 1, 2, 3$, where π_1, π_2, π_3 are the module actions of V on W_1, W_2 and W_3, respectively. A concept related to this is the notion of *intertwining map from* $W_1 \otimes W_2 \otimes W_3$ *to a module* W_4, a natural analogue of (1.1), defined to be a linear map

$$F : W_1 \otimes W_2 \otimes W_3 \longrightarrow W_4 \tag{1.8}$$

such that

$$\pi_4(v)F(w_{(1)} \otimes w_{(2)} \otimes w_{(3)}) = F\big(\pi_1(v)w_{(1)} \otimes w_{(2)} \otimes w_{(3)}\big) + F\big(w_{(1)} \otimes \pi_2(v)w_{(2)} \otimes w_{(3)}\big) + F\big(w_{(1)} \otimes w_{(2)} \otimes \pi_{(3)}(v)w_3\big), \tag{1.9}$$

with the obvious notation. The relation between (1.7) and (1.9) comes directly from the natural linear isomorphism

$$\mathrm{Hom}(W_1 \otimes W_2 \otimes W_3, W_4) \xrightarrow{\sim} \mathrm{Hom}\big(W_4^*, (W_1 \otimes W_2 \otimes W_3)^*\big); \tag{1.10}$$

given F, we have

$$\begin{aligned} W_4^* &\longrightarrow (W_1 \otimes W_2 \otimes W_3)^* \\ \nu &\mapsto \nu \circ F. \end{aligned} \tag{1.11}$$

Under this natural linear isomorphism, the intertwining maps correspond precisely to the V-module maps from W_4^* to $(W_1 \otimes W_2 \otimes W_3)^*$. In the situation for vertex algebras, as was the case for tensor products of two rather than three modules, there are analogues of all of the notions and comments discussed in this paragraph *except that we will not put V-module structure onto the vector space $W_1 \otimes W_2 \otimes W_3$*; as we have emphasized, we will instead base the theory on intertwining maps.

Two important ways of constructing maps of the type (1.8) are as follows: For modules W_1, W_2, W_3, W_4, M_1 and intertwining maps I_1 and I_2 of types $\binom{W_4}{W_1\,M_1}$ and $\binom{M_1}{W_2\,W_3}$, respectively, by definition the composition $I_1 \circ (1_{W_1} \otimes I_2)$ is an intertwining map from $W_1 \otimes W_2 \otimes W_3$ to W_4. Analogously, for intertwining maps I^1, I^2 of types $\binom{W_4}{M_2\,W_3}$ and $\binom{M_2}{W_1\,W_2}$, respectively, with M_2 also a module, the composition $I^1 \circ (I^2 \otimes 1_{W_3})$ is an intertwining map from $W_1 \otimes W_2 \otimes W_3$ to W_4. Hence we have two V-module homomorphisms

$$\begin{aligned} W_4^* &\longrightarrow (W_1 \otimes W_2 \otimes W_3)^* \\ \nu &\mapsto \nu \circ F_1, \end{aligned} \tag{1.12}$$

where F_1 is the intertwining map $I_1 \circ (1_{W_1} \otimes I_2)$; and

$$\begin{aligned} W_4^* &\longrightarrow (W_1 \otimes W_2 \otimes W_3)^* \\ \nu &\mapsto \nu \circ F_2, \end{aligned} \tag{1.13}$$

where F_2 is the intertwining map $I^1 \circ (I^2 \circ 1_{W_3})$.

The special cases in which the modules W_4 are two iterated tensor product modules and the "intermediate" modules M_1 and M_2 are two tensor product modules are particularly interesting: When $W_4 = W_1 \boxtimes (W_2 \boxtimes W_3)$ and $M_1 = W_2 \boxtimes W_3$, and I_1 and I_2 are the corresponding canonical intertwining maps, (1.12) gives the natural

V-module homomorphism

$$\begin{aligned} W_1 \boxbackslash (W_2 \boxtimes W_3) &\longrightarrow (W_1 \otimes W_2 \otimes W_3)^* \\ \nu &\mapsto \big(w_{(1)} \otimes w_{(2)} \otimes w_{(3)} \mapsto \nu\big(w_{(1)} \boxtimes (w_{(2)} \boxtimes w_{(3)})\big)\big); \end{aligned} \tag{1.14}$$

when $W_4 = (W_1 \boxtimes W_2) \boxtimes W_3$ and $M_2 = W_1 \boxtimes W_2$, and I^1 and I^2 are the corresponding canonical intertwining maps, (1.13) gives the natural V-module homomorphism

$$\begin{aligned} (W_1 \boxtimes W_2) \boxbackslash W_3 &\longrightarrow (W_1 \otimes W_2 \otimes W_3)^* \\ \nu &\mapsto \big(w_{(1)} \otimes w_{(2)} \otimes w_{(3)} \mapsto \nu\big((w_{(1)} \boxtimes w_{(2)}) \boxtimes w_{(3)}\big)\big). \end{aligned} \tag{1.15}$$

Clearly, in our Lie algebra case, both of the maps (1.14) and (1.15) are isomorphisms, since they both in fact amount to the identity map on $(W_1 \otimes W_2 \otimes W_3)^*$. However, in the vertex algebra case the analogues of these two maps are only injective homomorphisms, and typically not isomorphisms. (Recall the analogous situation, mentioned above, for double rather than triple tensor products.) These two maps enable us to identify both $W_1 \boxbackslash (W_2 \boxtimes W_3)$ and $(W_1 \boxtimes W_2) \boxbackslash W_3$ with subspaces of $(W_1 \otimes W_2 \otimes W_3)^*$. In the vertex algebra case we will have certain "compatibility conditions" and "local grading restriction conditions" on elements of $(W_1 \otimes W_2 \otimes W_3)^*$ to describe each of the two subspaces. In either the Lie algebra or the vertex algebra case, the construction of our desired natural associativity isomorphism between the two modules $(W_1 \boxtimes W_2) \boxtimes W_3$ and $W_1 \boxtimes (W_2 \boxtimes W_3)$ follows from showing that the ranges of homomorphisms (1.14) and (1.15) are equal to each other, which is of course obvious in the Lie algebra case since both (1.14) and (1.15) are isomorphisms to $(W_1 \otimes W_2 \otimes W_3)^*$. It turns out that, under this associativity isomorphism, (1.2) holds in both the Lie algebra case and the vertex algebra case; in the Lie algebra case, this is obvious because all the maps are the "tautological" ones.

Now we give the reader a preview of how, in the vertex algebra case, these compatibility and local grading restriction conditions on elements of $(W_1 \otimes W_2 \otimes W_3)^*$ will arise. As we have mentioned, in the Lie algebra case, an intertwining map from $W_1 \otimes W_2 \otimes W_3$ to W_4 corresponds to a module map from W_4^* to $(W_1 \otimes W_2 \otimes W_3)^*$. As was discussed in [53], for the vertex operator algebra analogue, the image of any $w'_{(4)} \in W'_4$ under such an analogous map satisfies certain "compatibility" and "local grading restriction" conditions, and so these conditions must be satisfied by those elements of $(W_1 \otimes W_2 \otimes W_3)^*$ lying in the ranges of the vertex-operator-algebra analogues of either of the maps (1.14) and (1.15) (or the maps (1.12) and (1.13)).

Besides these two conditions, satisfied by the elements of the ranges of the maps of both types (1.14) and (1.15), the elements of the ranges of the analogues of the homomorphisms (1.14) and (1.15) have their own separate properties. First note that any $\lambda \in (W_1 \otimes W_2 \otimes W_3)^*$ induces the two maps

$$\begin{aligned} \mu_\lambda^{(1)} : W_1 &\to (W_2 \otimes W_3)^* \\ w_{(1)} &\mapsto \lambda(w_{(1)} \otimes \cdot \otimes \cdot) \end{aligned} \tag{1.16}$$

and

$$\begin{aligned} \mu_\lambda^{(2)} : W_3 &\to (W_1 \otimes W_2)^* \\ w_{(3)} &\mapsto \lambda(\cdot \otimes \cdot \otimes w_{(3)}). \end{aligned} \tag{1.17}$$

In the vertex operator algebra analogue [53], if λ lies in the range of (1.14), then it must satisfy the condition that the elements $\mu_\lambda^{(1)}(w_{(1)})$ all lie in a suitable completion of the subspace $W_2 \boxtimes W_3$ of $(W_2 \otimes W_3)^*$, and if λ lies in the range of (1.15), then it must satisfy the condition that the elements $\mu_\lambda^{(2)}(w_{(3)})$ all lie in a suitable completion of the subspace $W_1 \boxtimes W_2$ of $(W_1 \otimes W_2)^*$. (Of course in the Lie algebra case, these statements are tautological.) In [53], these important conditions, that $\mu_\lambda^{(1)}(W_1)$ lies in a suitable completion of $W_2 \boxtimes W_3$ and that $\mu_\lambda^{(2)}(W_3)$ lies in a suitable completion of $W_1 \boxtimes W_2$, are understood as "local grading restriction conditions" with respect to the two different ways of composing intertwining maps.

In the construction of our desired natural associativity isomorphism, since we want the ranges of (1.14) and (1.15) to be the same submodule of $(W_1 \otimes W_2 \otimes W_3)^*$, the ranges of both (1.14) and (1.15) should satisfy both of these conditions. This amounts to a certain "expansion condition" in the vertex algebra case. When all these conditions are satisfied, it can in fact be proved [53] that the associativity isomorphism does indeed exist and that in addition, the "associativity of intertwining maps" holds; that is, the "product" of two suitable intertwining maps can be written, in a certain sense, as the "iterate" of two suitable intertwining maps, and conversely. This equality of products with iterates, highly nontrivial in the vertex algebra case, amounts in the Lie algebra case to the easy statement that in the notation above, any intertwining map of the form $I_1 \circ (1_{W_1} \otimes I_2)$ can also be written as an intertwining map of the form $I^1 \circ (I^2 \otimes 1_{W_3})$, for a suitable "intermediate module" M_2 and suitable intertwining maps I^1 and I^2, and conversely. The reason why this statement is easy in the Lie algebra case is that in fact *any* intertwining map F of the type (1.8) can be "factored" in either of these two ways; for example, to write F in the form $I_1 \circ (1_{W_1} \otimes I_2)$, take M_1 to be $W_2 \otimes W_3$, I_2 to be the canonical (identity) map and I_1 to be F itself (with the appropriate identifications having been made).

We are now ready to discuss the vertex algebra case.

1.4 The Vertex Algebra Case

In this section, which should be carefully compared with the previous one, we shall lay out our "road map" of the constructions of the tensor product functors and the associativity isomorphisms for a suitable class of vertex algebras, considerably generalizing, but also following the ideas of, the corresponding theory developed in [72–74] and [53] for vertex operator algebras. Without yet specifying the precise class of vertex algebras that we shall be using in the body of this work, except to say that our vertex algebras will be $\mathbb{Z}$-graded and our modules will be $\mathbb{C}$-graded at first and then $\mathbb{R}$-graded for the more substantial results, we now discuss the vertex algebra case. What follows applies to both the theory of [53, 72–74] and the present new logarithmic theory. In Remark 1.7 below, we comment on the substantial new features of the logarithmic generality.

In the vertex algebra case, the concept of intertwining map involves the moduli space of Riemann spheres with one negatively oriented puncture and two positively oriented punctures and with local coordinates around each puncture; the details of the geometric structures needed in this theory are presented in [52] and [56]. For each element of this moduli space there is a notion of intertwining map adapted to the particular element. Let z be a nonzero complex number and let $P(z)$ be the Riemann sphere $\hat{\mathbb{C}}$ with one negatively oriented puncture at ∞ and two positively oriented punctures at z and 0, with local coordinates $1/w$, $w-z$ and w at these three punctures, respectively.

Let V be a vertex algebra (on which appropriate assumptions, including the existence of a suitable $\mathbb{Z}$-grading, will be made later), and let $Y(\cdot, x)$ be the vertex operator map defining the algebra structure (see Sect. 2 below for a brief summary of basic notions and notation, including the formal delta function). Let W_1, W_2 and W_3 be modules for V, and let $Y_1(\cdot, x)$, $Y_2(\cdot, x)$ and $Y_3(\cdot, x)$ be the corresponding vertex operator maps. (The cases in which some of the W_i are V itself, and some of the Y_i are, correspondingly, Y, are important, but the most interesting cases are those where all three modules are different from V.) A "$P(z)$-intertwining map of type $\binom{W_3}{W_1\, W_2}$" is a linear map

$$I : W_1 \otimes W_2 \longrightarrow \overline{W_3}, \tag{1.18}$$

where $\overline{W_3}$ is a certain completion of W_3, related to its $\mathbb{C}$-grading, such that

$$\begin{aligned} &x_0^{-1}\delta\left(\frac{x_1 - z}{x_0}\right) Y_3(v, x_1) I(w_{(1)} \otimes w_{(2)}) \\ &\quad = z^{-1}\delta\left(\frac{x_1 - x_0}{z}\right) I\big(Y_1(v, x_0) w_{(1)} \otimes w_{(2)}\big) \\ &\qquad + x_0^{-1}\delta\left(\frac{z - x_1}{-x_0}\right) I\big(w_{(1)} \otimes Y_2(v, x_1) w_{(2)}\big) \end{aligned} \tag{1.19}$$

for $v \in V$, $w_{(1)} \in W_1$, $w_{(2)} \in W_2$, where x_0, x_1 and x_2 are commuting independent formal variables. This notion is motivated in detail in [71, 72] and [74]; we shall recall the motivation below.

Remark 1.3 In this theory, it is crucial to distinguish between formal variables and complex variables. Thus we shall use the following notational convention: *Throughout this work, unless we specify otherwise, the symbols $x, x_0, x_1, x_2, \ldots, y, y_0, y_1, y_2, \ldots$ will denote commuting independent formal variables, and by contrast, the symbols $z, z_0, z_1, z_2, \ldots$ will denote complex numbers in specified domains, not formal variables.*

Remark 1.4 Recall from [38] the definition of the notion of intertwining operator $\mathcal{Y}(\cdot, x)$ in the theory of vertex (operator) algebras. Given (W_1, Y_1), (W_2, Y_2) and (W_3, Y_3) as above, an intertwining operator of type $\binom{W_3}{W_1\,W_2}$ can be viewed as a certain type of linear map $\mathcal{Y}(\cdot, x)\cdot$ from $W_1 \otimes W_2$ to the vector space of formal series in x of the form $\sum_{n\in\mathbb{C}} w(n)x^n$, where the coefficients $w(n)$ lie in W_3, and where we are allowing arbitrary complex powers of x, suitably "truncated from below" in this sum. The main property of an intertwining operator is the following "Jacobi identity":

$$\begin{aligned}
&x_0^{-1}\delta\left(\frac{x_1-x_2}{x_0}\right)Y_3(v,x_1)\mathcal{Y}(w_{(1)},x_2)w_{(2)}\\
&\quad - x_0^{-1}\delta\left(\frac{x_2-x_1}{-x_0}\right)\mathcal{Y}(w_{(1)},x_2)Y_2(v,x_1)w_{(2)}\\
&= x_2^{-1}\delta\left(\frac{x_1-x_0}{x_2}\right)\mathcal{Y}\big(Y_1(v,x_0)w_{(1)},x_2\big)w_{(2)}
\end{aligned} \tag{1.20}$$

for $v \in V$, $w_{(1)} \in W_1$ and $w_{(2)} \in W_2$. (When all three modules W_i are V itself and all four operators Y_i and $\mathcal{Y}$ are Y itself, (1.20) becomes the usual Jacobi identity in the definition of the notion of vertex algebra. When W_1 is V, $W_2 = W_3$ and $\mathcal{Y} = Y_2 = Y_3$, (1.20) becomes the usual Jacobi identity in the definition of the notion of V-module.) The point is that by "substituting z for x_2" in (1.20), we obtain (1.19), where we make the identification

$$I(w_{(1)} \otimes w_{(2)}) = \mathcal{Y}(w_{(1)}, z)w_{(2)}; \tag{1.21}$$

the resulting complex powers of the complex number z are made precise by a choice of branch of the log function. The nonzero complex number z in the notion of $P(z)$-intertwining map thus "comes from" the substitution of z for x_2 in the Jacobi identity in the definition of the notion of intertwining operator. In fact, this correspondence (given a choice of branch of log) actually defines an isomorphism between the space of $P(z)$-intertwining maps and the space of intertwining operators of the same type ([72, 74]); this will be discussed.

There is a natural linear injection

$$\mathrm{Hom}(W_1 \otimes W_2, \overline{W}_3) \longrightarrow \mathrm{Hom}\big(W_3', (W_1 \otimes W_2)^*\big), \tag{1.22}$$

where here and below we denote by W' the (suitably defined) contragredient module of a V-module W; we have $W'' = W$. Under this injection, a map $I \in \mathrm{Hom}(W_1 \otimes W_2, \overline{W}_3)$ amounts to a map $I' : W_3' \longrightarrow (W_1 \otimes W_2)^*$:

$$w'_{(3)} \mapsto \big\langle w'_{(3)}, I(\cdot \otimes \cdot)\big\rangle, \tag{1.23}$$

where $\langle \cdot, \cdot \rangle$ denotes the natural pairing between the contragredient of a module and its completion. If I is a $P(z)$-intertwining map, then as in the Lie algebra case (see above), where such a map is a module map, the map (1.23) intertwines two natural V-actions on W_3' and $(W_1 \otimes W_2)^*$. We will see that in the present (vertex algebra) case, $(W_1 \otimes W_2)^*$ is typically not a V-module. The images of all the elements $w_{(3)}' \in W_3'$ under this map satisfy certain conditions, called the "$P(z)$-compatibility condition" and the "$P(z)$-local grading restriction condition," as formulated in [72] and [74]; we shall be discussing these.

Given a category of V-modules and two modules W_1 and W_2 in this category, as in the Lie algebra case, the "$P(z)$-tensor product of W_1 and W_2" is then defined to be a pair (W_0, I_0), where W_0 is a module in the category and I_0 is a $P(z)$-intertwining map of type $\binom{W_0}{W_1\, W_2}$, such that for any pair (W, I) with W a module in the category and I a $P(z)$-intertwining map of type $\binom{W}{W_1\, W_2}$, there is a unique morphism $\eta : W_0 \longrightarrow W$ such that $I = \bar{\eta} \circ I_0$; here and throughout this work we denote by $\bar{\chi}$ the linear map naturally extending a suitable linear map χ from a graded space to its appropriate completion. This universal property characterizes (W_0, I_0) up to canonical isomorphism, *if it exists*. We will denote the $P(z)$-tensor product of W_1 and W_2, if it exists, by $(W_1 \boxtimes_{P(z)} W_2, \boxtimes_{P(z)})$, and we will denote the image of $w_{(1)} \otimes w_{(2)}$ under $\boxtimes_{P(z)}$ by $w_{(1)} \boxtimes_{P(z)} w_{(2)}$, which is an element of $\overline{W_1 \boxtimes_{P(z)} W_2}$, not of $W_1 \boxtimes_{P(z)} W_2$.

From this definition and the natural map (1.22), we will see that if the $P(z)$-tensor product of W_1 and W_2 exists, then its contragredient module can be realized as the union of ranges of all maps of the form (1.23) as W_3' and I vary. Even if the $P(z)$-tensor product of W_1 and W_2 does not exist, we denote this union (which is always a subspace stable under a natural action of V) by $W_1 \boxbslash_{P(z)} W_2$. If the tensor product does exist, then

$$W_1 \boxtimes_{P(z)} W_2 = (W_1 \boxbslash_{P(z)} W_2)', \tag{1.24}$$

$$W_1 \boxbslash_{P(z)} W_2 = (W_1 \boxtimes_{P(z)} W_2)'; \tag{1.25}$$

examining (1.24) will show the reader why the notation $\boxbslash$ was chosen in the earlier papers ($\boxtimes = \boxbslash'$!). Several critical facts about $W_1 \boxbslash_{P(z)} W_2$ were proved in [72, 73] and [74], notably, $W_1 \boxbslash_{P(z)} W_2$ is equal to the subspace of $(W_1 \otimes W_2)^*$ consisting of all the elements satisfying the $P(z)$-compatibility condition and the $P(z)$-local grading restriction condition, and in particular, this subspace is V-stable; and the condition that $W_1 \boxbslash_{P(z)} W_2$ is a module is equivalent to the existence of the $P(z)$-tensor product $W_1 \boxtimes_{P(z)} W_2$. All these facts will be proved.

In order to construct vertex tensor category structure, we need to construct appropriate natural associativity isomorphisms. Assuming the existence of the relevant tensor products, we in fact need to construct an appropriate natural isomorphism from $(W_1 \boxtimes_{P(z_1 - z_2)} W_2) \boxtimes_{P(z_2)} W_3$ to $W_1 \boxtimes_{P(z_1)} (W_2 \boxtimes_{P(z_2)} W_3)$ for complex numbers z_1, z_2 satisfying $|z_1| > |z_2| > |z_1 - z_2| > 0$. Note that we are using two distinct nonzero complex numbers, and that certain inequalities hold. This situation corresponds to the fact that a Riemann sphere with one negatively oriented puncture and three positively oriented punctures can be seen in two different ways as the "product" of two Riemann spheres each with one negatively oriented puncture and two positively oriented punctures; the detailed geometric motivation is presented in [52, 56, 71] and [53].

To construct this natural isomorphism, we first consider compositions of certain intertwining maps. As we have mentioned, a $P(z)$-intertwining map I of type $\binom{W_3}{W_1 W_2}$ maps into $\overline{W}_3$ rather than W_3. Thus the existence of compositions of suitable intertwining maps always entails certain convergence. In particular, the existence of the composition $w_{(1)} \boxtimes_{P(z_1)} (w_{(2)} \boxtimes_{P(z_2)} w_{(3)})$ when $|z_1| > |z_2| > 0$ and the existence of the composition $(w_{(1)} \boxtimes_{P(z_1-z_2)} w_{(2)}) \boxtimes_{P(z_2)} w_{(3)}$ when $|z_2| > |z_1 - z_2| > 0$, for general elements $w_{(i)}$ of W_i, $i = 1, 2, 3$, requires the proof of certain convergence conditions. These conditions will be discussed in detail.

Let us now assume these convergence conditions and let z_1, z_2 satisfy $|z_1| > |z_2| > |z_1 - z_2| > 0$. To construct the desired associativity isomorphism from $(W_1 \boxtimes_{P(z_1-z_2)} W_2) \boxtimes_{P(z_2)} W_3$ to $W_1 \boxtimes_{P(z_1)} (W_2 \boxtimes_{P(z_2)} W_3)$, it is equivalent (by duality) to give a suitable natural isomorphism from $W_1 ⧄_{P(z_1)} (W_2 \boxtimes_{P(z_2)} W_3)$ to $(W_1 \boxtimes_{P(z_1-z_2)} W_2) ⧄_{P(z_2)} W_3$. As we mentioned in the previous section, instead of constructing this isomorphism directly, we shall embed both of these spaces, separately, into the single space $(W_1 \otimes W_2 \otimes W_3)^*$.

We will see that $(W_1 \otimes W_2 \otimes W_3)^*$ carries a natural V-action analogous to the contragredient of the diagonal action in the Lie algebra case (recall the similar action of V on $(W_1 \otimes W_2)^*$ mentioned above). Also, for four V-modules W_1, W_2, W_3 and W_4, we have a canonical notion of "$P(z_1, z_2)$-intertwining map from $W_1 \otimes W_2 \otimes W_3$ to $\overline{W}_4$" given by a vertex-algebraic analogue of (1.9); for this notion, we need only that z_1 and z_2 are nonzero and distinct. The relation between these two concepts comes from the natural linear injection

$$\begin{aligned} \mathrm{Hom}(W_1 \otimes W_2 \otimes W_3, \overline{W}_4) &\longrightarrow \mathrm{Hom}\big(W_4', (W_1 \otimes W_2 \otimes W_3)^*\big) \\ F &\mapsto F', \end{aligned} \tag{1.26}$$

where $F' : W_4' \longrightarrow (W_1 \otimes W_2 \otimes W_3)^*$ is given by

$$\nu \mapsto \nu \circ F, \tag{1.27}$$

which is indeed well defined. Under this natural map, the $P(z_1, z_2)$-intertwining maps correspond precisely to the maps from W_4' to $(W_1 \otimes W_2 \otimes W_3)^*$ that intertwine the two natural V-actions on W_4' and $(W_1 \otimes W_2 \otimes W_3)^*$.

Now for modules W_1, W_2, W_3, W_4, M_1, and a $P(z_1)$-intertwining map I_1 and a $P(z_2)$-intertwining map I_2 of types $\binom{W_4}{W_1 M_1}$ and $\binom{M_1}{W_2 W_3}$, respectively, it turns out that the composition $I_1 \circ (1_{W_1} \otimes I_2)$ exists and is a $P(z_1, z_2)$-intertwining map when $|z_1| > |z_2| > 0$. Analogously, for a $P(z_2)$-intertwining map I^1 and a $P(z_1 - z_2)$-intertwining map I^2 of types $\binom{W_4}{M_2 W_3}$ and $\binom{M_2}{W_1 W_2}$, respectively, where M_2 is also a module, the composition $I^1 \circ (I^2 \otimes 1_{W_3})$ is a $P(z_1, z_2)$-intertwining map when $|z_2| > |z_1 - z_2| > 0$. Hence we have two maps intertwining the V-actions:

$$\begin{aligned} W_4' &\longrightarrow (W_1 \otimes W_2 \otimes W_3)^* \\ \nu &\mapsto \nu \circ F_1, \end{aligned} \tag{1.28}$$

where F_1 is the intertwining map $I_1 \circ (1_{W_1} \otimes I_2)$, and

$$\begin{aligned} W_4' &\longrightarrow (W_1 \otimes W_2 \otimes W_3)^* \\ \nu &\mapsto \nu \circ F_2, \end{aligned} \tag{1.29}$$

where F_2 is the intertwining map $I^1 \circ (I^2 \circ 1_{W_3})$.

It is important to note that we can express these compositions $I_1 \circ (1_{W_1} \otimes I_2)$ and $I^1 \circ (I^2 \otimes 1_{W_3})$ in terms of intertwining operators, as discussed in Remark 1.4. Let $\mathcal{Y}_1, \mathcal{Y}_2, \mathcal{Y}^1$ and $\mathcal{Y}^2$ be the intertwining operators corresponding to I_1, I_2, I^1 and I^2, respectively. Then the compositions $I_1 \circ (1_{W_1} \otimes I_2)$ and $I^1 \circ (I^2 \otimes 1_{W_3})$ correspond to the "product" $\mathcal{Y}_1(\cdot, x_1)\mathcal{Y}_2(\cdot, x_2)\cdot$ and "iterate" $\mathcal{Y}^1(\mathcal{Y}^2(\cdot, x_0)\cdot, x_2)\cdot$ of intertwining operators, respectively, and we make the "substitutions" (in the sense of Remark 1.4) $x_1 \mapsto z_1$, $x_2 \mapsto z_2$ and $x_0 \mapsto z_1 - z_2$ in order to express the two compositions of intertwining maps as the "product" $\mathcal{Y}_1(\cdot, z_1)\mathcal{Y}_2(\cdot, z_2)\cdot$ and "iterate" $\mathcal{Y}^1(\mathcal{Y}^2(\cdot, z_1 - z_2)\cdot, z_2)\cdot$ of intertwining maps, respectively. (These products and iterates involve a branch of the log function and also certain convergence.)

Just as in the Lie algebra case, the special cases in which the modules W_4 are two iterated tensor product modules and the "intermediate" modules M_1 and M_2 are two tensor product modules are particularly interesting: When $W_4 = W_1 \boxtimes_{P(z_1)} (W_2 \boxtimes_{P(z_2)} W_3)$ and $M_1 = W_2 \boxtimes_{P(z_2)} W_3$, and I_1 and I_2 are the corresponding canonical intertwining maps, (1.28) gives the natural V-homomorphism

$$\begin{aligned} W_1 \boxslash_{P(z_1)} (W_2 \boxtimes_{P(z_2)} W_3) &\longrightarrow (W_1 \otimes W_2 \otimes W_3)^* \\ \nu &\mapsto \big(w_{(1)} \otimes w_{(2)} \otimes w_{(3)} \mapsto \\ &\qquad \nu\big(w_{(1)} \boxtimes_{P(z_1)} (w_{(2)} \boxtimes_{P(z_2)} w_{(3)})\big)\big); \end{aligned} \tag{1.30}$$

when $W_4 = (W_1 \boxtimes_{P(z_1-z_2)} W_2) \boxtimes_{P(z_2)} W_3$ and $M_2 = W_1 \boxtimes_{P(z_1-z_2)} W_2$, and I^1 and I^2 are the corresponding canonical intertwining maps, (1.29) gives the natural V-homomorphism

$$\begin{aligned} (W_1 \boxtimes_{P(z_1-z_2)} W_2) \boxslash_{P(z_2)} W_3 &\longrightarrow (W_1 \otimes W_2 \otimes W_3)^* \\ \nu &\mapsto \big(w_{(1)} \otimes w_{(2)} \otimes w_{(3)} \mapsto \\ &\qquad \nu\big((w_{(1)} \boxtimes_{P(z_1-z_2)} w_{(2)}) \boxtimes_{P(z_2)} w_{(3)}\big)\big). \end{aligned} \tag{1.31}$$

It turns out that both of these maps are injections, as in [53] (as we shall prove), so that we are embedding the spaces $W_1 \boxslash_{P(z_1)} (W_2 \boxtimes_{P(z_2)} W_3)$ and $(W_1 \boxtimes_{P(z_1-z_2)} W_2) \boxslash_{P(z_2)} W_3$ into the space $(W_1 \otimes W_2 \otimes W_3)^*$. Following the ideas in [53], we shall give a precise description of the ranges of these two maps, and under suitable conditions, prove that the two ranges are the same; this will establish the associativity isomorphism.

More precisely, as in [53], we prove that for any $P(z_1, z_2)$-intertwining map F, the image of any $\nu \in W_4'$ under F' (recall (1.27)) satisfies certain conditions that we call the "$P(z_1, z_2)$-compatibility condition" and the "$P(z_1, z_2)$-local grading restriction condition." Hence, as special cases, the elements of $(W_1 \otimes W_2 \otimes W_3)^*$ in the ranges of either of the maps (1.28) or (1.29), and in particular, of (1.30) or (1.31), satisfy these conditions.

In addition, any $\lambda \in (W_1 \otimes W_2 \otimes W_3)^*$ induces two maps $\mu_\lambda^{(1)}$ and $\mu_\lambda^{(2)}$ as in (1.16) and (1.17). We will see that any element λ of the range of (1.28), and in particular, of (1.30), must satisfy the condition that the elements $\mu_\lambda^{(1)}(w_{(1)})$ all lie, roughly speaking, in a suitable completion of the subspace $W_2 \boxslash_{P(z_2)} W_3$ of $(W_2 \otimes W_3)^*$, and any element λ of the range of (1.29), and in particular, of (1.31), must satisfy the condition that the elements $\mu_\lambda^{(2)}(w_{(3)})$ all lie, again roughly speaking, in a suitable completion of the subspace $W_1 \boxslash_{P(z_1-z_2)} W_2$ of $(W_1 \otimes W_2)^*$. These conditions will be called the "$P^{(1)}(z)$-local grading restriction condition" and the "$P^{(2)}(z)$-local grading restriction condition," respectively.

It turns out that the construction of the desired natural associativity isomorphism follows from showing that the ranges of both of (1.30) and (1.31) satisfy both of these conditions. This amounts to a certain "expansion condition" on our module category. When this expansion condition and a suitable convergence condition are satisfied, we show that the desired associativity isomorphisms do exist, and that in addition, the associativity of intertwining maps holds. That is, let z_1 and z_2 be complex numbers satisfying the inequalities $|z_1| > |z_2| > |z_1 - z_2| > 0$. Then for any $P(z_1)$-intertwining map I_1 and $P(z_2)$-intertwining map I_2 of types $\binom{W_4}{W_1 M_1}$ and $\binom{M_1}{W_2 W_3}$, respectively, there is a suitable module M_2, and a $P(z_2)$-intertwining map I^1 and a $P(z_1 - z_2)$-intertwining map I^2 of types $\binom{W_4}{M_2 W_3}$ and $\binom{M_2}{W_1 W_2}$, respectively, such that

$$\langle w'_{(4)}, I_1(w_{(1)} \otimes I_2(w_{(2)} \otimes w_{(3)}))\rangle = \langle w'_{(4)}, I^1(I^2(w_{(1)} \otimes w_{(2)}) \otimes w_{(3)})\rangle \quad (1.32)$$

for $w_{(1)} \in W_1$, $w_{(2)} \in W_2$, $w_{(3)} \in W_3$ and $w'_{(4)} \in W'_4$; and conversely, given I^1 and I^2 as indicated, there exist a suitable module M_1 and maps I_1 and I_2 with the indicated properties. In terms of intertwining operators (recall the comments above), the equality (1.32) reads

$$\begin{aligned}\langle w'_{(4)}, &\mathcal{Y}_1(w_{(1)}, x_1)\mathcal{Y}_2(w_{(2)}, x_2)w_{(3)}\rangle\big|_{x_1=z_1, x_2=z_2} \\ &= \langle w'_{(4)}, \mathcal{Y}^1(\mathcal{Y}^2(w_{(1)}, x_0)w_{(2)}, x_2)w_{(3)}\rangle\big|_{x_0=z_1-z_2, x_2=z_2}, \end{aligned} \quad (1.33)$$

where $\mathcal{Y}_1$, $\mathcal{Y}_2$, $\mathcal{Y}^1$ and $\mathcal{Y}^2$ are the intertwining operators corresponding to I_1, I_2, I^1 and I^2, respectively. (As we have been mentioning, the substitution of complex numbers for formal variables involves a branch of the log function and also certain convergence.) In this sense, the associativity asserts that the "product" of two suitable intertwining maps can be written as the "iterate" of two suitable intertwining maps, and conversely.

From this construction of the natural associativity isomorphisms we will see, by analogy with (1.2), that $(w_{(1)} \boxtimes_{P(z_1-z_2)} w_{(2)}) \boxtimes_{P(z_2)} w_{(3)}$ is mapped naturally to $w_{(1)} \boxtimes_{P(z_1)} (w_{(2)} \boxtimes_{P(z_2)} w_{(3)})$ under the natural extension of the corresponding associativity isomorphism (these elements in general lying in the algebraic completions of the corresponding tensor product modules). In fact, this property

$$(w_{(1)} \boxtimes_{P(z_1-z_2)} w_{(2)}) \boxtimes_{P(z_2)} w_{(3)} \mapsto w_{(1)} \boxtimes_{P(z_1)} (w_{(2)} \boxtimes_{P(z_2)} w_{(3)}) \tag{1.34}$$

for $w_{(1)} \in W_1$, $w_{(2)} \in W_2$ and $w_{(3)} \in W_3$ characterizes the associativity isomorphism

$$(W_1 \boxtimes_{P(z_1-z_2)} W_2) \boxtimes_{P(z_2)} W_3 \to W_1 \boxtimes_{P(z_1)} (W_2 \boxtimes_{P(z_2)} W_3) \tag{1.35}$$

(cf. (1.2)). The coherence property of the associativity isomorphisms will follow from this fact. We will of course have mutually inverse associativity isomorphisms.

Remark 1.5 Note that (1.33) can be written as

$$\mathcal{Y}_1(w_{(1)}, z_1)\mathcal{Y}_2(w_{(2)}, z_2) = \mathcal{Y}^1\big(\mathcal{Y}^2(w_{(1)}, z_1 - z_2)w_{(2)}, z_2\big), \tag{1.36}$$

with the appearance of the complex numbers being understood as substitutions in the sense mentioned above, and with the "generic" vectors $w_{(3)}$ and $w'_{(4)}$ being implicit. This (rigorous) equation amounts to the "operator product expansion" in the physics literature on conformal field theory; indeed, in our language, if we expand the right-hand side of (1.36) in powers of $z_1 - z_2$, we find that a product of intertwining maps is expressed as an expansion in powers of $z_1 - z_2$, with coefficients that are again intertwining maps, of the form $\mathcal{Y}^1(w, z_2)$. When all three modules are the vertex algebra itself, and all the intertwining operators are the canonical vertex operator $Y(\cdot, x)$ itself, this "operator product expansion" follows easily from the Jacobi identity. But for intertwining operators in general, it is a deep matter to prove the operator product expansion, that is, to prove the assertions involving (1.32) and (1.33) above. This was proved in [53] in the finitely reductive setting and is considerably generalized in the present work to the logarithmic setting.

Remark 1.6 The constructions of the tensor product modules and of the associativity isomorphisms previewed above for suitably general vertex algebras follow those in [72–74] and [53]. Alternative constructions are certainly possible. For example, an alternative construction of the tensor product modules was given in [100]. However, no matter what construction is used for the tensor product modules of suitably general vertex algebras, one cannot avoid constructing structures and proving results equivalent to what is carried out in this work. The constructions in this work of the tensor product functors and of the natural associativity isomorphisms are crucial in the deeper part of the theory of vertex tensor categories.

Remark 1.7 We have outlined the construction of the tensor product functors and the associativity isomorphisms without getting into the technical details. On the other hand, though the general ideas of the constructions are the same for both the semisimple theory developed in [72–74] and [53] and the nonsemisimple logarithmic theory carried out in the present work, many of the proofs of the results in the present work involve substantial new ideas and techniques, making the nonsemisimple logarithmic theory vastly more difficult technically than the semisimple theory.

First, we have had to further develop formal calculus beyond what had been developed in [37, 38, 53, 72–74, 99] and many other works. We have had to study new kinds of combinations of formal delta function expressions in several *formal and complex* variables. Second, we have extended formal calculus to include logarithms of formal variables. In formal calculus, logarithms of formal variables are in fact additional independent formal variables. We develop our "logarithmic formal calculus" in a much more general setting than what we need for the main results in this work. In particular, we at first allow the formal series in a formal variable and its logarithm to involve *infinitely many arbitrary complex powers* of the logarithm. This study of logarithmic formal calculus has surprising connections with various classes of combinatorial identities and has been extended and exploited by Robinson [129–131]. Third, to construct the natural associativity isomorphisms and other data for the tensor categories and to prove the coherence property, it is necessary to use complex analysis. We wanted to carry out our theory under the most general natural sets of assumptions that would indeed *yield* a theory. This required us to work with series involving *arbitrary real* powers of the complex variables, *with the powers not even being lower bounded*. We have in fact extended a number of classical results in complex analysis to results that can be applied to such series. In particular, we have had to prove many results that allow us to switch orders of infinite sums, by either proving the multiconvergence of the corresponding multisums or by using Taylor expansion for analytic functions. Fourth, since our theory also involves logarithms of complex variables, we have also had to extend those same classical results in complex analysis to results that can be applied still further to series involving logarithms of the complex variables. In particular, we prove that when the powers of the logarithm of a complex variable are bounded above in a series involving *arbitrary real powers* of the variable and nonnegative integral powers of its logarithm, the convergence of suitable *iterated* sums implies absolute convergence of the corresponding *double* sums. We also prove what we call the "unique expansion property" for the set $\mathbb{R} \times \{0, \dots, N\}$ (see Proposition 7.8), which says that the coefficients of an absolutely convergent series of the form just indicated are determined uniquely by its sum. One important difference from the logarithmic *formal* calculus is that when we use complex analysis, it is necessary for the powers of the logarithms to be bounded from above, essentially because a complex variable z can also be expressed as the sum of the series $z = \sum_{n\in\mathbb{N}} \frac{(\log z)^n}{n!}$. Fifth, we have had to combine our results on formal calculus, on logarithmic formal calculus, and on complex analysis for series with both arbitrary real powers and also logarithms to prove our main results on the construction of the tensor category structures. In many proofs, we encounter expressions involving both formal variables and complex variables, and thus we have had to develop new and delicate methods exploiting both the formal and complex analysis methods that we have just mentioned. The proofs, which are not short (and cannot be), accomplish the necessary interchanges of order of summations.

Remark 1.8 The operator product expansion and resulting braided tensor category structure constructed by the theory in [53, 72–74] were originally structures whose existence was conjectured: It was in their important study of conformal field theory that Moore and Seiberg [111, 112] first discovered a set of polynomial equations from a suitable axiom system for a "rational conformal field theory." Inspired by a comment of Witten, they observed an analogy between the theory of these polynomial equations and the theory of tensor categories. The structures given by these Moore-Seiberg equations were called "modular tensor categories" by I. Frenkel. However, in the work of Moore and Seiberg, as they commented, neither tensor product structure nor other related structures were either formulated or constructed mathematically. Later, Turaev formulated a precise notion of modular tensor category in [136] and [137] and gave examples of such tensor categories from representations of quantum groups at roots of unity, based on results obtained by many people on quantum groups and their representations, especially those in the pioneering work [127] and [128] by Reshetikhin and Turaev on the construction of knot and 3-manifold invariants from representations of quantum groups. On the other hand, on the "rational conformal field theory" side, a modular tensor category structure in this sense on certain module categories for affine Lie algebras, and much more generally, on certain module categories for "chiral algebras" associated with rational conformal field theories, was then believed to exist by both physicists and mathematicians, but such structure was not in fact constructed at that time. Moore and Seiberg observed the analogy mentioned above based on the assumption of the existence of a suitable tensor product functor (including a tensor product module) and derived their polynomial equations based on the assumption of the existence of a suitable operator product expansion for chiral vertex operators, which is essentially equivalent to assuming the associativity of intertwining maps, as we have expressed it above. As we have discussed, the desired tensor product modules and functors were constructed under suitable conditions in the series of papers [72, 73] and [74], and in [53] the appropriate natural associativity isomorphisms among tensor products of triples of modules were constructed, and it was shown that this is equivalent to the desired associativity of intertwining maps (and thus the existence of a suitable operator product expansion). In particular, this work [72–74] and [53] served to construct the desired braided tensor category structure in the generality of suitable vertex operator algebras, including those associated with affine Lie algebras and the Virasoro algebra as a very special case; see [75] and [54], respectively. (For a discussion of the remaining parts of the modular tensor category structure in this generality, see below and [66].) The results in these papers will be generalized in this work. In the special case of affine Lie algebras and also in the special case of Virasoro-algebraic structures, using the work of Tsuchiya-Ueno-Yamada [135] and Beilinson-Feigin-Mazur [13] combined with a formulation of braided tensor category structure by Deligne [18], one can obtain the braided tensor category structure discussed above (but not the modular tensor category structure).

1.5 Some Recent Applications and Related Literature

We begin with a discussion concerning the "rational" case, with semisimple module categories. We also refer the reader to the recent review by Fuchs, Runkel and

Schweigert [41] on rational conformal field theory, which also in fact briefly discusses nonrational conformal field theories, including in particular logarithmic conformal field theories.

After the important work [111] and [112] of Moore and Seiberg, it was widely believed that the category of modules for a suitable vertex operator algebra must have a structure of braided tensor category satisfying additional properties related to the modular invariance of the vertex operator algebra. As is mentioned in Remark 1.8, for a suitable vertex operator algebra, the work [72–74] and [53] constructed a structure of braided tensor category on the category of modules for the vertex operator algebra; see also [54] and [75]. On the other hand, the precise and conceptual formulation of the notion of modular tensor category by Turaev [136] led to a mathematical conjecture that the category of modules for a suitable vertex operator algebra can be endowed in a natural way with modular tensor category structure in this sense. It was in 2005 that this conjecture was finally proved by the first author in [64] (see also the announcement [61] and the exposition [62]). The hardest part of the proof of this conjecture was the proof of the rigidity property of the braided tensor category constructed in [72–74] and [53].

Even in the case of a vertex operator algebra associated to an affine Lie algebra or the Virasoro algebra, there was no proof of rigidity for the braided tensor category of modules in the literature, before the proof discovered in [64]. The works of Tsuchiya-Ueno-Yamada [135] and Beilinson-Feigin-Mazur [13] can be used to construct a structure of braided tensor category on the category of modules for such a vertex operator algebra, but neither the rigidity property nor the other main axiom for modular tensor category structure, called the nondegeneracy property, of these braided tensor categories has ever been proved using the results or methods in those works. Under the assumption that the braided tensor category structure on the category of integrable highest weight (standard) modules of a fixed positive integral level for an affine Lie algebra was already known to have the rigidity property, Finkelberg [28, 29] showed that this braided tensor category structure could be recovered by transporting to this category the corresponding rigid braided tensor category structure previously constructed for negative levels by Kazhdan and Lusztig [91–95]. But since the rigidity was an assumption needed in the proof, the work [28, 29] did not actually serve to give a construction of the braided tensor category structure at positive integral level. The book [12] asserted that one had a construction of the structure of modular tensor category on the category of modules for a vertex operator algebra associated to an affine Lie algebra at positive integral level, and while a construction of the structure of braided tensor category was indeed given, there was no proof of the rigidity property, so that even in the cases of affine Lie algebras and the Virasoro algebra, the construction of the corresponding modular tensor category structures was still an unsolved open problem before 2005.

Under the assumption of the rigidity for positive integral level, the work [28, 29] of Finkelberg combined with the work [91–95] of Kazhdan and Lusztig established the important equivalence between the braided tensor category of a semisimple subquotient of the category of modules for a quantum group at a root of unity and the braided tensor category of integrable highest weight modules of a positive integral level for an affine Lie algebra. The proof of the rigidity of the braided tensor category of integrable highest weight modules of a positive integral level for an affine Lie algebra, as a special case in [64], based on the braided tensor category structure constructed in [75], as a special case in [72–74] and [53], thus in fact provided the completion of the proof of the equivalence theorem that was the goal in [28] and [29] above. As we have mentioned, the only known proof of this rigidity requires the work [72–74] and [53], and in particular, in the affine Lie algebra case, uses the work [75].

The proof of the rigidity in [64] is highly nontrivial. The reason why the rigidity was so hard is that one needed to prove the Verlinde conjecture for suitable vertex operator algebras in order to prove the rigidity, and the Verlinde conjecture requires the consideration of genus-one as opposed to genus-zero conformal field theory. The nondegeneracy property of the modular tensor category also follows from the truth of the Verlinde conjecture. The Verlinde conjecture was discovered by E. Verlinde [138] in 1987, and as was demonstrated by Moore and Seiberg [111, 112] in 1988, the validity of the conjecture follows from their axiom system for a rational conformal field theory. However, the construction of rational conformal field theories is much harder than the construction of modular tensor categories, and this in turn requires the proof of the Verlinde conjecture without the assumption of the axioms for a rational conformal field theory. The Verlinde conjecture for suitable vertex operator algebras was proved in 2004 by the first author in [63] (without the assumption of the axioms for a rational conformal field theory), and its proof in turn depended on the aspects of the theory of intertwining operators (the genus-zero theory) developed in [59] and on the aspects of the theory of q-traces of products or iterates of intertwining operators and their modular invariance (the genus-one theory) developed in [60]. (These works in turn depended on [72–74] and [53].) The modular invariance theorem proved in the pioneering work [143, 144] of Zhu actually turned out to be only a very special case of the stronger necessary result proved in [60], and was far from enough for the purpose of establishing either the required rigidity property or the required nondegeneracy property of the modular tensor category structure. The paper [60] established the most general modular invariance result in the semisimple case and also constructed all genus-one correlation functions of the corresponding chiral rational conformal field theories. After Zhu's modular invariance was proved in 1990, the modular invariance for products or iterates of more than one intertwining operator was an open problem for a long time. In the case of products or iterates of at most one intertwining operator and any number of vertex operators for modules, a straightforward generalization of Zhu's result using his same method gives the modular invariance (see [107]). But for products or iterates of more than one intertwining operator, Zhu's method is not sufficient because the commutator formula that he used to derive his recurrence formula in his proof has no generalization for intertwining operators. This was one of the main reasons that for about 15 years after 1990, there had been not much progress toward the proof of the rigidity and nondegeneracy properties. In [60], this difficulty was overcome by means of a proof that q-traces of products or iterates of intertwining operators satisfy modular invariant differential equations with regular singular points; the need for a recurrence formula was thus bypassed.

We have been discussing the case of rational conformal field theories. The present work includes as a special case a complete treatment of the work [72–74] and [53], with much stronger results added as well; this work is required for the results that we have just discussed. The main theme of the present work being the logarithmic generalization of this theory, allowing categories of modules that are not completely reducible, we would now like to comment on some recent applications and related literature in the (much greater) logarithmic generality, and also, in this generality we are in addition able to replace vertex operator algebras by much more general vertex algebras equipped with a suitable additional grading by an abelian group. (Allowing logarithmic structures and allowing vertex algebras with a grading by an abelian group are "unrelated" generalizations of the context of [72–74] and [53]; in the present work we are able to carry out both generalizations simultaneously.)

The triplet $\mathcal{W}$-algebras $\mathcal{W}(1, p)$, mentioned above, are a class of vertex operator algebras of central charge $1 - 6\frac{(p-1)^2}{p}$ which in recent years have attracted a lot of attention from physicists and mathematicians. Introduced by Kausch [89], they have been studied extensively by Flohr [30, 31], Gaberdiel-Kausch [43, 45], Kausch [90], Fuchs-Hwang-Semikhatov-Tipunin [40], Abe [1], Feigin-Gaĭnutdinov-Semikhatov-Tipunin [26, 27], Carqueville-Flohr [17], Flohr-Gaberdiel [32], Fuchs [39], Adamović-Milas [6, 9, 11], Flohr-Grabow-Koehn [34], Flohr-Knuth [33], Gaberdiel-Runkel [46, 47], Gaĭnutdinov-Tipunin [49], Pearce-Rasmussen-Ruelle [116, 117], Nagatomo-Tsuchiya [114] and Rasmussen [122]. A triplet $\mathcal{W}$-algebra $V = \coprod_{n\in\mathbb{Z}} V_{(n)}$ satisfies the positive energy condition ($V_{(0)} = \mathbb{C}\mathbf{1}$ and $V_{(n)} = 0$ for $n < 0$) and the C_2-cofiniteness condition (the quotient space $V/C_2(V)$ is finite dimensional, where $C_2(V)$ is the subspace of V spanned by the elements of the form $u_{-2}v$ for $u, v \in V$). The C_2-cofiniteness condition was proved by Abe [1] in the simplest $p = 2$ case and by Carqueville-Flohr [17] and Adamović-Milas [6] in the general case.

In [65], the first author proved that for a vertex operator algebra V satisfying the positive energy condition and the C_2-cofiniteness condition, the category of grading-restricted generalized V-modules satisfies the assumptions needed to invoke the theory carried out in the present work. The present work, combined with [65] (for proving the assumptions of the present work), thus establishes the logarithmic operator product expansion and constructs the logarithmic tensor category theory for any vertex operator algebra satisfying the positive energy condition and the C_2-cofiniteness condition. For example, the logarithmic tensor products used heavily in the papers [108, 109] and [110] of Miyamoto are in fact constructed in the present work together with [65]. In particular, for a triplet $\mathcal{W}$-algebra V discussed above, the category of grading-restricted generalized V-modules indeed has the natural braided tensor category structure constructed in the present work. Many of the assertions involving a logarithmic operator product expansion and a logarithmic tensor category theory in the works on triplet $\mathcal{W}$-algebras mentioned above are mathematically formulated and established in the present work together with the paper [65], so that now, they do not have to be taken as unproved assumptions in those works.

Based on the results of Feigin-Gaĭnutdinov-Semikhatov-Tipunin [26] and of Fuchs-Hwang-Semikhatov-Tipunin [40], Feigin, Gaĭnutdinov, Semikhatov and

Tipunin conjectured [27] an equivalence between the braided finite tensor category of grading-restricted generalized modules for a triplet $\mathcal{W}$-algebra and the braided finite tensor category of suitable modules for a restricted quantum group. Their formulation of the conjecture also includes the statement that the categories of grading-restricted generalized modules for the triplet $\mathcal{W}$-algebras considered in their paper are indeed braided tensor categories. Assuming the existence of the braided tensor category structure on the triplet $\mathcal{W}$-algebra with $p = 2$, Feigin, Gaĭnutdinov, Semikhatov and Tipunin gave a proof of their conjecture. However, in the case $p \neq 2$, Kondo and Saito [97] showed that the tensor category of modules for the corresponding restricted quantum group is not braided. Thus, the conjecture in the case $p \neq 2$ cannot be true as it is stated, although the equivalence between the abelian categories was proved in [114] for all p. It is believed that the correct formulation of the conjecture and the proof will be possible only after the conformal-field-theoretic aspects of the representations of triplet $\mathcal{W}$-algebras are studied thoroughly. As we mentioned above, the present work, the paper [65] and the papers [1, 17] and [6] provide a proof of the assumption in their conjecture that the categories of grading-restricted generalized modules for the triplet $\mathcal{W}$-algebras are indeed braided tensor categories. We expect that further studies of the tensor-categorical structures and conformal-field-theoretic properties for triplet $\mathcal{W}$-algebras will provide a correct formulation and proof of suitable equivalence between categories of suitable modules for triplet $\mathcal{W}$-algebras and for restricted quantum groups.

In [67], the first author introduced a notion of generalized twisted module associated to a general automorphism of a vertex operator algebra, including an automorphism of infinite order. The first author in [67] also gave a construction of such generalized twisted modules associated to the automorphisms obtained by exponentiating weight 1 elements of the vertex operator algebra. If the automorphism of the vertex operator algebra does not act semisimply, the twisted vertex operators for these generalized twisted modules must involve the logarithm of a formal or complex variable, and we need additional $\mathbb{C}/\mathbb{Z}$- or $\mathbb{C}$-gradings on these generalized twisted modules. As was noticed by Milas, the triplet $\mathcal{W}$-algebras are fixed-point subalgebras of suitable vertex operator algebras constructed from a one-dimensional lattice under an automorphism obtained by exponentiating a weight 1 element. In particular, some logarithmic intertwining operators constructed in [7] are in fact twisted vertex operators. Thus the paper [67] provided an orbifold approach to the representation theory of triplet $\mathcal{W}$-algebras. (This orbifold point of view is one of the analogues of the orbifold point of view for vertex operator algebras introduced in [37].) Since the automorphisms involved indeed do not act on the vertex operator algebra semisimply, the twisted vertex operators for the generalized twisted modules associated to these automorphisms must involve the logarithm of the variables, and we also need additional $\mathbb{C}/\mathbb{Z}$- or $\mathbb{C}$-gradings on these generalized twisted modules. Here $\mathbb{C}/\mathbb{Z}$ or $\mathbb{C}$ are instances of the additional grading abelian group in the present work. Thus we need the general framework and results in the present work, including both the logarithmic generality and also the additional abelian-group gradings, for the study of these generalized twisted modules.

Many of the results on the representation theory of triplet $\mathcal{W}$-algebras have also been generalized to the more general case of $\mathcal{W}(p, q)$-algebras [25] of central charge $1 - 6\frac{(p-q)^2}{pq}$, $q > p > 0$ coprime (see for example [5, 10, 48, 119–121, 123, 125, 134] and [140]), and to $N = 1$ triplet vertex operator superalgebras (see [8, 9] and [11]). Results have also been obtained for the vertex operator subalgebras of the algebras $\mathcal{W}(p, q)$ generated by the Virasoro algebra (see [16, 24, 44, 103, 115, 124, 126] and [50]). The C_2-cofiniteness of the $\mathcal{W}(2, q)$-algebras has been proved by Adamović and Milas in [10]. Thus using the results obtained in [67], the theory developed in the present work applies to these $\mathcal{W}(2, q)$-algebras, yielding braided tensor categories. The $N = 1$ triplet vertex operator superalgebras introduced by Adamović and Milas in [8] are also proved by these authors in [9] to be C_2-cofinite. As was mentioned above in Sect. 1.2, the theory developed in this work also applies to vertex superalgebras. The same remarks apply to the results in [67]. Thus the theory developed in the present work applies to these $N = 1$ triplet vertex operator superalgebras, producing the corresponding braided tensor categories.

Finally, we would like to emphasize that it is interesting that the methods developed and used in the present work, even in the special case of categories of modules for an affine Lie algebra at negative levels, are very different from those developed and used by Kazhdan and Lusztig in [91–95], and are much more general. The methods used in [91–95], closely related to algebraic geometry, depended heavily on the Knizhnik-Zamolodchikov equations. In the present work, we use and develop the general theory of vertex (operator) algebras (and generalizations), requiring both formal calculus theory and complex analysis, and we do not use algebraic geometry. Also, in the present work and in the work [141] and [142], which verified the assumptions needed for the application of the present theory, although we need to show that products of intertwining operators satisfy certain differential equations with regular singular points, no explicit form of the equations, such as the explicit form of the Knizhnik-Zamolodchikov equations, is needed. In fact, because for a general vertex (operator) algebra satisfying those assumptions in the present work or in [65] no explicit form of the differential equations such as the form of the Knizhnik-Zamolodchikov equations exists, it was crucial that in the work [53, 59, 72–74], the present work and [65], we have developed methods that are independent of the explicit form of the differential equations. Another interesting difference between the present general theory and this work of Kazhdan and Lusztig is that logarithmic structures (necessarily) pervade our theory, starting from the vertex-algebraic foundations, while the logarithmic nature of solutions of the Knizhnik-Zamolodchikov equations involved in [91–95] did not have to be emphasized there.

1.6 Main Results of the Present Work

In this section, we state the main results of the present work, numbered as in the main text. The reader is referred to the relevant sections for definitions, notations and details.

Let A be an abelian group and $\tilde{A}$ an abelian group containing A as a subgroup. Let V be a strongly A-graded Möbius or conformal vertex algebra, as defined in

Sect. 2. Let $\mathcal{C}$ be a full subcategory of the category $\mathcal{M}_{sg}$ of strongly $\tilde{A}$-graded (ordinary) V-modules or the category $\mathcal{GM}_{sg}$ of strongly $\tilde{A}$-graded generalized V-modules, closed under the contragredient functor and under taking finite direct sums; see Sect. 2 and Assumptions 4.1 and 5.30.

In Sect. 4, the notions of $P(z)$- and $Q(z)$-tensor product functor are defined in terms of $P(z)$- and $Q(z)$-intertwining maps and $P(z)$- and $Q(z)$-products; intertwining maps are related to logarithmic intertwining operators, defined and studied in Sect. 3. The symbols $P(z)$ and $Q(z)$ refer to the moduli space elements described in Remarks 4.3 and 4.37, respectively. In Sect. 5, we give a construction of the $P(z)$-tensor product of two objects of $\mathcal{C}$, when this structure exists. For $W_1, W_2 \in \operatorname{ob}\mathcal{C}$, define the subset

$$W_1 \boxbackslash_{P(z)} W_2 \subset (W_1 \otimes W_2)^*$$

of $(W_1 \otimes W_2)^*$ to be the union, or equivalently, the sum, of the images

$$I'(W') \subset (W_1 \otimes W_2)^*$$

as $(W; I)$ ranges through all the $P(z)$-products of W_1 and W_2 with $W \in \operatorname{ob}\mathcal{C}$, where I' is a map corresponding naturally to the $P(z)$-intertwining map I and where W' is the contragredient (generalized) module of W.

The following two results give the construction of the $P(z)$-tensor product:

Proposition 5.37 *Let $W_1, W_2 \in \operatorname{ob}\mathcal{C}$. If $(W_1 \boxbackslash_{P(z)} W_2, Y'_{P(z)})$ is an object of $\mathcal{C}$ (where $Y'_{P(z)}$ is the natural action of V), denote by $(W_1 \boxtimes_{P(z)} W_2, Y_{P(z)})$ its contragredient (generalized) module:*

$$W_1 \boxtimes_{P(z)} W_2 = (W_1 \boxbackslash_{P(z)} W_2)'.$$

Then the $P(z)$-tensor product of W_1 and W_2 in $\mathcal{C}$ exists and is

$$\left(W_1 \boxtimes_{P(z)} W_2, Y_{P(z)}; i'\right),$$

where i is the natural inclusion from $W_1 \boxbackslash_{P(z)} W_2$ to $(W_1 \otimes W_2)^*$. Conversely, let us assume that $\mathcal{C}$ is closed under images. If the $P(z)$-tensor product of W_1 and W_2 in $\mathcal{C}$ exists, then $(W_1 \boxbackslash_{P(z)} W_2, Y'_{P(z)})$ is an object of $\mathcal{C}$.

For

$$\lambda \in (W_1 \otimes W_2)^*,$$

let W_λ be the smallest doubly graded subspace of $((W_1 \otimes W_2)^*)_{[\mathbb{C}]}^{(\tilde{A})}$ (the direct sum of the homogeneous subspaces with respect to the gradings both by conformal generalized weights and by $\tilde{A}$) containing λ and stable under the component operators of the operators $Y'_{P(z)}(v, x)$ for $v \in V$, $m \in \mathbb{Z}$, and under the operators $L'_{P(z)}(-1)$, $L'_{P(z)}(0)$ and $L'_{P(z)}(1)$ (to handle the Möbius but non-conformal case). Let

$$\mathrm{COMP}_{P(z)}\big((W_1 \otimes W_2)^*\big),$$
$$\mathrm{LGR}_{[\mathbb{C}];\,P(z)}\big((W_1 \otimes W_2)^*\big)$$

and

$$\mathrm{LGR}_{(\mathbb{C});\,P(z)}\big((W_1 \otimes W_2)^*\big)$$

be the spaces of elements of $(W_1 \otimes W_2)^*$ satisfying the $P(z)$-compatibility condition, the $P(z)$-local grading restriction condition and the $L(0)$-semisimple $P(z)$-local grading restriction condition, respectively, as defined in Sect. 4; the subscript $(\mathbb{C})$ refers to the semisimplicity of the action of $L(0)$ in this case, so that generalized weights are weights.

Theorem 5.50 *Suppose that for every element*

$$\lambda \in \mathrm{COMP}_{P(z)}\big((W_1 \otimes W_2)^*\big) \cap \mathrm{LGR}_{[\mathbb{C}];\,P(z)}\big((W_1 \otimes W_2)^*\big)$$

the space W_λ (which is a (strongly-graded) generalized module) is a generalized submodule of some object of $\mathcal{C}$ included in $(W_1 \otimes W_2)^*$ (this holds vacuously if $\mathcal{C} = \mathcal{GM}_{sg}$). Then

$$W_1 \boxtimes_{P(z)} W_2 = \mathrm{COMP}_{P(z)}\big((W_1 \otimes W_2)^*\big) \cap \mathrm{LGR}_{[\mathbb{C}];\,P(z)}\big((W_1 \otimes W_2)^*\big).$$

Suppose that $\mathcal{C}$ is a category of strongly-graded V-modules (that is, $\mathcal{C} \subset \mathcal{M}_{sg}$) and that for every element

$$\lambda \in \mathrm{COMP}_{P(z)}\big((W_1 \otimes W_2)^*\big) \cap \mathrm{LGR}_{(\mathbb{C});\,P(z)}\big((W_1 \otimes W_2)^*\big)$$

the space W_λ (which is a (strongly-graded) V-module) is a submodule of some object of $\mathcal{C}$ included in $(W_1 \otimes W_2)^*$ (which holds vacuously if $\mathcal{C} = \mathcal{M}_{sg}$). Then

$$W_1 \boxtimes_{P(z)} W_2 = \mathrm{COMP}_{P(z)}\big((W_1 \otimes W_2)^*\big) \cap \mathrm{LGR}_{(\mathbb{C});\,P(z)}\big((W_1 \otimes W_2)^*\big).$$

The hard parts of the proof of Theorem 5.50 are given in Sect. 6.

We also give an analogous construction of $Q(z)$-tensor products in these sections.

For the construction of the natural associativity isomorphism between suitable pairs of triple tensor product functors, we assume that for any object of $\mathcal{C}$, all the (generalized) weights are real numbers and in addition there exists $K \in \mathbb{Z}_+$ such that

$$\big(L(0) - L(0)_s\big)^K = 0$$

on the module, $L(0)_s$ being the semisimple part of $L(0)$ (the latter condition holding vacuously when $\mathcal{C}$ is in $\mathcal{M}_{sg}$); see Assumption 7.11.

The main hard parts of the construction of the associativity isomorphisms are presented in Sect. 9, after necessary preparation in Sect. 8. To discuss these results, we need the important $P^{(1)}(z)$- and $P^{(2)}(z)$-local grading restriction conditions on

$$\lambda \in (W_1 \otimes W_2 \otimes W_3)^*$$

(where W_1, W_2, and W_3 are objects of $\mathcal{C}$) and their $L(0)$-semisimple versions. Here we state the (two-part) $P^{(2)}(z)$-local grading restriction condition, the other conditions being analogous:

The $P^{(2)}(z)$-Local Grading Restriction Condition

(a) The $P^{(2)}(z)$-*grading condition*: For any $w_{(3)} \in W_3$, there exists a formal series $\sum_{n\in\mathbb{R}} \lambda_n^{(2)}$ with

$$\lambda_n^{(2)} \in \coprod_{\beta\in\tilde{A}} \left((W_1 \otimes W_2)^*\right)_{[n]}^{(\beta)}$$

for $n \in \mathbb{R}$, an open neighborhood of $z' = 0$, and $N \in \mathbb{N}$ such that for $w_{(1)} \in W_1$ and $w_{(2)} \in W_2$, the series

$$\sum_{n\in\mathbb{R}} \left(e^{z' L'_{P(z)}(0)} \lambda_n^{(2)}\right)(w_{(1)} \otimes w_{(2)})$$

has the following properties:

(i) It can be written as the iterated series

$$\sum_{n\in\mathbb{R}} e^{nz'} \left(\left(\sum_{i=0}^{N} \frac{(z')^i}{i!} \left(L'_{P(z)}(0) - n\right)^i \lambda_n^{(2)}\right)(w_{(1)} \otimes w_{(2)})\right).$$

(ii) It is absolutely convergent for $z' \in \mathbb{C}$ in the neighborhood of $z' = 0$ above.

(iii) It is absolutely convergent to $\mu^{(2)}_{\lambda, w_{(3)}}(w_{(1)} \otimes w_{(2)})$ when $z' = 0$:

$$\sum_{n\in\mathbb{R}} \lambda_n^{(2)}(w_{(1)} \otimes w_{(2)}) = \mu^{(2)}_{\lambda, w_{(3)}}(w_{(1)} \otimes w_{(2)}) = \lambda(w_{(1)} \otimes w_{(2)} \otimes w_{(3)})$$

(the last equality being the definition of $\mu^{(2)}_{\lambda, w_{(3)}}$).

(b) For any $w_{(3)} \in W_3$, let $W^{(2)}_{\lambda, w_{(3)}}$ be the smallest doubly graded subspace of $((W_1 \otimes W_2)^*)_{[\mathbb{R}]}^{(\tilde{A})}$ containing all the terms $\lambda_n^{(2)}$ in the formal series in (a) and stable under the component operators of the operators $Y'_{P(z)}(v, x)$ for $v \in V$, $m \in \mathbb{Z}$, and under the operators $L'_{P(z)}(-1)$, $L'_{P(z)}(0)$ and $L'_{P(z)}(1)$. Then $W^{(2)}_{\lambda, w_{(3)}}$ has the properties

$$\dim\left(W^{(2)}_{\lambda, w_{(3)}}\right)_{[n]}^{(\beta)} < \infty,$$

$$\left(W^{(2)}_{\lambda, w_{(3)}}\right)_{[n+k]}^{(\beta)} = 0 \quad \text{for } k \in \mathbb{Z} \text{ sufficiently negative}$$

for any $n \in \mathbb{R}$ and $\beta \in \tilde{A}$, where the subscripts denote the $\mathbb{R}$-grading by $L'_{P(z)}(0)$-(generalized) eigenvalues and the superscripts denote the $\tilde{A}$-grading.

The following result gives, among other things, the deep fact that when λ is obtained from a suitable product of intertwining maps, the elements $\lambda_n^{(2)}$ for $n \in \mathbb{R}$ in the assumed $P^{(2)}(z)$-local grading restriction condition for suitable $z \in \mathbb{C}^\times$ satisfy the $P(z)$-compatibility condition:

Theorem 9.17 *Assume that the convergence condition for intertwining maps in $\mathcal{C}$ (see Sect.* 7*) holds and that*

$$|z_1| > |z_2| > |z_1 - z_2| > 0.$$

Let W_1, W_2, W_3, W_4, M_1 and M_2 be objects of $\mathcal{C}$ and let I_1, I_2, I^1 and I^2 be $P(z_1)$-, $P(z_2)$-, $P(z_2)$- and $P(z_1 - z_2)$-intertwining maps of types $\binom{W_4}{W_1 M_1}$, $\binom{M_1}{W_2 W_3}$, $\binom{W_4}{M_2 W_3}$ and $\binom{M_2}{W_1 W_2}$, respectively. Let $w'_{(4)} \in W'_4$.

1. *Suppose that $(I_1 \circ (1_{W_1} \otimes I_2))'(w'_{(4)})$ satisfies Part* (a) *of the $P^{(2)}(z_1 - z_2)$-local grading restriction condition, that is, the $P^{(2)}(z_1 - z_2)$-grading condition (or the $L(0)$-semisimple $P^{(2)}(z_1 - z_2)$-grading condition when $\mathcal{C}$ is in $\mathcal{M}_{sg}$). For any $w_{(3)} \in W_3$, let $\sum_{n\in\mathbb{R}} \lambda_n^{(2)}$ be a series weakly absolutely convergent to*

$$\mu^{(2)}_{(I_1\circ(1_{W_1}\otimes I_2))'(w'_{(4)}),w_{(3)}} \in (W_1 \otimes W_2)^*$$

as indicated in the $P^{(2)}(z_1 - z_2)$-grading condition (or the $L(0)$-semisimple $P^{(2)}(z_1 - z_2)$-grading condition), and suppose in addition that the elements $\lambda_n^{(2)} \in (W_1 \otimes W_2)^*$, $n \in \mathbb{R}$, satisfy the $P(z_1 - z_2)$-lower truncation condition (Part (a) *of the $P(z_1 - z_2)$-compatibility condition in Sect.* 5*). Then each $\lambda_n^{(2)}$ satisfies the (full) $P(z_1 - z_2)$-compatibility condition. Moreover, the corresponding space*

$$W^{(2)}_{(I_1\circ(1_{W_1}\otimes I_2))'(w'_{(4)}),w_{(3)}} \subset (W_1 \otimes W_2)^*,$$

equipped with the vertex operator map given by $Y'_{P(z_1-z_2)}$ and the operators $L'_{P(z_1-z_2)}(j)$ for $j = -1, 0, 1$, is a doubly-graded generalized V-module, and when $\mathcal{C}$ is in $\mathcal{M}_{sg}$, a doubly-graded V-module. In particular, if $(I_1 \circ (1_{W_1} \otimes I_2))'(w'_{(4)})$ satisfies the full $P^{(2)}(z_1 - z_2)$-local grading restriction condition (or the $L(0)$-semisimple $P^{(2)}(z_1 - z_2)$-local grading restriction condition when $\mathcal{C}$ is in $\mathcal{M}_{sg}$), then $W^{(2)}_{(I_1\circ(1_{W_1}\otimes I_2))'(w'_{(4)}),w_{(3)}}$ is an object of $\mathcal{GM}_{sg}$ (or $\mathcal{M}_{sg}$ when $\mathcal{C}$ is in $\mathcal{M}_{sg}$); in this case, the assumption that each $\lambda_n^{(2)}$ satisfies the $P(z_1 - z_2)$-lower truncation condition is redundant.
2. *Analogously, suppose that $(I^1 \circ (I^2 \otimes 1_{W_3}))'(w'_{(4)})$ satisfies Part* (a) *of the $P^{(1)}(z_2)$-local grading restriction condition, that is, the $P^{(1)}(z_2)$-grading condition (or the $L(0)$-semisimple $P^{(1)}(z_2)$-grading condition when $\mathcal{C}$ is in $\mathcal{M}_{sg}$).*

For any $w_{(1)} \in W_1$, *let* $\sum_{n\in\mathbb{R}} \lambda_n^{(1)}$ *be a series weakly absolutely convergent to*

$$\mu^{(1)}_{(I^1\circ(I^2\otimes 1_{W_3}))'(w'_{(4)}),w_{(1)}} \in (W_2 \otimes W_3)^*$$

as indicated in the $P^{(1)}(z_2)$*-grading condition (or the* $L(0)$*-semisimple* $P^{(1)}(z_2)$*-grading condition), and suppose in addition that the elements* $\lambda_n^{(1)} \in (W_2 \otimes W_3)^*$, $n \in \mathbb{R}$, *satisfy the* $P(z_2)$*-lower truncation condition (Part* (a) *of the* $P(z_2)$*-compatibility condition). Then each* $\lambda_n^{(1)}$ *satisfies the (full)* $P(z_2)$*-compatibility condition. Moreover, the corresponding space*

$$W^{(1)}_{(I^1\circ(I^2\otimes 1_{W_3}))'(w'_{(4)}),w_{(1)}} \subset (W_2 \otimes W_3)^*,$$

equipped with the vertex operator map given by $Y'_{P(z_2)}$ *and the operators* $L'_{P(z_2)}(j)$ *for* $j = -1, 0, 1$, *is a doubly-graded generalized* V*-module, and when* $\mathcal{C}$ *is in* $\mathcal{M}_{sg}$, *a doubly-graded* V*-module. In particular, if* $(I^1 \circ (I^2 \otimes 1_{W_3}))'(w'_{(4)})$ *satisfies the full* $P^{(1)}(z_2)$*-local grading restriction condition (or the* $L(0)$*-semisimple* $P^{(1)}(z_2)$*-local grading restriction condition when* $\mathcal{C}$ *is in* $\mathcal{M}_{sg}$*), then* $W^{(1)}_{(I^1\circ(I^2\otimes 1_{W_3}))'(w'_{(4)}),w_{(1)}}$ *is an object of* $\mathcal{GM}_{sg}$ *(or* $\mathcal{M}_{sg}$ *when* $\mathcal{C}$ *is in* $\mathcal{M}_{sg}$*); in this case, the assumption that each* $\lambda_n^{(1)}$ *satisfies the* $P(z_2)$*-lower truncation condition is redundant.*

The following result, based heavily on the previous theorem, establishes the associativity of intertwining maps:

Theorem 9.23 *Assume that* $\mathcal{C}$ *is closed under images, that the convergence condition for intertwining maps in* $\mathcal{C}$ *holds and that*

$$|z_1| > |z_2| > |z_1 - z_2| > 0.$$

Let W_1, W_2, W_3, W_4, M_1 *and* M_2 *be objects of* $\mathcal{C}$. *Assume also that* $W_1 \boxtimes_{P(z_1-z_2)} W_2$ *and* $W_2 \boxtimes_{P(z_2)} W_3$ *exist in* $\mathcal{C}$.

1. *Let* I_1 *and* I_2 *be* $P(z_1)$*- and* $P(z_2)$*-intertwining maps of types* $\binom{W_4}{W_1 M_1}$ *and* $\binom{M_1}{W_2 W_3}$, *respectively. Suppose that for each* $w'_{(4)} \in W'_4$,

$$\lambda = \big(I_1 \circ (1_{W_1} \otimes I_2)\big)'\big(w'_{(4)}\big) \in (W_1 \otimes W_2 \otimes W_3)^*$$

satisfies the $P^{(2)}(z_1 - z_2)$*-local grading restriction condition (or the* $L(0)$*-semisimple* $P^{(2)}(z_1 - z_2)$*-local grading restriction condition when* $\mathcal{C}$ *is in* $\mathcal{M}_{sg}$*). For* $w'_{(4)} \in W'_4$ *and* $w_{(3)} \in W_3$, *let* $\sum_{n\in\mathbb{R}} \lambda_n^{(2)}$ *be the (unique) series weakly absolutely convergent to* $\mu^{(2)}_{\lambda, w_{(3)}}$ *as indicated in the* $P^{(2)}(z_1 - z_2)$*-grading condition (or the* $L(0)$*-semisimple* $P^{(2)}(z_1 - z_2)$*-grading condition). Suppose also that for* *each* $n \in \mathbb{R}$, $w'_{(4)} \in W'_4$ *and* $w_{(3)} \in W_3$, *the generalized* V*-submodule of* $W^{(2)}_{\lambda, w_{(3)}}$ *generated by* $\lambda^{(2)}_n$ *is a generalized* V*-submodule of some object of* $\mathcal{C}$ *included in* $(W_1 \otimes W_2)^*$. *Then the product*

$$I_1 \circ (1_{W_1} \otimes I_2)$$

can be expressed as an iterate, and in fact, there exists a unique $P(z_2)$*-intertwining map* I^1 *of type* $\binom{W_4}{W_1 \boxtimes_{P(z_1-z_2)} W_2 \; W_3}$ *such that*

$$\langle w'_{(4)}, I_1(w_{(1)} \otimes I_2(w_{(2)} \otimes w_{(3)}))\rangle = \langle w'_{(4)}, I^1((w_{(1)} \boxtimes_{P(z_1-z_2)} w_{(2)}) \otimes w_{(3)})\rangle$$

for all $w_{(1)} \in W_1$, $w_{(2)} \in W_2$, $w_{(3)} \in W_3$ *and* $w'_{(4)} \in W'_4$.

2. *Analogously, let* I^1 *and* I^2 *be* $P(z_2)$*- and* $P(z_1 - z_2)$*-intertwining maps of types* $\binom{W_4}{M_2 W_3}$ *and* $\binom{M_2}{W_1 W_2}$, *respectively. Suppose that for each* $w'_{(4)} \in W'_4$,

$$\lambda = (I^1 \circ (I^2 \otimes 1_{W_3}))'(w'_{(4)}) \in (W_1 \otimes W_2 \otimes W_3)^*$$

satisfies the $P^{(1)}(z_2)$*-local grading restriction condition (or the* $L(0)$*-semisimple* $P^{(1)}(z_2)$*-local grading restriction condition when* $\mathcal{C}$ *is in* $\mathcal{M}_{sg}$*). For* $w'_{(4)} \in W'_4$ *and* $w_{(1)} \in W_1$, *let* $\sum_{n\in\mathbb{R}} \lambda^{(1)}_n$ *be the (unique) series weakly absolutely convergent to* $\mu^{(1)}_{\lambda, w_{(1)}}$ *as indicated in the* $P^{(1)}(z_2)$*-grading condition (or the* $L(0)$*-semisimple* $P^{(1)}(z_2)$*-grading condition). Suppose also that for each* $n \in \mathbb{R}$, $w'_{(4)} \in W'_4$ *and* $w_{(1)} \in W_1$, *the generalized* V*-submodule of* $W^{(1)}_{\lambda, w_{(1)}}$ *generated by* $\lambda^{(1)}_n$ *is a generalized* V*-submodule of some object of* $\mathcal{C}$ *included in* $(W_2 \otimes W_3)^*$. *Then the iterate*

$$I^1 \circ (I^2 \otimes 1_{W_3})$$

can be expressed as a product, and in fact, there exists a unique $P(z_1)$*-intertwining map* I_1 *of type* $\binom{W_4}{W_1 \; W_2 \boxtimes_{P(z_2)} W_3}$ *such that*

$$\langle w'_{(4)}, I^1(I^2(w_{(1)} \otimes w_{(2)}) \otimes w_{(3)})\rangle = \langle w'_{(4)}, I_1(w_{(1)} \otimes (w_{(2)} \boxtimes_{P(z_2)} w_{(3)}))\rangle$$

for all $w_{(1)} \in W_1$, $w_{(2)} \in W_2$, $w_{(3)} \in W_3$ *and* $w'_{(4)} \in W'_4$.

The hard part of the proof of this theorem is the proof of Lemma 9.22. This associativity of intertwining maps immediately gives the following important associativity of logarithmic intertwining operators, which is a strong version of logarithmic operator product expansion:

Corollary 9.24 *Assume that* $\mathcal{C}$ *is closed under images, that the convergence condition for intertwining maps in* $\mathcal{C}$ *holds and that*

$$|z_1| > |z_2| > |z_1 - z_2| > 0.$$

Let W_1, W_2, W_3, W_4, M_1 and M_2 be objects of $\mathcal{C}$. Assume also that $W_1 \boxtimes_{P(z_1-z_2)} W_2$ and $W_2 \boxtimes_{P(z_2)} W_3$ exist in $\mathcal{C}$.

1. *Let $\mathcal{Y}_1$ and $\mathcal{Y}_2$ be logarithmic intertwining operators (ordinary intertwining operators in the case that $\mathcal{C}$ is in $\mathcal{M}_{sg}$) of types $\binom{W_4}{W_1\,M_1}$ and $\binom{M_1}{W_2\,W_3}$, respectively. Suppose that for each $w'_{(4)} \in W'_4$, the element $\lambda \in (W_1 \otimes W_2 \otimes W_3)^*$ given by*

$$\lambda(w_{(1)} \otimes w_{(2)} \otimes w_{(3)}) = \langle w'_{(4)}, \mathcal{Y}_1(w_{(1)}, x_1)\mathcal{Y}_2(w_{(2)}, x_2)w_{(3)}\rangle\big|_{x_1=z_1,x_2=z_2}$$

for $w_{(1)} \in W_1$, $w_{(2)} \in W_2$ and $w_{(3)} \in W_3$ satisfies the $P^{(2)}(z_1 - z_2)$-local grading restriction condition (or the $L(0)$-semisimple $P^{(2)}(z_1 - z_2)$-local grading restriction condition when $\mathcal{C}$ is in $\mathcal{M}_{sg}$). For $w'_{(4)} \in W'_4$ and $w_{(3)} \in W_3$, let $\sum_{n\in\mathbb{R}} \lambda_n^{(2)}$ be the (unique) series weakly absolutely convergent to $\mu^{(2)}_{\lambda,w_{(3)}}$ as indicated in the $P^{(2)}(z_1 - z_2)$-grading condition (or the $L(0)$-semisimple $P^{(2)}(z_1 - z_2)$-grading condition). Suppose also that for each $n \in \mathbb{R}$, $w'_{(4)} \in W'_4$ and $w_{(3)} \in W_3$, the generalized V-submodule of $W^{(2)}_{\lambda,w_{(3)}}$ generated by $\lambda_n^{(2)}$ is a generalized V-submodule of some object of $\mathcal{C}$ included in $(W_1 \otimes W_2)^*$. Then there exists a unique logarithmic intertwining operator (a unique ordinary intertwining operator in the case that $\mathcal{C}$ is in $\mathcal{M}_{sg}$) $\mathcal{Y}^1$ of type $\binom{W_4}{W_1\boxtimes_{P(z_1-z_2)} W_2\;\; W_3}$ such that

$$\begin{aligned}&\langle w'_{(4)}, \mathcal{Y}_1(w_{(1)}, x_1)\mathcal{Y}_2(w_{(2)}, x_2)w_{(3)}\rangle\big|_{x_1=z_1,x_2=z_2}\\ &\quad= \langle w'_{(4)}, \mathcal{Y}^1(\mathcal{Y}_{\boxtimes_{P(z_1-z_2)},0}(w_{(1)}, x_0)w_{(2)}, x_2)w_{(3)}\rangle\big|_{x_0=z_1-z_2,x_2=z_2}\end{aligned}$$

for all $w_{(1)} \in W_1$, $w_{(2)} \in W_2$, $w_{(3)} \in W_3$ and $w'_{(4)} \in W'_4$. In particular, the product of the logarithmic intertwining operators (ordinary intertwining operators in the case that $\mathcal{C}$ is in $\mathcal{M}_{sg}$) $\mathcal{Y}_1$ and $\mathcal{Y}_2$ evaluated at z_1 and z_2, respectively, can be expressed as an iterate (with the intermediate generalized V-module $W_1 \boxtimes_{P(z_1-z_2)} W_2$) of logarithmic intertwining operators (ordinary intertwining operators in the case that $\mathcal{C}$ is in $\mathcal{M}_{sg}$) evaluated at z_2 and $z_1 - z_2$.
2. *Analogously, let $\mathcal{Y}^1$ and $\mathcal{Y}^2$ be logarithmic intertwining operators (ordinary intertwining operators in the case that $\mathcal{C}$ is in $\mathcal{M}_{sg}$) of types $\binom{W_4}{M_2\,W_3}$ and $\binom{M_2}{W_1\,W_2}$, respectively. Suppose that for each $w'_{(4)} \in W'_4$, the element $\lambda \in (W_1 \otimes W_2 \otimes W_3)^*$ given by*

$$\lambda(w_{(1)} \otimes w_{(2)} \otimes w_{(3)}) = \langle w'_{(4)}, \mathcal{Y}^1(\mathcal{Y}^2(w_{(1)}, x_0)w_{(2)}, x_2)w_{(3)}\rangle\big|_{x_0=z_1-z_2,x_2=z_2}$$

satisfies the $P^{(1)}(z_2)$-local grading restriction condition (or the $L(0)$-semisimple $P^{(1)}(z_2)$-local grading restriction condition when $\mathcal{C}$ is in $\mathcal{M}_{sg}$). For $w'_{(4)} \in W'_4$ and $w_{(1)} \in W_1$, let $\sum_{n\in\mathbb{R}} \lambda_n^{(1)}$ be the (unique) series weakly absolutely convergent to $\mu^{(1)}_{\lambda,w_{(1)}}$ as indicated in the $P^{(1)}(z_2)$-grading condition (or the $L(0)$-semisimple $P^{(1)}(z_2)$-grading condition). Suppose also that for each $n \in \mathbb{R}$,

$w'_{(4)} \in W'_4$ and $w_{(1)} \in W_1$, the generalized V-submodule of $W^{(1)}_{\lambda, w_{(1)}}$ generated by $\lambda^{(1)}_n$ is a generalized V-submodule of some object of $\mathcal{C}$ included in $(W_2 \otimes W_3)^*$. Then there exists a unique logarithmic intertwining operator (a unique ordinary intertwining operator in the case that $\mathcal{C}$ is in $\mathcal{M}_{sg}$) $\mathcal{Y}_1$ of type $\binom{W_4}{W_1\ W_2 \boxtimes_{P(z_2)} W_3}$ such that

$$\langle w'_{(4)}, \mathcal{Y}^1(\mathcal{Y}^2(w_{(1)}, x_0)w_{(2)}, x_2)w_{(3)}\rangle\big|_{x_0=z_1-z_2, x_2=z_2}$$
$$= \langle w'_{(4)}, \mathcal{Y}_1(w_{(1)}, x_1)\mathcal{Y}_{\boxtimes_{P(z_2)},0}(w_{(2)}, x_2)w_{(3)}\rangle\big|_{x_1=z_1, x_2=z_2}$$

for all $w_{(1)} \in W_1$, $w_{(2)} \in W_2$, $w_{(3)} \in W_3$ and $w'_{(4)} \in W'_4$. In particular, the iterate of the logarithmic intertwining operators (ordinary intertwining operators in the case that $\mathcal{C}$ is in $\mathcal{M}_{sg}$) $\mathcal{Y}^1$ and $\mathcal{Y}^2$ evaluated at z_2 and $z_1 - z_2$, respectively, can be expressed as a product (with the intermediate generalized V-module $W_2 \boxtimes_{P(z_2)} W_3$) of logarithmic intertwining operators (ordinary intertwining operators in the case that $\mathcal{C}$ is in $\mathcal{M}_{sg}$) evaluated at z_1 and z_2.

In Sect. 10, we construct the associativity isomorphisms, under certain assumptions: In addition to the assumptions above, we assume that $\mathcal{C}$ is closed under images and that for some $z \in \mathbb{C}^\times$ (and hence for every $z \in \mathbb{C}^\times$), $\mathcal{C}$ is closed under $P(z)$-tensor products; see Assumption 10.1. Besides the convergence condition (Sect. 7), at the end of Sect. 9 we introduce what we call the "expansion condition," which, roughly speaking, states that an element of $(W_1 \otimes W_2 \otimes W_3)^*$ obtained from a product or an iterate of intertwining maps satisfies the $P^{(2)}(z)$- or $P^{(1)}(z)$-local grading restriction condition, respectively, for suitable $z \in \mathbb{C}^\times$, along with certain other "minor" conditions. Then we have:

Theorem 10.3 *Assume that the convergence condition and the expansion condition for intertwining maps in $\mathcal{C}$ both hold. Let z_1, z_2 be complex numbers satisfying*

$$|z_1| > |z_2| > |z_1 - z_2| > 0$$

(so that in particular $z_1 \neq 0$, $z_2 \neq 0$ and $z_1 \neq z_2$). Then there exists a unique natural isomorphism

$$\mathcal{A}^{P(z_1-z_2),P(z_2)}_{P(z_1),P(z_2)} : \boxtimes_{P(z_1)} \circ (1 \times \boxtimes_{P(z_2)}) \to \boxtimes_{P(z_2)} \circ (\boxtimes_{P(z_1-z_2)} \times 1)$$

such that for all $w_{(1)} \in W_1$, $w_{(2)} \in W_2$ and $w_{(3)} \in W_3$, with W_j objects of $\mathcal{C}$,

$$\overline{\mathcal{A}^{P(z_1-z_2),P(z_2)}_{P(z_1),P(z_2)}}(w_{(1)} \boxtimes_{P(z_1)} (w_{(2)} \boxtimes_{P(z_2)} w_{(3)})) = (w_{(1)} \boxtimes_{P(z_1-z_2)} w_{(2)}) \boxtimes_{P(z_2)} w_{(3)},$$

where for simplicity we use the same notation $\mathcal{A}^{P(z_1-z_2),P(z_2)}_{P(z_1),P(z_2)}$ to denote the isomorphism of generalized modules

$$\mathcal{A}^{P(z_1-z_2),P(z_2)}_{P(z_1),P(z_2)} : W_1 \boxtimes_{P(z_1)} (W_2 \boxtimes_{P(z_2)} W_3) \longrightarrow (W_1 \boxtimes_{P(z_1-z_2)} W_2) \boxtimes_{P(z_2)} W_3.$$

Here we are using the notation

$$\overline{\eta}: \overline{W_1} \to \overline{W_2}$$

to denote the natural extension of a map $\eta: W_1 \to W_2$ of generalized modules to the (suitably defined) formal completions; such natural extensions enter into many of the constructions in this work.

In Sect. 11, we give results which will allow us to verify the convergence and expansion conditions. We need the "convergence and extension property" for products or iterates and the "convergence and extension property without logarithms" for products or iterates. Here we only give the convergence and extension property for products:

Given objects W_1, W_2, W_3, W_4 and M_1 of the category $\mathcal{C}$, let $\mathcal{Y}_1$ and $\mathcal{Y}_2$ be logarithmic intertwining operators of types $\binom{W_4}{W_1 M_1}$ and $\binom{M_1}{W_2 W_3}$, respectively.

Convergence and Extension Property for Products For any $\beta \in \tilde{A}$, there exists an integer N_β depending only on $\mathcal{Y}_1$, $\mathcal{Y}_2$ and β, and for any weight-homogeneous elements $w_{(1)} \in (W_1)^{(\beta_1)}$ and $w_{(2)} \in (W_2)^{(\beta_2)}$ $(\beta_1, \beta_2 \in \tilde{A})$ and any $w_{(3)} \in W_3$ and $w'_{(4)} \in W'_4$ such that

$$\beta_1 + \beta_2 = -\beta,$$

there exist $M \in \mathbb{N}$, $r_k, s_k \in \mathbb{R}$, $i_k, j_k \in \mathbb{N}$, $k = 1, \ldots, M$; $K \in \mathbb{Z}_+$ independent of $w_{(1)}$ and $w_{(2)}$ such that each $i_k < K$; and analytic functions $f_k(z)$ on $|z| < 1$, $k = 1, \ldots, M$, satisfying

$$\mathrm{wt}\, w_{(1)} + \mathrm{wt}\, w_{(2)} + s_k > N_\beta, \quad k = 1, \ldots, M,$$

such that

$$\langle w'_{(4)}, \mathcal{Y}_1(w_{(1)}, x_2)\mathcal{Y}_2(w_{(2)}, x_2)w_{(3)}\rangle_{W_4}\Big|_{x_1=z_1, x_2=z_2}$$

is absolutely convergent when $|z_1| > |z_2| > 0$ and can be analytically extended to the multivalued analytic function

$$\sum_{k=1}^{M} z_2^{r_k}(z_1 - z_2)^{s_k}(\log z_2)^{i_k}\left(\log(z_1 - z_2)\right)^{j_k} f_k\left(\frac{z_1 - z_2}{z_2}\right)$$

(here $\log(z_1 - z_2)$ and $\log z_2$, and in particular, the powers of the variables, mean the multivalued functions, not the particular branch we have been using) in the region $|z_2| > |z_1 - z_2| > 0$.

Theorem 11.4 *Suppose that the following two conditions are satisfied:*

1. *Every finitely-generated lower bounded doubly-graded (as defined in Sect.* 11*) generalized V-module is an object of $\mathcal{C}$ (or every finitely-generated lower bounded doubly-graded V-module is an object of $\mathcal{C}$, when $\mathcal{C}$ is in $\mathcal{M}_{sg}$).*

2. *The convergence and extension property for either products or iterates holds in $\mathcal{C}$ (or the convergence and extension property without logarithms for either products or iterates holds in $\mathcal{C}$, when $\mathcal{C}$ is in $\mathcal{M}_{sg}$).*

Then the convergence and expansion conditions for intertwining maps in $\mathcal{C}$ both hold.

In the following two results, we assume that the grading abelian groups A and $\tilde{A}$ are trivial. Set

$$V_+ = \coprod_{n>0} V_{(n)}.$$

Let W be a generalized V-module and let

$$C_1(W) = \mathrm{span}\{u_{-1}w \mid u \in V_+, w \in W\}.$$

If $W/C_1(W)$ is finite dimensional, we say that W is C_1-cofinite or satisfies the C_1-cofiniteness condition. If for any $N \in \mathbb{R}$, $\coprod_{n<N} W_{[n]}$ is finite dimensional, we say that W is quasi-finite dimensional or satisfies the quasi-finite-dimensionality condition. The following result in Sect. 11 allows us to verify the convergence and extension properties and thus the convergence and expansion conditions:

Theorem 11.6 *Let W_i for $i = 0, \ldots, n+1$ be generalized V-modules satisfying the C_1-cofiniteness condition and the quasi-finite-dimensionality condition. Then for any $w'_{(0)} \in W'_0$, $w_{(1)} \in W_1, \ldots, w_{(n+1)} \in W_{n+1}$, there exist*

$$a_{k,l}(z_1, \ldots, z_n) \in \mathbb{C}\big[z_1^{\pm 1}, \ldots, z_n^{\pm 1}, (z_1 - z_2)^{-1}, (z_1 - z_3)^{-1}, \ldots, (z_{n-1} - z_n)^{-1}\big],$$

for $k = 1, \ldots, m$ and $l = 1, \ldots, n$, such that the following holds: For any generalized V-modules $\widetilde{W}_1, \ldots, \widetilde{W}_{n-1}$, and any logarithmic intertwining operators

$$\mathcal{Y}_1, \mathcal{Y}_2, \ldots, \mathcal{Y}_{n-1}, \mathcal{Y}_n$$

of types

$$\binom{W_0}{W_1\,\widetilde{W}_1}, \binom{\widetilde{W}_1}{W_2\,\widetilde{W}_2}, \ldots, \binom{\widetilde{W}_{n-2}}{W_{n-1}\,\widetilde{W}_{n-1}}, \binom{\widetilde{W}_{n-1}}{W_n\,W_{n+1}},$$

respectively, the series

$$\langle w'_{(0)}, \mathcal{Y}_1(w_{(1)}, z_1) \cdots \mathcal{Y}_n(w_{(n)}, z_n) w_{(n+1)} \rangle$$

satisfies the system of differential equations

$$\frac{\partial^m \varphi}{\partial z_l^m} + \sum_{k=1}^{m} \iota_{|z_1|>\cdots>|z_n|>0}\big(a_{k,l}(z_1, \ldots, z_n)\big) \frac{\partial^{m-k} \varphi}{\partial z_l^{m-k}} = 0, \quad l = 1, \ldots, n$$

in the region $|z_1| > \cdots > |z_n| > 0$, *where*

$$\iota_{|z_1|>\cdots>|z_n|>0}\big(a_{k,l}(z_1, \ldots, z_n)\big)$$

for $k = 1, \ldots, m$ *and* $l = 1, \ldots, n$ *are the (unique) Laurent expansions of* $a_{k,l}(z_1, \ldots, z_n)$ *in the region* $|z_1| > \cdots > |z_n| > 0$. *Moreover, for any set of possible singular points of the system*

$$\frac{\partial^m \varphi}{\partial z_l^m} + \sum_{k=1}^{m} a_{k,l}(z_1, \ldots, z_n) \frac{\partial^{m-k} \varphi}{\partial z_l^{m-k}} = 0, \quad l = 1, \ldots, n$$

such that either $z_i = 0$ *or* $z_i = \infty$ *for some* i *or* $z_i = z_j$ *for some* $i \neq j$, *the* $a_{k,l}(z_1, \ldots, z_n)$ *can be chosen for* $k = 1, \ldots, m$ *and* $l = 1, \ldots, n$ *so that these singular points are regular.*

Using this result, we prove the following:

Theorem 11.8 *Suppose that all generalized* V*-modules in* $\mathcal{C}$ *satisfy the* C_1*-cofiniteness condition and the quasi-finite-dimensionality condition. Then:*

1. *The convergence and extension properties for products and iterates hold in* $\mathcal{C}$. *If* $\mathcal{C}$ *is in* $\mathcal{M}_{sg}$ *and if every object of* $\mathcal{C}$ *is a direct sum of irreducible objects of* $\mathcal{C}$ *and there are only finitely many irreducible objects of* $\mathcal{C}$ *(up to equivalence), then the convergence and extension properties without logarithms for products and iterates hold in* $\mathcal{C}$.
2. *For any* $n \in \mathbb{Z}_+$, *any objects* $W_1, \ldots, W_{n+1}$ *and* $\widetilde{W}_1, \ldots, \widetilde{W}_{n-1}$ *of* $\mathcal{C}$, *any logarithmic intertwining operators*

$$\mathcal{Y}_1, \mathcal{Y}_2, \ldots, \mathcal{Y}_{n-1}, \mathcal{Y}_n$$

of types

$$\binom{W_0}{W_1\,\widetilde{W}_1}, \binom{\widetilde{W}_1}{W_2\,\widetilde{W}_2}, \ldots, \binom{\widetilde{W}_{n-2}}{W_{n-1}\,\widetilde{W}_{n-1}}, \binom{\widetilde{W}_{n-1}}{W_n\,W_{n+1}},$$

respectively, and any $w'_{(0)} \in W'_0$, $w_{(1)} \in W_1, \ldots, w_{(n+1)} \in W_{n+1}$, *the series*

$$\langle w'_{(0)}, \mathcal{Y}_1(w_{(1)}, z_1) \cdots \mathcal{Y}_n(w_{(n)}, z_n) w_{(n+1)} \rangle \tag{11.37}$$

is absolutely convergent in the region $|z_1| > \cdots > |z_n| > 0$ *and its sum can be analytically extended to a multivalued analytic function on the region given by* $z_i \neq 0$, $i = 1, \ldots, n$, $z_i \neq z_j$, $i \neq j$, *such that for any set of possible singular points with either* $z_i = 0$, $z_i = \infty$ *or* $z_i = z_j$ *for* $i \neq j$, *this multivalued analytic function can be expanded near the singularity as a series having the same form as the expansion near the singular points of a solution of a system of differential equations with regular singular points.*

We now return to the assumptions before Theorem 11.6, that is, we do not assume that A and $\tilde{A}$ are trivial. To construct the braided tensor category structure, we need more assumptions in addition to those mentioned above, which are collected in Assumption 10.1. We assume in addition that the Möbius or conformal vertex algebra V, viewed as a V-module, is an object of $\mathcal{C}$; and also that the product of three logarithmic intertwining operators is absolutely convergent in a suitable region and can be analytically extended to a multivalued analytic function, admitting suitable expansions as series in powers of the variables and their logarithms near its singularities (expansions that hold for solutions of systems of differential equations with regular singularities), on a suitable largest possible region containing the original region for the convergence of the product. See Assumptions 12.1 and 12.2 for the precise statements. Under these assumptions, we construct, in addition to the tensor product bifunctor $\boxtimes = \boxtimes_{P(1)}$, a braiding isomorphism $\mathcal{R}$, an associativity isomorphism $\mathcal{A}$ (for the braided tensor category structure, different from the associativity isomorphisms above), a left unit isomorphism l and a right unit isomorphism r. The following main results of this work are given in Sect. 12:

Theorem 12.15 *Let V be a Möbius or conformal vertex algebra and $\mathcal{C}$ a full subcategory of $\mathcal{M}_{sg}$ or $\mathcal{GM}_{sg}$ satisfying Assumptions* 10.1, 12.1 *and* 12.2. *Then the category $\mathcal{C}$, equipped with the tensor product bifunctor $\boxtimes$, the unit object V, the braiding isomorphism $\mathcal{R}$, the associativity isomorphism $\mathcal{A}$, and the left and right unit isomorphisms l and r, is an additive braided monoidal category.*

Corollary 12.16 *If the category $\mathcal{C}$ is an abelian category, then $\mathcal{C}$, equipped with the tensor product bifunctor $\boxtimes$, the unit object V, the braiding isomorphism $\mathcal{R}$, the associativity isomorphism $\mathcal{A}$, and the left and right unit isomorphisms l and r, is a braided tensor category.*

2 The Setting: Strongly Graded Conformal and Möbius Vertex Algebras and Their Generalized Modules

In this section we define and discuss the basic structures and introduce some notation that will be used in this work. More specifically, we first introduce the notions of "conformal vertex algebra" and "Möbius vertex algebra." A conformal vertex algebra is just a vertex algebra equipped with a conformal vector satisfying the usual axioms; a Möbius vertex algebra is a variant of a "quasi-vertex operator algebra" as in [38], with the difference that the two grading restriction conditions in the definition of vertex operator algebra are not required. We then define the notion of module for each of these types of vertex algebra. Relaxing the $L(0)$-semisimplicity in the definition of module we obtain the notion of "generalized module." Finally, we notice that in order to have a contragredient functor on the module category under consideration, we need to impose a stronger grading condition. This leads to the notions of "strong gradedness" of Möbius vertex algebras and their generalized modules. In this work we are mainly interested in certain full subcategories of the category of strongly graded generalized modules for certain strongly graded Möbius vertex algebras.

Throughout the work we shall assume some familiarity with the material in [15, 21, 37, 38] and [99].

In particular, we recall the necessary basic material on "formal calculus," starting with the "formal delta function." Formal calculus will be needed throughout this work, and in fact, the theory of formal calculus will be considerably developed, whenever new formal-calculus ideas are needed for the formulations and for the proofs of the results.

Throughout, we shall use the notation $\mathbb{N}$ for the nonnegative integers and $\mathbb{Z}_+$ for the positive integers.

We shall continue to use the notational convention concerning formal variables and complex variables given in Remark 1.3. Recall from [37, 38] or [99] that the formal delta function is defined as the formal series

$$\delta(x) = \sum_{n\in\mathbb{Z}} x^n$$

in the formal variable x. We will consistently use the *binomial expansion convention*: For any complex number λ, $(x+y)^\lambda$ is to be expanded as a formal series in nonnegative integral powers of the second variable, i.e.,

$$(x+y)^\lambda = \sum_{n\in\mathbb{N}} \binom{\lambda}{n} x^{\lambda-n} y^n.$$

Here x or y might be something other than a formal variable (or a nonzero complex multiple of a formal variable); for instance, x or y (but not both; this expansion is understood to be formal) might be a nonzero complex number, or x or y might be some more complicated object. The use of the binomial expansion convention will be clear in context.

Objects like $\delta(x)$ and $(x+y)^\lambda$ lie in spaces of formal series. Some of the spaces that we will use are, with W a vector space (over $\mathbb{C}$) and x a formal variable:

$$W[x] = \left\{ \sum_{n\in\mathbb{N}} a_n x^n \middle| a_n \in W, \text{ all but finitely many } a_n = 0 \right\}$$

(the space of formal polynomials with coefficients in W),

$$W[x, x^{-1}] = \left\{ \sum_{n\in\mathbb{Z}} a_n x^n \middle| a_n \in W, \text{ all but finitely many } a_n = 0 \right\}$$

(the formal Laurent polynomials),

$$W[[x]] = \left\{ \sum_{n\in\mathbb{N}} a_n x^n \middle| a_n \in W \text{ (with possibly infinitely many } a_n \text{ not } 0) \right\}$$

(the formal power series),

$$W((x)) = \left\{ \sum_{n\in\mathbb{Z}} a_n x^n \middle| a_n \in W, a_n = 0 \text{ for sufficiently small } n \right\}$$

(the truncated formal Laurent series), and

$$W[[x, x^{-1}]] = \left\{ \sum_{n\in\mathbb{Z}} a_n x^n \middle| a_n \in W \text{ (with possibly infinitely many } a_n \text{ not 0)} \right\}$$

(the formal Laurent series). We will also need the space

$$W\{x\} = \left\{ \sum_{n\in\mathbb{C}} a_n x^n \middle| a_n \in W \text{ for } n \in \mathbb{C} \right\} \tag{2.1}$$

as in [37]; here the powers of the formal variable are complex, and the coefficients may all be nonzero. We will also use analogues of these spaces involving two or more formal variables. Note that for us, a "formal power series" involves only nonnegative integral powers of the formal variable(s), and a "formal Laurent series" can involve all the integral powers of the formal variable(s).

The following formal version of Taylor's theorem is easily verified by direct expansion (see Proposition 8.3.1 of [37]): For $f(x) \in W\{x\}$,

$$e^{y(d/dx)} f(x) = f(x + y), \tag{2.2}$$

where the exponential denotes the formal exponential series, and where we are using the binomial expansion convention on the right-hand side. It is important to note that this formula holds for arbitrary formal series $f(x)$ with complex powers of x, where $f(x)$ need not be an expansion in any sense of an analytic function (again, see Proposition 8.3.1 of [37]).

The formal delta function $\delta(x)$ has the following simple and fundamental property: For any $f(x) \in W[x, x^{-1}]$,

$$f(x)\delta(x) = f(1)\delta(x). \tag{2.3}$$

(Here we are taking the liberty of writing complex numbers to the right of vectors in W.) This is proved immediately by observing its truth for $f(x) = x^n$ and then using linearity. This property has many important variants; in general, whenever an expression is multiplied by the formal delta function, we may formally set the argument appearing in the delta function equal to 1, provided that the relevant algebraic expressions make sense. For example, for any

$$X(x_1, x_2) \in (\text{End } W)[[x_1, x_1^{-1}, x_2, x_2^{-1}]]$$

such that

$$\lim_{x_1 \to x_2} X(x_1, x_2) = X(x_1, x_2)\big|_{x_1 = x_2} \tag{2.4}$$

exists, we have

$$X(x_1, x_2)\delta\left(\frac{x_1}{x_2}\right) = X(x_2, x_2)\delta\left(\frac{x_1}{x_2}\right). \tag{2.5}$$

The existence of the "algebraic limit" defined in (2.4) means that for an arbitrary vector $w \in W$, the coefficient of each power of x_2 in the formal expansion $X(x_1, x_2)w|_{x_1=x_2}$ is a finite sum. In general, the existence of such "algebraic limits," and also such products of formal sums, always means that the coefficient of each monomial in the relevant formal variables gives a finite sum. Often, proving the existence of the relevant algebraic limits (or products) is a much more subtle matter than computing such limits (or products), just as in analysis. (In this work, we will typically use "substitution notation" like $|_{x_1=x_2}$ or $X(x_2, x_2)$ rather than the formal limit notation on the left-hand side of (2.4).) Below, we will give a more sophisticated analogue of the delta-function substitution principle (2.5), an analogue that we will need in this work.

This analogue, and in fact, many fundamental principles of vertex operator algebra theory, are based on certain delta-function expressions of the following type, involving three (commuting and independent, as usual) formal variables:

$$x_0^{-1}\delta\left(\frac{x_1 - x_2}{x_0}\right) = \sum_{n\in\mathbb{Z}} \frac{(x_1 - x_2)^n}{x_0^{n+1}} = \sum_{m\in\mathbb{N}, n\in\mathbb{Z}} (-1)^m \binom{n}{m} x_0^{-n-1} x_1^{n-m} x_2^m;$$

here the binomial expansion convention is of course being used.

The following important identities involving such three-variable delta-function expressions are easily proved (see [37] or [99], where extensive motivation for these formulas is also given):

$$x_2^{-1}\delta\left(\frac{x_1 - x_0}{x_2}\right) = x_1^{-1}\delta\left(\frac{x_2 + x_0}{x_1}\right), \tag{2.6}$$

$$x_0^{-1}\delta\left(\frac{x_1 - x_2}{x_0}\right) - x_0^{-1}\delta\left(\frac{x_2 - x_1}{-x_0}\right) = x_2^{-1}\delta\left(\frac{x_1 - x_0}{x_2}\right). \tag{2.7}$$

Note that the three terms in (2.7) involve nonnegative integral powers of x_2, x_1 and x_0, respectively. In particular, the two terms on the left-hand side of (2.7) are unequal formal Laurent series in three variables, even though they might appear equal at first glance. We shall use these two identities extensively.

Remark 2.1 Here is the useful analogue, mentioned above, of the delta-function substitution principle (2.5): Let

$$f(x_1, x_2, y) \in (\operatorname{End} W)[[x_1, x_1^{-1}, x_2, x_2^{-1}, y, y^{-1}]] \tag{2.8}$$

be such that

$$\lim_{x_1 \to x_2} f(x_1, x_2, y) \text{ exists} \tag{2.9}$$

and such that for any $w \in W$,

$$f(x_1, x_2, y)w \in W\big[\big[x_1, x_1^{-1}, x_2, x_2^{-1}\big]\big]((y)). \tag{2.10}$$

Then

$$x_1^{-1}\delta\left(\frac{x_2 - y}{x_1}\right) f(x_1, x_2, y) = x_1^{-1}\delta\left(\frac{x_2 - y}{x_1}\right) f(x_2 - y, x_2, y). \tag{2.11}$$

For this principle, see Remark 2.3.25 of [99], where the proof is also presented.

The following formal residue notation will be useful: For

$$f(x) = \sum_{n \in \mathbb{C}} a_n x^n \in W\{x\}$$

(note that the powers of x need not be integral),

$$\mathrm{Res}_x\, f(x) = a_{-1}.$$

For instance, for the expression in (2.6),

$$\mathrm{Res}_{x_2}\, x_2^{-1}\delta\left(\frac{x_1 - x_0}{x_2}\right) = 1. \tag{2.12}$$

For a vector space W, we will denote its vector space dual by W^* ($= \mathrm{Hom}_{\mathbb{C}}(W, \mathbb{C})$), and we will use the notation $\langle \cdot, \cdot \rangle_W$, or $\langle \cdot, \cdot \rangle$ if the underlying space W is clear, for the canonical pairing between W^* and W.

We will use the following version of the notion of "conformal vertex algebra": A conformal vertex algebra is a vertex algebra (in the sense of Borcherds [15]; see [99]) equipped with a $\mathbb{Z}$-grading and with a conformal vector satisfying the usual compatibility conditions. Specifically:

Definition 2.2 A *conformal vertex algebra* is a $\mathbb{Z}$-graded vector space

$$V = \coprod_{n \in \mathbb{Z}} V_{(n)} \tag{2.13}$$

(for $v \in V_{(n)}$, we say the *weight* of v is n and we write $\mathrm{wt}\, v = n$) equipped with a linear map $V \otimes V \to V[[x, x^{-1}]]$, or equivalently,

$$\begin{aligned} V &\to (\mathrm{End}\, V)\big[\big[x, x^{-1}\big]\big] \\ v &\mapsto Y(v, x) = \sum_{n \in \mathbb{Z}} v_n x^{-n-1} \quad (\text{where } v_n \in \mathrm{End}\, V), \end{aligned} \tag{2.14}$$

$Y(v, x)$ denoting the *vertex operator associated with* v, and equipped also with two distinguished vectors $\mathbf{1} \in V_{(0)}$ (the *vacuum vector*) and $\omega \in V_{(2)}$ (the *conformal vector*), satisfying the following conditions for $u, v \in V$: the *lower truncation condition*:

$$u_n v = 0 \quad \text{for } n \text{ sufficiently large} \tag{2.15}$$

(or equivalently, $Y(u,x)v \in V((x))$); the *vacuum property*:

$$Y(\mathbf{1}, x) = 1_V; \tag{2.16}$$

the *creation property*:

$$Y(v,x)\mathbf{1} \in V[[x]] \quad \text{and} \quad \lim_{x \to 0} Y(v,x)\mathbf{1} = v \tag{2.17}$$

(that is, $Y(v,x)\mathbf{1}$ involves only nonnegative integral powers of x and the constant term is v); the *Jacobi identity* (the main axiom):

$$x_0^{-1}\delta\left(\frac{x_1 - x_2}{x_0}\right)Y(u,x_1)Y(v,x_2) - x_0^{-1}\delta\left(\frac{x_2 - x_1}{-x_0}\right)Y(v,x_2)Y(u,x_1)$$
$$= x_2^{-1}\delta\left(\frac{x_1 - x_0}{x_2}\right)Y\big(Y(u,x_0)v, x_2\big) \tag{2.18}$$

(note that when each expression in (2.18) is applied to any element of V, the coefficient of each monomial in the formal variables is a finite sum; on the right-hand side, the notation $Y(\cdot, x_2)$ is understood to be extended in the obvious way to $V[[x_0, x_0^{-1}]]$); the *Virasoro algebra relations*:

$$\big[L(m), L(n)\big] = (m-n)L(m+n) + \frac{1}{12}\big(m^3 - m\big)\delta_{n+m,0}c \tag{2.19}$$

for $m, n \in \mathbb{Z}$, where

$$L(n) = \omega_{n+1} \quad \text{for } n \in \mathbb{Z}, \text{ i.e., } Y(\omega, x) = \sum_{n \in \mathbb{Z}} L(n)x^{-n-2}, \tag{2.20}$$

$$c \in \mathbb{C} \tag{2.21}$$

(the *central charge* or *rank* of V);

$$\frac{d}{dx}Y(v,x) = Y\big(L(-1)v, x\big) \tag{2.22}$$

(the $L(-1)$-*derivative property*); and

$$L(0)v = nv = (\operatorname{wt} v)v \quad \text{for } n \in \mathbb{Z} \text{ and } v \in V_{(n)}. \tag{2.23}$$

This completes the definition of the notion of conformal vertex algebra. We will denote such a conformal vertex algebra by $(V, Y, \mathbf{1}, \omega)$ or simply by V.

The only difference between the definition of conformal vertex algebra and the definition of *vertex operator algebra* (in the sense of [37] and [38]) is that a vertex operator algebra V also satisfies the two *grading restriction conditions*

$$V_{(n)} = 0 \quad \text{for } n \text{ sufficiently negative}, \tag{2.24}$$

and

$$\dim V_{(n)} < \infty \quad \text{for } n \in \mathbb{Z}. \tag{2.25}$$

(As we mentioned above, a *vertex algebra* is the same thing as a conformal vertex algebra but without the assumptions of a grading or a conformal vector, or, of course, the $L(n)$'s.)

Remark 2.3 Of course, not every vertex algebra is conformal. For example, it is well known [15] that any commutative associative algebra A with unit 1, together with a derivation $D : A \to A$ can be equipped with a vertex algebra structure, by:

$$Y(\cdot, x)\cdot : A \times A \to A[[x]], \quad Y(a, x)b = \left(e^{xD}a\right)b,$$

and $\mathbf{1} = 1$. In particular, $u_n = 0$ for any $u \in A$ and $n \geq 0$. If ω is a conformal vector for such a vertex algebra, then for any $u \in A$, $Du = u_{-2}\mathbf{1} = L(-1)u$ from (2.17) and (2.22), so $D = L(-1) = \omega_0$, which equals 0 because $\omega = L(0)\omega/2 = \omega_1\omega/2 = 0$. Thus a vertex algebra constructed from a commutative associative algebra with nonzero derivation in this way cannot be conformal.

Remark 2.4 The theory of vertex tensor categories inherently uses the whole moduli space of spheres with two positively oriented punctures and one negatively oriented puncture (and in fact, more generally, with arbitrary numbers of positively oriented punctures and one negatively oriented puncture) equipped with general (analytic) local coordinates vanishing at the punctures. Because of the analytic local coordinates, our constructions require certain conditions on the Virasoro algebra operators. However, recalling the definition of the moduli space elements $P(z)$ from Sect. 1.4, we point out that if we restrict our attention to elements of the moduli space of only the type $P(z)$, then the relevant operations of sewing and subsequently decomposing Riemann spheres continue to yield spheres of the same type, and rather than general conformal transformations around the punctures, only Möbius (projective) transformations around the punctures are needed. This makes it possible to develop the essential structure of our tensor product theory by working entirely with spheres of this special type; the general vertex tensor category theory then follows from the structure thus developed. This is why, in the present work, we are focusing on the theory of $P(z)$-tensor products. Correspondingly, it turns out that it is very natural for us to consider, along with the notion of conformal vertex algebra (Definition 2.2), a weaker notion of vertex algebra involving only the three-dimensional subalgebra of the Virasoro algebra corresponding to the group of Möbius transformations. That is, instead of requiring an action of the whole Virasoro algebra, we use only the action of the Lie algebra $\mathfrak{sl}(2)$ generated by $L(-1)$, $L(0)$ and $L(1)$. Thus we get a notion essentially identical to the notion of "quasi-vertex operator algebra" in [38]; the reason for focusing on this notion here is the same as the reason why it was considered in [38]. Here we designate this notion by the term "Möbius vertex algebra"; the only difference between the definition of Möbius vertex algebra and the definition of quasi-vertex operator algebra [38] is that a quasi-vertex operator algebra V also satisfies the two grading restriction conditions (2.24) and (2.25).

Thus we formulate:

Definition 2.5 The notion of *Möbius vertex algebra* is defined in the same way as that of conformal vertex algebra except that in addition to the data and axioms concerning V, Y and $\mathbf{1}$ (through (2.18) in Definition 2.2), we assume (in place of the existence of the conformal vector ω and the Virasoro algebra conditions (2.19), (2.20) and (2.21)) the following: We have a representation ρ of $\mathfrak{sl}(2)$ on V given by

$$L(j) = \rho(L_j), \quad j = 0, \pm 1, \tag{2.26}$$

where $\{L_{-1}, L_0, L_1\}$ is a basis of $\mathfrak{sl}(2)$ with Lie brackets

$$[L_0, L_{-1}] = L_{-1}, \qquad [L_0, L_1] = -L_1, \quad \text{and} \quad [L_{-1}, L_1] = -2L_0, \tag{2.27}$$

and the following conditions hold for $v \in V$:

$$\big[L(-1), Y(v, x)\big] = Y\big(L(-1)v, x\big), \tag{2.28}$$

$$\big[L(0), Y(v, x)\big] = Y\big(L(0)v, x\big) + xY\big(L(-1)v, x\big), \tag{2.29}$$

$$\begin{aligned}\big[L(1), Y(v, x)\big] &= Y\big(L(1)v, x\big) + 2xY\big(L(0)v, x\big)\\ &\quad + x^2 Y\big(L(-1)v, x\big),\end{aligned} \tag{2.30}$$

and also, (2.22) and (2.23). Of course, (2.28)–(2.30) can be written as

$$\begin{aligned}\big[L(j), Y(v, x)\big] &= \sum_{k=0}^{j+1} \binom{j+1}{k} x^k Y\big(L(j-k)v, x\big)\\ &= \sum_{k=0}^{j+1} \binom{j+1}{k} x^{j+1-k} Y\big(L(k-1)v, x\big)\end{aligned} \tag{2.31}$$

for $j = 0, \pm 1$.

We will denote such a Möbius vertex algebra by $(V, Y, \mathbf{1}, \rho)$ or simply by V. Note that there is no notion of central charge (or rank) for a Möbius vertex algebra. Also, a conformal vertex algebra can certainly be viewed as a Möbius vertex algebra in the obvious way. (Of course, a conformal vertex algebra could have other $\mathfrak{sl}(2)$-structures making it a Möbius vertex algebra in a different way.)

Remark 2.6 By (2.26) and (2.27) we have $[L(0), L(j)] = -jL(j)$ for $j = 0, \pm 1$. Hence

$$L(j)V_{(n)} \subset V_{(n-j)}, \quad \text{for } j = 0, \pm 1. \tag{2.32}$$

Moreover, from (2.28), (2.29) and (2.30) with $v = \mathbf{1}$ we get, by (2.16) and (2.17),

$$L(j)\mathbf{1} = 0 \quad \text{for } j = 0, \pm 1.$$

Remark 2.7 Not every Möbius vertex algebra is conformal. As an example, take the commutative associative algebra $\mathbb{C}[t]$ with derivation $D = -d/dt$, and form a vertex algebra as in Remark 2.3. By Remark 2.3, this vertex algebra is not conformal. However, define linear operators

$$L(-1) = D, \qquad L(0) = tD, \qquad L(1) = t^2 D$$

on $\mathbb{C}[t]$. Then it is straightforward to verify that $\mathbb{C}[t]$ becomes a Möbius vertex algebra with these operators giving a representation of $\mathfrak{sl}(2)$ having the desired properties and with the $\mathbb{Z}$-grading (by nonpositive integers) given by the eigenspace decomposition with respect to $L(0)$.

Remark 2.8 It is also easy to see that not every vertex algebra is Möbius. For example, take the two-dimensional commutative associative algebra $A = \mathbb{C}1 \oplus \mathbb{C}a$ with 1 as identity and $a^2 = 0$. The linear operator D defined by $D(1) = 0$, $D(a) = a$ is a nonzero derivation of A. Hence A has a vertex algebra structure by Remark 2.3. Now if it is a module for $\mathfrak{sl}(2)$ as in Definition 2.5, since A is two-dimensional and $L(0)1 = 0$, $L(0)$ must act as 0. But then $D = L(-1) = [L(0), L(-1)] = 0$, a contradiction.

A module for a conformal vertex algebra V is a module for V viewed as a vertex algebra such that the conformal element acts in the same way as in the definition of vertex operator algebra. More precisely:

Definition 2.9 Given a conformal vertex algebra $(V, Y, \mathbf{1}, \omega)$, a *module* for V is a $\mathbb{C}$-graded vector space

$$W = \coprod_{n\in\mathbb{C}} W_{(n)} \tag{2.33}$$

(graded by *weights*) equipped with a linear map $V \otimes W \to W[[x, x^{-1}]]$, or equivalently,

$$\begin{aligned} V &\to (\operatorname{End} W)[[x, x^{-1}]] \\ v &\mapsto Y(v, x) = \sum_{n\in\mathbb{Z}} v_n x^{-n-1} \quad (\text{where } v_n \in \operatorname{End} W) \end{aligned} \tag{2.34}$$

(note that the sum is over $\mathbb{Z}$, not $\mathbb{C}$), $Y(v, x)$ denoting the *vertex operator on* W *associated with* v, such that all the defining properties of a conformal vertex algebra that make sense hold. That is, the following conditions are satisfied: the lower truncation condition: for $v \in V$ and $w \in W$,

$$v_n w = 0 \quad \text{for } n \text{ sufficiently large} \tag{2.35}$$

(or equivalently, $Y(v, x)w \in W((x))$); the vacuum property:

$$Y(\mathbf{1}, x) = 1_W; \tag{2.36}$$

the Jacobi identity for vertex operators on W: for $u, v \in V$,

$$x_0^{-1}\delta\left(\frac{x_1 - x_2}{x_0}\right)Y(u, x_1)Y(v, x_2) - x_0^{-1}\delta\left(\frac{x_2 - x_1}{-x_0}\right)Y(v, x_2)Y(u, x_1)$$
$$= x_2^{-1}\delta\left(\frac{x_1 - x_0}{x_2}\right)Y\big(Y(u, x_0)v, x_2\big) \tag{2.37}$$

(note that on the right-hand side, $Y(u, x_0)$ is the operator on V associated with u); the Virasoro algebra relations on W with scalar c equal to the central charge of V:

$$\big[L(m), L(n)\big] = (m - n)L(m + n) + \frac{1}{12}\big(m^3 - m\big)\delta_{n+m,0}c \tag{2.38}$$

for $m, n \in \mathbb{Z}$, where

$$L(n) = \omega_{n+1} \quad \text{for } n \in \mathbb{Z}, \text{ i.e., } Y(\omega, x) = \sum_{n\in\mathbb{Z}} L(n)x^{-n-2}; \tag{2.39}$$

$$\frac{d}{dx}Y(v, x) = Y\big(L(-1)v, x\big) \tag{2.40}$$

(the $L(-1)$-derivative property); and

$$\big(L(0) - n\big)w = 0 \quad \text{for } n \in \mathbb{C} \text{ and } w \in W_{(n)}. \tag{2.41}$$

This completes the definition of the notion of module for a conformal vertex algebra.

Remark 2.10 The Virasoro algebra relations (2.38) for a module action follow from the corresponding relations (2.19) for V together with the Jacobi identities (2.18) and (2.37) and the $L(-1)$-derivative properties (2.22) and (2.40), as we recall from (for example) [38] or [99].

We also have:

Definition 2.11 The notion of *module* for a Möbius vertex algebra is defined in the same way as that of module for a conformal vertex algebra except that in addition to the data and axioms concerning W and Y (through (2.37) in Definition 2.9), we assume (in place of the Virasoro algebra conditions (2.38) and (2.39)) a representation ρ of $\mathfrak{sl}(2)$ on W given by (2.26) and the conditions (2.28), (2.29) and (2.30), for operators acting on W, and also, (2.40) and (2.41).

In addition to modules, we have the following notion of *generalized module* (or *logarithmic module*, as in, for example, [104]):

Definition 2.12 A *generalized module* for a conformal (respectively, Möbius) vertex algebra is defined in the same way as a module for a conformal (respectively,

Möbius) vertex algebra except that in the grading (2.33), each space $W_{(n)}$ is replaced by $W_{[n]}$, where $W_{[n]}$ is the generalized $L(0)$-eigenspace corresponding to the (generalized) eigenvalue $n \in \mathbb{C}$; that is, (2.33) and (2.41) in the definition are replaced by

$$W = \coprod_{n \in \mathbb{C}} W_{[n]} \tag{2.42}$$

and

$$\text{for } n \in \mathbb{C} \text{ and } w \in W_{[n]},$$
$$\big(L(0) - n\big)^m w = 0 \quad \text{for } m \in \mathbb{N} \text{ sufficiently large,} \tag{2.43}$$

respectively. For $w \in W_{[n]}$, we still write $\mathrm{wt}\, w = n$ for the (generalized) weight of w.

We will denote such a module or generalized module just defined by (W, Y), or sometimes by (W, Y_W) or simply by W. We will use the notation

$$\pi_n : W \to W_{[n]} \tag{2.44}$$

for the projection from W to its subspace of (generalized) weight n, and for its natural extensions to spaces of formal series with coefficients in W. In either the conformal or Möbius case, a module is of course a generalized module.

Remark 2.13 For any vector space U on which an operator, say, $L(0)$, acts in such a way that

$$U = \coprod_{n \in \mathbb{C}} U_{[n]} \tag{2.45}$$

where for $n \in \mathbb{C}$,

$$U_{[n]} = \big\{u \in U \,|\, \big(L(0) - n\big)^m u = 0 \text{ for } m \in \mathbb{N} \text{ sufficiently large}\big\},$$

we shall typically use the same projection notation

$$\pi_n : U \to U_{[n]} \tag{2.46}$$

as in (2.44). If instead of (2.45) we have only

$$U = \sum_{n \in \mathbb{C}} U_{[n]},$$

then in fact this sum is indeed direct, and for any $L(0)$-stable subspace T of U, we have

$$T = \coprod_{n \in \mathbb{C}} T_{[n]}$$

(as with ordinary rather than generalized eigenspaces).

Remark 2.14 A module for a conformal vertex algebra V is obviously again a module for V viewed as a Möbius vertex algebra, and conversely, a module for V viewed as a Möbius vertex algebra is a module for V viewed as a conformal vertex algebra, by Remark 2.10. Similarly, the generalized modules for a conformal vertex algebra V are exactly the generalized modules for V viewed as a Möbius vertex algebra.

Remark 2.15 A conformal or Möbius vertex algebra is a module for itself (and in particular, a generalized module for itself).

Remark 2.16 In either the conformal or Möbius vertex algebra case, we have the obvious notions of *V-module homomorphism*, *submodule*, *quotient module*, and so on; in particular, homomorphisms are understood to be grading-preserving. We sometimes write the vector space of (generalized-) module maps (homomorphisms) $W_1 \to W_2$ for (generalized) V-modules W_1 and W_2 as $\mathrm{Hom}_V(W_1, W_2)$.

Remark 2.17 We have chosen the name "generalized module" here because the vector space underlying the module is graded by generalized eigenvalues. (This notion is different from the notion of "generalized module" used in [72]. A generalized module for a vertex operator algebra V as defined in, for example, Definition 2.11 of [72] is precisely a module for V viewed as a conformal vertex algebra.)

We will use the following notion of (formal algebraic) completion of a generalized module:

Definition 2.18 Let $W = \coprod_{n\in\mathbb{C}} W_{[n]}$ be a generalized module for a Möbius (or conformal) vertex algebra. We denote by $\overline{W}$ the (formal) completion of W with respect to the $\mathbb{C}$-grading, that is,

$$\overline{W} = \prod_{n\in\mathbb{C}} W_{[n]}. \tag{2.47}$$

We will use the same notation $\overline{U}$ for any $\mathbb{C}$-graded subspace U of W. We will continue to use the notation π_n for the projection from $\overline{W}$ to $W_{[n]}$:

$$\pi_n : \overline{W} \to W_{[n]}.$$

We will also continue to use the notation $\langle\cdot,\cdot\rangle_W$, or $\langle\cdot,\cdot\rangle$ if the underlying space is clear, for the canonical pairing between the subspace $\coprod_{n\in\mathbb{C}}(W_{[n]})^*$ of W^*, and $\overline{W}$. We are of course viewing $(W_{[n]})^*$ as embedded in W^* in the natural way, that is, for $w^* \in (W_{[n]})^*$,

$$\langle w^*, w\rangle_W = \langle w^*, w_n\rangle_{W_{[n]}} \tag{2.48}$$

for any $w = \sum_{m\in\mathbb{C}} w_m$ (finite sum) in W, where $w_m \in W_{[m]}$.

The following weight formula holds for generalized modules, generalizing the corresponding formula in the module case (cf. [104]):

Proposition 2.19 *Let W be a generalized module for a Möbius (or conformal) vertex algebra V. Let both $v \in V$ and $w \in W$ be homogeneous. Then*

$$\mathrm{wt}(v_n w) = \mathrm{wt}\, v + \mathrm{wt}\, w - n - 1 \quad \textit{for any } n \in \mathbb{Z}, \tag{2.49}$$

$$\mathrm{wt}\big(L(j)w\big) = \mathrm{wt}\, w - j \quad \textit{for } j = 0, \pm 1. \tag{2.50}$$

Proof Applying the $L(-1)$-derivative property (2.40) to formula (2.29), with the operators acting on W, and extracting the coefficient of x^{-n-1}, we obtain:

$$\big[L(0), v_n\big] = \big(L(0)v\big)_n + (-n-1)v_n. \tag{2.51}$$

This can be written as

$$\big(L(0) - (\mathrm{wt}\, v - n - 1)\big)v_n = v_n L(0),$$

and so we have

$$\big(L(0) - (\mathrm{wt}\, v + m - n - 1)\big)v_n = v_n\big(L(0) - m\big)$$

for any $m \in \mathbb{C}$. Applying this repeatedly we get

$$\big(L(0) - (\mathrm{wt}\, v + m - n - 1)\big)^t v_n = v_n\big(L(0) - m\big)^t$$

for any $t \in \mathbb{N}$, $m \in \mathbb{C}$, and (2.49) follows.

For (2.50), since as operators acting on W we have

$$\big[L(0), L(j)\big] = -jL(j) \tag{2.52}$$

for $j = 0, \pm 1$, we get $(L(0) + j)L(j) = L(j)L(0)$ so that

$$\big(L(0) - m + j\big)L(j) = L(j)\big(L(0) - m\big)$$

for any $m \in \mathbb{C}$. Thus

$$\big(L(0) - m + j\big)^t L(j) = L(j)\big(L(0) - m\big)^t$$

for any $t \in \mathbb{N}$, $m \in \mathbb{C}$, and (2.50) follows. □

Remark 2.20 From Proposition 2.19 we see that a generalized V-module W decomposes into submodules corresponding to the congruence classes of its weights modulo $\mathbb{Z}$: For $\mu \in \mathbb{C}/\mathbb{Z}$, let

$$W_{[\mu]} = \coprod_{\bar{n} = \mu} W_{[n]}, \tag{2.53}$$

where $\bar{n}$ denotes the equivalence class of $n \in \mathbb{C}$ in $\mathbb{C}/\mathbb{Z}$. Then

$$W = \coprod_{\mu \in \mathbb{C}/\mathbb{Z}} W_{[\mu]} \tag{2.54}$$

and each $W_{[\mu]}$ is a V-submodule of W. Thus if a generalized module W is indecomposable (in particular, if it is irreducible), then all complex numbers n for which $W_{[n]} \neq 0$ are congruent modulo $\mathbb{Z}$ to each other.

Remark 2.21 Let W be a generalized module for a Möbius (or conformal) vertex algebra V. We consider the "semisimple part" $L(0)_s \in \operatorname{End} W$ of the operator $L(0)$:

$$L(0)_s w = nw \quad \text{for } w \in W_{[n]}, n \in \mathbb{C}.$$

Then on W we have

$$\big[L(0)_s, v_n\big] = \big[L(0), v_n\big] \quad \text{for all } v \in V \text{ and } n \in \mathbb{Z}; \tag{2.55}$$

$$\big[L(0)_s, L(j)\big] = \big[L(0), L(j)\big] \quad \text{for } j = 0, \pm 1. \tag{2.56}$$

Indeed, for homogeneous elements $v \in V$ and $w \in W$, (2.49) and (2.51) imply that

$$\begin{aligned}
\big[L(0)_s, v_n\big]w &= L(0)_s(v_n w) - v_n\big(L(0)_s w\big)\\
&= (\operatorname{wt} v + \operatorname{wt} w - n - 1)v_n w - (\operatorname{wt} w)v_n w\\
&= (\operatorname{wt} v)v_n w + (-n-1)v_n w\\
&= \big(L(0)v\big)_n w + (-n-1)v_n w\\
&= \big[L(0), v_n\big]w.
\end{aligned}$$

Similarly, for any homogeneous element $w \in W$ and $j = 0, \pm 1$, (2.50) and (2.52) imply that

$$\begin{aligned}
\big[L(0)_s, L(j)\big]w &= L(0)_s\big(L(j)w\big) - L(j)\big(L(0)_s w\big)\\
&= (\operatorname{wt} w - j)L(j)w - (\operatorname{wt} w)L(j)w\\
&= -jL(j)w\\
&= \big[L(0), L(j)\big]w.
\end{aligned}$$

Thus the "locally nilpotent part" $L(0) - L(0)_s$ of $L(0)$ commutes with the action of V and of $\mathfrak{sl}(2)$ on W. In other words, $L(0) - L(0)_s$ is a V-homomorphism from W to itself.

Now suppose that $L(1)$ acts locally nilpotently on a Möbius (or conformal) vertex algebra V, that is, for any $v \in V$, there is $m \in \mathbb{N}$ such that $L(1)^m v = 0$. Then generalizing formula (3.20) in [72] (the case of ordinary modules for a vertex operator algebra), we define the *opposite vertex operator* on a generalized V-module (W, Y_W) associated to $v \in V$ by

$$Y_W^o(v, x) = Y_W\big(e^{xL(1)}\big(-x^{-2}\big)^{L(0)} v, x^{-1}\big), \tag{2.57}$$

that is, for $k \in \mathbb{Z}$ and $v \in V_{(k)}$,

$$\begin{aligned} Y_W^o(v,x) &= \sum_{n\in\mathbb{Z}} v_n^o x^{-n-1} \\ &= \sum_{n\in\mathbb{Z}} \left((-1)^k \sum_{m\in\mathbb{N}} \frac{1}{m!} \big(L(1)^m v\big)_{-n-m-2+2k} \right) x^{-n-1}, \end{aligned} \tag{2.58}$$

as in [72]. (In the present work, we are replacing the symbol $*$ used in [72] for opposite vertex operators by the symbol o; see also Sect. 5.1 below.) Here we are defining the component operators

$$v_n^o = (-1)^k \sum_{m\in\mathbb{N}} \frac{1}{m!} \big(L(1)^m v\big)_{-n-m-2+2k} \tag{2.59}$$

for $v \in V_{(k)}$ and $n, k \in \mathbb{Z}$. Note that the $L(1)$-local nilpotence ensures well-definedness here. Clearly, $v \mapsto Y_W^o(v,x)$ is a linear map $V \to (\operatorname{End} W)[[x, x^{-1}]]$ such that $V \otimes W \to W((x^{-1}))$ $(v \otimes w \mapsto Y_W^o(v,x)w)$.

By (2.59), (2.32) and (2.49), we see that for $n, k \in \mathbb{Z}$ and $v \in V_{(k)}$, the operator v_n^o is of generalized weight $n + 1 - k$ $(= n + 1 - \operatorname{wt} v)$, in the sense that

$$v_n^o W_{[m]} \subset W_{[m+n+1-k]} \quad \text{for any } m \in \mathbb{C}. \tag{2.60}$$

As mentioned in [72] (see (3.23) in [72]), the proof of the Jacobi identity in Theorem 5.2.1 of [38] proves the following *opposite Jacobi identity* for Y_W^o in the case where V is a vertex operator algebra and W is a V-module:

$$\begin{aligned} & x_0^{-1}\delta\left(\frac{x_1 - x_2}{x_0}\right) Y_W^o(v, x_2) Y_W^o(u, x_1) \\ & \quad - x_0^{-1}\delta\left(\frac{x_2 - x_1}{-x_0}\right) Y_W^o(u, x_1) Y_W^o(v, x_2) \\ & = x_2^{-1}\delta\left(\frac{x_1 - x_0}{x_2}\right) Y_W^o\big(Y(u, x_0)v, x_2\big) \end{aligned} \tag{2.61}$$

for $u, v \in V$, and taking Res_{x_0} gives us the *opposite commutator formula*. Similarly, the proof of the $L(-1)$-derivative property in Theorem 5.2.1 of [38] proves the following $L(-1)$-*derivative property* for Y_W^o in the same case:

$$\frac{d}{dx} Y_W^o(v,x) = Y_W^o\big(L(-1)v, x\big). \tag{2.62}$$

The same proofs carry over and prove the opposite Jacobi identity and the $L(-1)$-derivative property for Y_W^o in the present case, where V is a Möbius (or conformal) vertex algebra with $L(1)$ acting locally nilpotently and where W is a generalized

V-module. In the case in which V is a conformal vertex algebra, we have

$$Y_W^o(\omega, x) = Y_W\left(x^{-4}\omega, x^{-1}\right) = \sum_{n\in\mathbb{Z}} L(n)x^{n-2} \tag{2.63}$$

since $L(1)\omega = 0$.

For opposite vertex operators, we have the following analogues of (2.28)–(2.31) in the Möbius case:

Lemma 2.22 *For $v \in V$,*

$$\left[Y_W^o(v, x), L(1)\right] = Y_W^o\left(L(-1)v, x\right), \tag{2.64}$$

$$\left[Y_W^o(v, x), L(0)\right] = Y_W^o\left(L(0)v, x\right) + xY_W^o\left(L(-1)v, x\right), \tag{2.65}$$

$$\begin{aligned}\left[Y_W^o(v, x), L(-1)\right] = {}& Y_W^o\left(L(1)v, x\right) + 2xY_W^o\left(L(0)v, x\right)\\ &+ x^2Y_W^o\left(L(-1)v, x\right).\end{aligned} \tag{2.66}$$

Equivalently,

$$\begin{aligned}\left[Y_W^o(v, x), L(-j)\right] &= \sum_{k=0}^{j+1}\binom{j+1}{k}x^kY_W^o\left(L(j-k)v, x\right)\\ &= \sum_{k=0}^{j+1}\binom{j+1}{k}x^{j+1-k}Y_W^o\left(L(k-1)v, x\right)\end{aligned} \tag{2.67}$$

for $j = 0, \pm 1$.

Proof For $j = 0, \pm 1$, by definition and (2.31) we have

$$\begin{aligned}&\left[Y_W^o(v, x), L(j)\right]\\ &\quad = -\left[L(j), Y_W\left(e^{xL(1)}\left(-x^{-2}\right)^{L(0)}v, x^{-1}\right)\right]\\ &\quad = -\sum_{k=0}^{j+1}\binom{j+1}{k}x^{-k}Y_W\left(L(j-k)e^{xL(1)}\left(-x^{-2}\right)^{L(0)}v, x^{-1}\right).\end{aligned} \tag{2.68}$$

By (5.2.14) in [38] and the fact that

$$x^{L(0)}L(j)x^{-L(0)} = x^{-j}L(j) \tag{2.69}$$

(easily proved by applying to a homogeneous vector),

$$\begin{aligned}&L(-1)e^{xL(1)}\left(-x^{-2}\right)^{L(0)}\\ &\quad = e^{xL(1)}L(-1)\left(-x^{-2}\right)^{L(0)} - 2xe^{xL(1)}L(0)\left(-x^{-2}\right)^{L(0)}\end{aligned}$$

$$
\begin{aligned}
&\quad + x^2 e^{xL(1)} L(1) \big(-x^{-2}\big)^{L(0)}\\
&= -x^2 e^{xL(1)} \big(-x^{-2}\big)^{L(0)} L(-1) - 2x e^{xL(1)} \big(-x^{-2}\big)^{L(0)} L(0)\\
&\quad - e^{xL(1)} \big(-x^{-2}\big)^{L(0)} L(1)\\
&= -e^{xL(1)} \big(-x^{-2}\big)^{L(0)} \big(x^2 L(-1) + 2x L(0) + L(1)\big).
\end{aligned} \tag{2.70}
$$

We also have

$$
\begin{aligned}
L(1) e^{xL(1)} \big(-x^{-2}\big)^{L(0)} &= e^{xL(1)} L(1) \big(-x^{-2}\big)^{L(0)}\\
&= -x^{-2} e^{xL(1)} \big(-x^{-2}\big)^{L(0)} L(1).
\end{aligned} \tag{2.71}
$$

By (2.70), (2.71), $L(0) = \frac{1}{2}[L(1), L(-1)]$ and $[L(1), L(0)] = L(1)$, we have

$$
\begin{aligned}
&L(0) e^{xL(1)} \big(-x^{-2}\big)^{L(0)}\\
&= \frac{1}{2} L(1) L(-1) e^{xL(1)} \big(-x^{-2}\big)^{L(0)} - \frac{1}{2} L(-1) L(1) e^{xL(1)} \big(-x^{-2}\big)^{L(0)}\\
&= -\frac{1}{2} L(1) e^{xL(1)} \big(-x^{-2}\big)^{L(0)} \big(x^2 L(-1) + 2x L(0) + L(1)\big)\\
&\quad + \frac{1}{2} x^{-2} L(-1) e^{xL(1)} \big(-x^{-2}\big)^{L(0)} L(1)\\
&= \frac{1}{2} x^{-2} e^{xL(1)} \big(-x^{-2}\big)^{L(0)} L(1) \big(x^2 L(-1) + 2x L(0) + L(1)\big)\\
&\quad - \frac{1}{2} x^{-2} e^{xL(1)} \big(-x^{-2}\big)^{L(0)} \big(x^2 L(-1) + 2x L(0) + L(1)\big) L(1)\\
&= e^{xL(1)} \big(-x^{-2}\big)^{L(0)} L(0) + x^{-1} e^{xL(1)} \big(-x^{-2}\big)^{L(0)} L(1)\\
&= e^{xL(1)} \big(-x^{-2}\big)^{L(0)} \big(L(0) + x^{-1} L(1)\big).
\end{aligned} \tag{2.72}
$$

Thus we obtain

$$
\begin{aligned}
&\big[Y_W^o(v, x), L(1)\big]\\
&= -\sum_{k=0}^{2} \binom{2}{k} x^{-k} Y_W\big(L(1-k) e^{xL(1)} \big(-x^{-2}\big)^{L(0)} v, x^{-1}\big)\\
&= -Y_W\big(L(1) e^{xL(1)} \big(-x^{-2}\big)^{L(0)} v, x^{-1}\big)\\
&\quad - 2x^{-1} Y_W\big(L(0) e^{xL(1)} \big(-x^{-2}\big)^{L(0)} v, x^{-1}\big)\\
&\quad - x^{-2} Y_W\big(L(-1) e^{xL(1)} \big(-x^{-2}\big)^{L(0)} v, x^{-1}\big)\\
&= x^{-2} Y_W\big(e^{xL(1)} \big(-x^{-2}\big)^{L(0)} L(1) v, x^{-1}\big)
\end{aligned}
$$

$$
\begin{aligned}
&\quad - 2x^{-1}Y_W\big(e^{xL(1)}\big(-x^{-2}\big)^{L(0)}\big(L(0)+x^{-1}L(1)\big)v, x^{-1}\big)\\
&\quad + x^{-2}Y_W\big(e^{xL(1)}\big(-x^{-2}\big)^{L(0)}\big(x^2L(-1)+2xL(0)+L(1)\big)v, x^{-1}\big)\\
&= Y_W\big(e^{xL(1)}\big(-x^{-2}\big)^{L(0)}L(-1)v, x^{-1}\big)\\
&= Y_W^o\big(L(-1)v, x\big),
\end{aligned}
$$

$$
\begin{aligned}
&\big[Y_W^o(v,x), L(0)\big]\\
&= -\sum_{k=0}^{1}\binom{1}{k}x^{-k}Y_W\big(L(-k)e^{xL(1)}\big(-x^{-2}\big)^{L(0)}v, x^{-1}\big)\\
&= -Y_W\big(L(0)e^{xL(1)}\big(-x^{-2}\big)^{L(0)}v, x^{-1}\big)\\
&\quad - x^{-1}Y_W\big(L(-1)e^{xL(1)}\big(-x^{-2}\big)^{L(0)}v, x^{-1}\big)\\
&= -Y_W\big(e^{xL(1)}\big(-x^{-2}\big)^{L(0)}\big(L(0)+x^{-1}L(1)\big)v, x^{-1}\big)\\
&\quad + x^{-1}Y_W\big(e^{xL(1)}\big(-x^{-2}\big)^{L(0)}\big(x^2L(-1)+2xL(0)+L(1)\big)v, x^{-1}\big)\\
&= Y_W\big(e^{xL(1)}\big(-x^{-2}\big)^{L(0)}\big(xL(-1)+L(0)\big)v, x^{-1}\big)\\
&= Y_W^o\big(L(0)v, x\big) + xY_W^o\big(L(-1)v, x\big)
\end{aligned}
$$

and

$$
\begin{aligned}
\big[Y_W^o(v,x), L(-1)\big] &= -Y_W\big(L(-1)e^{xL(1)}\big(-x^{-2}\big)^{L(0)}v, x^{-1}\big)\\
&= Y_W\big(e^{xL(1)}\big(-x^{-2}\big)^{L(0)}\big(x^2L(-1)+2xL(0)+L(1)\big)v, x^{-1}\big)\\
&= Y_W^o\big(L(1)v, x\big) + 2xY_W^o\big(L(0)v, x\big) + x^2Y_W^o\big(L(-1)v, x\big),
\end{aligned}
$$

proving the lemma. □

As in Sect. 5.2 of [38], we can define a V-action on W^* as follows:

$$\big\langle Y'(v,x)w', w\big\rangle = \big\langle w', Y_W^o(v,x)w\big\rangle \tag{2.73}$$

for $v \in V$, $w' \in W^*$ and $w \in W$; the correspondence $v \mapsto Y'(v,x)$ is a linear map from V to $(\text{End}\, W^*)[[x, x^{-1}]]$. Writing

$$Y'(v,x) = \sum_{n\in\mathbb{Z}} v_n x^{-n-1}$$

($v_n \in \text{End}\, W^*$), we have

$$\big\langle v_n w', w\big\rangle = \big\langle w', v_n^o w\big\rangle \tag{2.74}$$

for $v \in V$, $w' \in W^*$ and $w \in W$. (Actually, in [38] this V-action was defined on a space smaller than W^*, but this definition holds without change on all of W^*.) In the case in which V is a conformal vertex algebra we define the operators $L'(n)$ ($n \in \mathbb{Z}$) by

$$Y'(\omega, x) = \sum_{n\in\mathbb{Z}} L'(n)x^{-n-2};$$

then, by extracting the coefficient of x^{-n-2} in (2.73) with $v = \omega$ and using the fact that $L(1)\omega = 0$ we have

$$\big\langle L'(n)w', w\big\rangle = \big\langle w', L(-n)w\big\rangle \quad \text{for } n \in \mathbb{Z} \tag{2.75}$$

(see (2.63)), as in Sect. 5.2 of [38]. In the case where V is only a Möbius vertex algebra, we define operators $L'(-1)$, $L'(0)$ and $L'(1)$ on W^* by formula (2.75) for $n = 0, \pm 1$. It follows from (2.50) that

$$L'(j)(W_{[m]})^* \subset (W_{[m-j]})^* \tag{2.76}$$

for $m \in \mathbb{C}$ and $j = 0, \pm 1$. By combining (2.74) with (2.60) we get

$$v_n(W_{[m]})^* \subset (W_{[m+k-n-1]})^* \tag{2.77}$$

for any $n, k \in \mathbb{Z}$, $v \in V_{(k)}$ and $m \in \mathbb{C}$.

We have just seen that the $L(1)$-local nilpotence condition enables us to define a natural vertex operator action on the vector space dual of a generalized module for a Möbius (or conformal) vertex algebra. This condition is satisfied by all vertex operator algebras, due to (2.32) and the grading restriction condition (2.24). However, the functor $W \mapsto W^*$ is certainly not involutive, and W^* is not in general a generalized module. In this work we will need certain module categories equipped with an involutive "contragredient functor" $W \mapsto W'$ which generalizes the contragredient functor for the category of modules for vertex operator algebras. For this purpose, we introduce the following:

Definition 2.23 Let A be an abelian group. A Möbius (or conformal) vertex algebra

$$V = \coprod_{n\in\mathbb{Z}} V_{(n)}$$

is said to be *strongly graded with respect to A* (or *strongly A-graded*, or just *strongly graded* if the abelian group A is understood) if V is equipped with a second gradation, by A,

$$V = \coprod_{\alpha\in A} V^{(\alpha)},$$

such that the following conditions are satisfied: the two gradations are compatible, that is,

$$V^{(\alpha)} = \coprod_{n\in\mathbb{Z}} V^{(\alpha)}_{(n)} \quad \big(\text{where } V^{(\alpha)}_{(n)} = V_{(n)} \cap V^{(\alpha)}\big) \text{ for any } \alpha \in A;$$

for any $\alpha, \beta \in A$ and $n \in \mathbb{Z}$,

$$V_{(n)}^{(\alpha)} = 0 \quad \text{for } n \text{ sufficiently negative;} \tag{2.78}$$

$$\dim V_{(n)}^{(\alpha)} < \infty; \tag{2.79}$$

$$\mathbf{1} \in V_{(0)}^{(0)}; \tag{2.80}$$

$$v_l V^{(\beta)} \subset V^{(\alpha+\beta)} \quad \text{for any } v \in V^{(\alpha)}, l \in \mathbb{Z}; \tag{2.81}$$

and

$$L(j)V^{(\alpha)} \subset V^{(\alpha)} \quad \text{for } j = 0, \pm 1. \tag{2.82}$$

If V is in fact a conformal vertex algebra, we in addition require that

$$\omega \in V_{(2)}^{(0)}, \tag{2.83}$$

so that for all $j \in \mathbb{Z}$, (2.82) follows from (2.81).

Remark 2.24 Note that the notion of conformal vertex algebra strongly graded with respect to the trivial group is exactly the notion of vertex operator algebra. Also note that (2.32), (2.78) and (2.82) imply the local nilpotence of $L(1)$ acting on V, and hence we have the construction and properties of opposite vertex operators on a generalized module for a strongly graded Möbius (or conformal) vertex algebra.

For (generalized) modules for a strongly graded algebra we will also have a second grading by an abelian group, and it is natural to allow this group to be larger than the second grading group A for the algebra. (Note that this already occurs for the *first* grading group, which is $\mathbb{Z}$ for algebras and $\mathbb{C}$ for (generalized) modules.) We now define the notions of strongly graded module and generalized module, and also, at the end of this definition, the notions of lower bounded such structures.

Definition 2.25 Let A be an abelian group and V a strongly A-graded Möbius (or conformal) vertex algebra. Let $\tilde{A}$ be an abelian group containing A as a subgroup. A V-module (respectively, generalized V-module)

$$W = \coprod_{n \in \mathbb{C}} W_{(n)} \quad \left(\text{respectively, } W = \coprod_{n \in \mathbb{C}} W_{[n]}\right)$$

is said to be *strongly graded with respect to* $\tilde{A}$ (or *strongly* $\tilde{A}$*-graded*, or just *strongly graded*) if the abelian group $\tilde{A}$ is understood) if W is equipped with a second gradation, by $\tilde{A}$,

$$W = \coprod_{\beta \in \tilde{A}} W^{(\beta)}, \tag{2.84}$$

such that the following conditions are satisfied: the two gradations are compatible, that is, for any $\beta \in \tilde{A}$,

$$W^{(\beta)} = \coprod_{n\in\mathbb{C}} W^{(\beta)}_{(n)} \quad \big(\text{where } W^{(\beta)}_{(n)} = W_{(n)} \cap W^{(\beta)}\big)$$

$$\Big(\text{respectively, } W^{(\beta)} = \coprod_{n\in\mathbb{C}} W^{(\beta)}_{[n]} \ \big(\text{where } W^{(\beta)}_{[n]} = W_{[n]} \cap W^{(\beta)}\big)\Big);$$

for any $\alpha \in A$, $\beta \in \tilde{A}$ and $n \in \mathbb{C}$,

$$W^{(\beta)}_{(n+k)} = 0 \quad \big(\text{respectively, } W^{(\beta)}_{[n+k]} = 0\big) \text{ for } k \in \mathbb{Z} \text{ sufficiently negative;} \tag{2.85}$$

$$\dim W^{(\beta)}_{(n)} < \infty \quad \big(\text{respectively, } \dim W^{(\beta)}_{[n]} < \infty\big); \tag{2.86}$$

$$v_l W^{(\beta)} \subset W^{(\alpha+\beta)} \quad \text{for any } v \in V^{(\alpha)}, l \in \mathbb{Z}; \tag{2.87}$$

and

$$L(j)W^{(\beta)} \subset W^{(\beta)} \quad \text{for } j = 0, \pm 1. \tag{2.88}$$

(Note that if V is in fact a conformal vertex algebra, then for all $j \in \mathbb{Z}$, (2.88) follows from (2.83) and (2.87).) A strongly $\tilde{A}$-graded (generalized) V-module W is said to be *lower bounded* if instead of (2.85), it satisfies the stronger condition that for any $\beta \in \tilde{A}$,

$$W^{(\beta)}_{(n)} = 0 \quad \big(\text{respectively, } W^{(\beta)}_{[n]} = 0\big) \text{ for } \Re(n) \text{ sufficiently negative} \tag{2.89}$$

($n \in \mathbb{C}$).

Remark 2.26 A strongly A-graded conformal or Möbius vertex algebra is a strongly A-graded module for itself (and in particular, a strongly A-graded generalized module for itself), and is in fact lower bounded.

Remark 2.27 Let V be a vertex operator algebra, viewed (equivalently) as a conformal vertex algebra strongly graded with respect to the trivial group (recall Remark 2.24). Then the V-modules that are strongly graded with respect to the trivial group (in the sense of Definition 2.25) are exactly the ($\mathbb{C}$-graded) modules for V as a vertex operator algebra, with the grading restrictions as follows: For $n \in \mathbb{C}$,

$$W_{(n+k)} = 0 \quad \text{for } k \in \mathbb{Z} \text{ sufficiently negative} \tag{2.90}$$

and

$$\dim W_{(n)} < \infty, \tag{2.91}$$

and the lower bounded such structures have (2.90) replaced by:

$$W_{(n)} = 0 \quad \text{for } \Re(n) \text{ sufficiently negative.} \tag{2.92}$$

Also, the generalized V-modules that are strongly graded with respect to the trivial group are exactly the generalized V-modules (in the sense of Definition 2.12) such that for $n \in \mathbb{C}$,

$$W_{[n+k]} = 0 \quad \text{for } k \in \mathbb{Z} \text{ sufficiently negative} \tag{2.93}$$

and

$$\dim W_{[n]} < \infty, \tag{2.94}$$

and the lower bounded ones have (2.90) replaced by:

$$W_{[n]} = 0 \quad \text{for } \Re(n) \text{ sufficiently negative.} \tag{2.95}$$

Remark 2.28 In the strongly graded case, algebra and module homomorphisms are of course understood to preserve the grading by A or $\tilde{A}$.

Example 2.29 An important source of examples of strongly graded conformal vertex algebras and modules comes from the vertex algebras and modules associated with even lattices. Let L be an even lattice, i.e., a finite-rank free abelian group equipped with a nondegenerate symmetric bilinear form $\langle \cdot, \cdot \rangle$, not necessarily positive definite, such that $\langle \alpha, \alpha \rangle \in 2\mathbb{Z}$ for all $\alpha \in L$. Then there is a natural structure of conformal vertex algebra on a certain vector space V_L; see [15] and Chap. 8 of [37]. If the form $\langle \cdot, \cdot \rangle$ on L is also positive definite, then V_L is a vertex operator algebra (that is, the grading restrictions hold). If L is not necessarily positive definite, then V_L is equipped with a natural second grading given by L itself, making V_L a strongly L-graded conformal vertex algebra in the sense of Definition 2.23. Any (rational) sublattice M of the "dual lattice" L° of L containing L gives rise to a lower bounded strongly M-graded module for the strongly L-graded conformal vertex algebra (see Chap. 8 of [37]; cf. [99]).

In the next two remarks, we mention certain important properties of compositions of two or more vertex operators, properties that will also be important in the further generality of logarithmic intertwining operators in the future.

Remark 2.30 As mentioned in Remark 2.24, strong gradedness for a Möbius (or conformal) vertex algebra V implies the local nilpotence of $L(1)$ acting on V. In fact, strong gradedness implies much more that will be important for us: From (2.78), (2.79), (2.81) and (2.82) (and (2.83) in the conformal vertex algebra case), it is clear that strong gradedness for V implies the following *local grading restriction condition on* V (see [58]):

(i) for any $m > 0$ and $v_{(1)}, \ldots, v_{(m)} \in V$, there exists $r \in \mathbb{Z}$ such that the coefficient of each monomial in $x_1, \ldots, x_{m-1}$ in the formal series

$$Y(v_{(1)}, x_1) \cdots Y(v_{(m-1)}, x_{m-1}) v_{(m)}$$

lies in $\coprod_{n>r} V_{(n)}$;

(ii) in the conformal vertex algebra case: for any element of the conformal vertex algebra V homogeneous with respect to the weight grading, the Virasoro-algebra submodule $M = \coprod_{n\in\mathbb{Z}} M_{(n)}$ (where $M_{(n)} = M \cap V_{(n)}$) of V generated by this element satisfies the following grading restriction conditions: $M_{(n)} = 0$ when n is sufficiently negative and $\dim M_{(n)} < \infty$ for $n \in \mathbb{Z}$

or

(ii′) in the Möbius vertex algebra case: for any element of the Möbius vertex algebra V homogeneous with respect to the weight grading, the $\mathfrak{sl}(2)$-submodule $M = \coprod_{n\in\mathbb{Z}} M_{(n)}$ (where $M_{(n)} = M \cap V_{(n)}$) of V generated by this element satisfies the following grading restriction conditions: $M_{(n)} = 0$ when n is sufficiently negative and $\dim M_{(n)} < \infty$ for $n \in \mathbb{Z}$.

As was pointed out in [58], Condition (i) above was first stated in [21] (see formula (9.39), Proposition 9.17 and Theorem 12.33 in [21]) for generalized vertex algebras and abelian intertwining algebras (certain generalizations of vertex algebras); it guarantees the convergence, rationality and commutativity properties of the matrix coefficients of products of more than two vertex operators. Conditions (i) and (ii) (or (ii′)) together ensure that all the essential results involving the Virasoro operators and the geometry of vertex operator algebras in [52] and [56] still hold for these algebras.

Remark 2.31 Similarly, from (2.85), (2.86), (2.87) and (2.88) (and (2.83) in the conformal vertex algebra case), it is clear that strong gradedness for (generalized) modules implies the following *local grading restriction condition on a (generalized) module* W for a strongly graded Möbius (or conformal) vertex algebra V:

(i) for any $m > 0$, $v_{(1)}, \ldots, v_{(m-1)} \in V$, $n \in \mathbb{C}$ and $w \in W_{[n]}$, there exists $r \in \mathbb{Z}$ such that the coefficient of each monomial in $x_1, \ldots, x_{m-1}$ in the formal series

$$Y(v_{(1)}, x_1) \cdots Y(v_{(m-1)}, x_{m-1})w$$

lies in $\coprod_{k>r} W_{[n+k]}$;

(ii) in the conformal vertex algebra case: for any $w \in W_{[n]}$ ($n \in \mathbb{C}$), the Virasoro-algebra submodule $M = \coprod_{k\in\mathbb{Z}} M_{[n+k]}$ (where $M_{[n+k]} = M \cap W_{[n+k]}$) of W generated by w satisfies the following grading restriction conditions: $M_{[n+k]} = 0$ when k is sufficiently negative and $\dim M_{[n+k]} < \infty$ for $k \in \mathbb{Z}$

or

(ii′) in the Möbius vertex algebra case: for any $w \in W_{[n]}$ ($n \in \mathbb{C}$), the $\mathfrak{sl}(2)$-submodule $M = \coprod_{k\in\mathbb{Z}} M_{[n+k]}$ (where $M_{[n+k]} = M \cap W_{[n+k]}$) of W generated by w satisfies the following grading restriction conditions: $M_{[n+k]} = 0$ when k is sufficiently negative and $\dim M_{[n+k]} < \infty$ for $k \in \mathbb{Z}$.

Note that in the case of ordinary (as opposed to generalized) modules, all the generalized weight spaces such as $W_{[n]}$ mentioned here are ordinary weight spaces $W_{(n)}$. Analogous statements of course hold for lower bounded (generalized) modules.

With the strong gradedness condition on a (generalized) module, we can now define the corresponding notion of contragredient module. First we give:

Definition 2.32 Let $W = \coprod_{\beta\in\tilde{A},n\in\mathbb{C}} W_{[n]}^{(\beta)}$ be a strongly $\tilde{A}$-graded generalized module for a strongly A-graded Möbius (or conformal) vertex algebra. For each $\beta \in \tilde{A}$ and $n \in \mathbb{C}$, let us identify $(W_{[n]}^{(\beta)})^*$ with the subspace of W^* consisting of the linear functionals on W vanishing on each $W_{[m]}^{(\gamma)}$ with $\gamma \neq \beta$ or $m \neq n$ (cf. (2.48)). We define W' to be the $(\tilde{A} \times \mathbb{C})$-graded vector subspace of W^* given by

$$W' = \coprod_{\beta\in\tilde{A},n\in\mathbb{C}} (W')_{[n]}^{(\beta)}, \quad \text{where } (W')_{[n]}^{(\beta)} = \left(W_{[n]}^{(-\beta)}\right)^*; \tag{2.96}$$

we also use the notations

$$\left(W'\right)^{(\beta)} = \coprod_{n\in\mathbb{C}} \left(W_{[n]}^{(-\beta)}\right)^* \subset \left(W^{(-\beta)}\right)^* \subset W^* \tag{2.97}$$

(where $(W^{(\beta)})^*$ consists of the linear functionals on W vanishing on all $W^{(\gamma)}$ with $\gamma \neq \beta$) and

$$\left(W'\right)_{[n]} = \coprod_{\beta\in\tilde{A}} \left(W_{[n]}^{(-\beta)}\right)^* \subset (W_{[n]})^* \subset W^* \tag{2.98}$$

for the homogeneous subspaces of W' with respect to the $\tilde{A}$- and $\mathbb{C}$-grading, respectively. (The reason for the minus signs here will become clear below.) We will still use the notation $\langle\cdot,\cdot\rangle_W$, or $\langle\cdot,\cdot\rangle$ when the underlying space is clear, for the canonical pairing between W' and

$$\overline{W} \subset \prod_{\beta\in\tilde{A},n\in\mathbb{C}} W_{[n]}^{(\beta)}$$

(recall (2.47)).

Remark 2.33 In the case of ordinary rather than generalized modules, Definition 2.32 still applies, and all of the generalized weight subspaces $W_{[n]}$ of W are ordinary weight spaces $W_{(n)}$. In this case, we can write $(W')_{(n)}$ rather than $(W')_{[n]}$ for the corresponding subspace of W'.

Let W be a strongly graded (generalized) module for a strongly graded Möbius (or conformal) vertex algebra V. Recall that we have the action (2.73) of V on W^* and that (2.77) holds. Furthermore, (2.59), (2.74) and (2.87) imply for any $n, k \in \mathbb{Z}$, $\alpha \in A$, $\beta \in \tilde{A}$, $v \in V_{(k)}^{(\alpha)}$ and $m \in \mathbb{C}$,

$$v_n\left(\left(W'\right)_{[m]}^{(\beta)}\right) = v_n\left(\left(W_{[m]}^{(-\beta)}\right)^*\right) \subset \left(W_{[m+k-n-1]}^{(-\alpha-\beta)}\right)^* = \left(W'\right)_{[m+k-n-1]}^{(\alpha+\beta)}. \tag{2.99}$$

Thus v_n preserves W' for $v \in V$, $n \in \mathbb{Z}$. Similarly (in the Möbius case), (2.75), (2.76) and (2.88) imply that W' is stable under the operators $L'(-1)$, $L'(0)$ and $L'(1)$, and in fact

$$L'(j)(W')^{(\beta)}_{[n]} \subset (W')^{(\beta)}_{[n-j]}$$

for any $j = 0, \pm 1$, $\beta \in \tilde{A}$ and $n \in \mathbb{C}$. In the case or ordinary rather than generalized modules, the symbols $(W')^{(\beta)}_{[n]}$, etc., can be replaced by $(W')^{(\beta)}_{(n)}$, etc.

For any fixed $\beta \in \tilde{A}$ and $n \in \mathbb{C}$, by (2.43) and the finite-dimensionality (2.86) of $W^{(-\beta)}_{[n]}$, there exists $N \in \mathbb{N}$ such that $(L(0) - n)^N W^{(-\beta)}_{[n]} = 0$. But then for any $w' \in (W')^{(\beta)}_{[n]}$,

$$\langle (L'(0) - n)^N w', w \rangle = \langle w', (L(0) - n)^N w \rangle = 0 \tag{2.100}$$

for all $w \in W$. Thus $(L'(0) - n)^N w' = 0$. So (2.43) holds with W replaced by W'. In the case of ordinary modules, we of course take $N = 1$.

By (2.85) and (2.99) we have the lower truncation condition for the action Y' of V on W':

$$\text{For any } v \in V \text{ and } w' \in W', \quad v_n w' = 0 \quad \text{for } n \text{ sufficiently large.} \tag{2.101}$$

As a consequence, the Jacobi identity can now be formulated on W'. In fact, by the above, and using the same proofs as those of Theorems 5.2.1 and 5.3.1 in [38], together with Lemma 2.22, we obtain:

Theorem 2.34 *Let $\tilde{A}$ be an abelian group containing A as a subgroup and V a strongly A-graded Möbius (or conformal) vertex algebra. Let (W, Y) be a strongly $\tilde{A}$-graded V-module (respectively, generalized V-module). Then the pair (W', Y') carries a strongly $\tilde{A}$-graded V-module (respectively, generalized V-module) structure, and*

$$(W'', Y'') = (W, Y).$$

If W is lower bounded, then so is W'.

Definition 2.35 The pair (W', Y') in Theorem 2.34 will be called the *contragredient module* of (W, Y).

Let W_1 and W_2 be strongly $\tilde{A}$-graded (generalized) V-modules and let $f : W_1 \to W_2$ be a module homomorphism (which is of course understood to preserve both the $\mathbb{C}$-grading and the $\tilde{A}$-grading, and to preserve the action of $\mathfrak{sl}(2)$ in the Möbius case). Then by (2.74) and (2.75), the linear map

$$f' : W_2' \to W_1'$$

given by

$$\langle f'(w'_{(2)}), w_{(1)} \rangle = \langle w'_{(2)}, f(w_{(1)}) \rangle \tag{2.102}$$

for any $w_{(1)} \in W_1$ and $w'_{(2)} \in W'_2$ is well defined and is clearly a module homomorphism from W'_2 to W'_1.

Notation 2.36 In this work we will be especially interested in the case where V is strongly A-graded, and we will be focusing on the category of all strongly $\tilde{A}$-graded (ordinary) V-modules, for which we will use the notation

$$\mathcal{M}_{sg},$$

or the category of all strongly $\tilde{A}$-graded generalized V-modules, which we will call

$$\mathcal{GM}_{sg}.$$

From the above we see that in the strongly graded case we have contravariant functors

$$(\cdot)' : (W, Y) \mapsto (W', Y'),$$

the *contragredient functors*, from $\mathcal{M}_{sg}$ to itself and from $\mathcal{GM}_{sg}$ to itself, and also from the full subcategories of lower bounded such structures to themselves. We also know that V itself is a (lower bounded) object of $\mathcal{M}_{sg}$ (and thus of $\mathcal{GM}_{sg}$ as well); recall Remark 2.26. Our main objects of study will be certain full subcategories $\mathcal{C}$ of $\mathcal{M}_{sg}$ or $\mathcal{GM}_{sg}$ that are closed under the contragredient functor and such that $V \in \mathrm{ob}\,\mathcal{C}$.

Remark 2.37 In order to formulate certain results in this work, even in the case when our Möbius or conformal vertex algebra V is strongly graded we will in fact sometimes use the category whose objects are *all* the modules for V and whose morphisms are all the V-module homomorphisms, and also the category of *all* the generalized modules for V. (If V is conformal, then the category of all the V-modules is the same whether V is viewed as either conformal or Möbius, by Remark 2.14, and similarly for the category of all the generalized V-modules.) Note that in view of Remark 2.28, the categories $\mathcal{M}_{sg}$ and $\mathcal{GM}_{sg}$ are not full subcategories of these categories of *all* modules and generalized modules.

We now recall from [21, 37, 38] and [99] the well-known principles that vertex operator algebras (which are exactly conformal vertex algebras strongly graded with respect to the trivial group; recall Remark 2.24) and their modules have important "rationality," "commutativity" and "associativity" properties, and that these properties can in fact be used as axioms replacing the Jacobi identity in the definition of the notion of vertex operator algebra. (These principles in fact generalize to all vertex algebras, as in [99].)

In the propositions below,

$$\mathbb{C}[x_1, x_2]_S$$

is the ring of formal rational functions obtained by inverting (localizing with respect to) the products of (zero or more) elements of the set S of nonzero homogeneous linear polynomials in x_1 and x_2. Also, ι_{12} (which might also be written as $\iota_{x_1x_2}$) is the operation of expanding an element of $\mathbb{C}[x_1, x_2]_S$, that is, a polynomial in x_1 and x_2 divided by a product of homogeneous linear polynomials in x_1 and x_2, as a formal series containing at most finitely many negative powers of x_2 (using binomial expansions for negative powers of linear polynomials involving both x_1 and x_2); similarly for ι_{21} and so on. (The distinction between rational functions and formal Laurent series is crucial.)

Let V be a vertex operator algebra. For W a ($\mathbb{C}$-graded) V-module (including possibly V itself), the space W' is just the "restricted dual space"

$$W' = \coprod_{n\in\mathbb{C}} W^*_{(n)}. \tag{2.103}$$

Proposition 2.38 *We have*:

(a) (*Rationality of products*) *For $v, v_1, v_2 \in V$ and $v' \in V'$, the formal series*

$$\langle v', Y(v_1, x_1)Y(v_2, x_2)v\rangle, \tag{2.104}$$

which involves only finitely many negative powers of x_2 and only finitely many positive powers of x_1, lies in the image of the map ι_{12}:

$$\langle v', Y(v_1, x_1)Y(v_2, x_2)v\rangle = \iota_{12} f(x_1, x_2), \tag{2.105}$$

where the (uniquely determined) element $f \in \mathbb{C}[x_1, x_2]_S$ is of the form

$$f(x_1, x_2) = \frac{g(x_1, x_2)}{x_1^r x_2^s (x_1 - x_2)^t} \tag{2.106}$$

for some $g \in \mathbb{C}[x_1, x_2]$ and $r, s, t \in \mathbb{Z}$.

(b) (*Commutativity*) *We also have*

$$\langle v', Y(v_2, x_2)Y(v_1, x_1)v\rangle = \iota_{21} f(x_1, x_2). \tag{2.107}$$

Proposition 2.39 *We have*:

(a) (*Rationality of iterates*) *For $v, v_1, v_2 \in V$ and $v' \in V'$, the formal series*

$$\langle v', Y(Y(v_1, x_0)v_2, x_2)v\rangle, \tag{2.108}$$

which involves only finitely many negative powers of x_0 and only finitely many positive powers of x_2, lies in the image of the map ι_{20}:

$$\langle v', Y(Y(v_1, x_0)v_2, x_2)v\rangle = \iota_{20} h(x_0, x_2), \tag{2.109}$$

where the (uniquely determined) element $h \in \mathbb{C}[x_0, x_2]_S$ is of the form

$$h(x_0, x_2) = \frac{k(x_0, x_2)}{x_0^r x_2^s (x_0 + x_2)^t} \tag{2.110}$$

for some $k \in \mathbb{C}[x_0, x_2]$ and $r, s, t \in \mathbb{Z}$.

(b) *The formal series* $\langle v', Y(v_1, x_0 + x_2)Y(v_2, x_2)v\rangle$, *which involves only finitely many negative powers of* x_2 *and only finitely many positive powers of* x_0, *lies in the image of* ι_{02}, *and in fact*

$$\big\langle v', Y(v_1, x_0 + x_2)Y(v_2, x_2)v\big\rangle = \iota_{02}h(x_0, x_2). \tag{2.111}$$

Proposition 2.40 (Associativity) *We have the following equality of formal rational functions*:

$$\iota_{12}^{-1}\big\langle v', Y(v_1, x_1)Y(v_2, x_2)v\big\rangle = \big(\iota_{20}^{-1}\big\langle v', Y\big(Y(v_1, x_0)v_2, x_2\big)v\big\rangle\big)\big|_{x_0=x_1-x_2}, \tag{2.112}$$

that is,

$$f(x_1, x_2) = h(x_1 - x_2, x_2).$$

Proposition 2.41 *In the presence of the other axioms for the notion of vertex operator algebra, the Jacobi identity follows from the rationality of products and iterates, and commutativity and associativity. In particular, in the definition of vertex operator algebra, the Jacobi identity may be replaced by these properties.*

The rationality, commutativity and associativity properties immediately imply the following result, in which the formal variables x_1 and x_2 are specialized to nonzero complex numbers in suitable domains:

Corollary 2.42 *The formal series obtained by specializing* x_1 *and* x_2 *to (nonzero) complex numbers* z_1 *and* z_2, *respectively, in* (2.104) *converges to a rational function of* z_1 *and* z_2 *in the domain*

$$|z_1| > |z_2| > 0 \tag{2.113}$$

and the analogous formal series obtained by specializing x_1 *and* x_2 *to* z_1 *and* z_2, *respectively, in* (2.107) *converges to the same rational function of* z_1 *and* z_2 *in the (disjoint) domain*

$$|z_2| > |z_1| > 0. \tag{2.114}$$

Moreover, the formal series obtained by specializing x_0 *and* x_2 *to* $z_1 - z_2$ *and* z_2, *respectively, in* (2.108) *converges to this same rational function of* z_1 *and* z_2 *in the domain*

$$|z_2| > |z_1 - z_2| > 0. \tag{2.115}$$

In particular, in the common domain

$$|z_1| > |z_2| > |z_1 - z_2| > 0, \tag{2.116}$$

we have the equality

$$\big\langle v', Y(v_1, z_1)Y(v_2, z_2)v\big\rangle = \big\langle v', Y\big(Y(v_1, z_1 - z_2)v_2, z_2\big)v\big\rangle \tag{2.117}$$

of rational functions of z_1 *and* z_2.

Remark 2.43 These last five results also hold for modules for a vertex operator algebra V; in the statements, one replaces the vectors v and v' by elements w and w' of a V-module W and its restricted dual W', respectively, and Proposition 2.41becomes: Given a vertex operator algebra V, in the presence of the other axioms for the notion of V-module, the Jacobi identity follows from the rationality of products and iterates, and commutativity and associativity. In particular, in the definition of V-module, the Jacobi identity may be replaced by these properties.

For either vertex operator algebras or modules, it is sometimes convenient to express the equalities of rational functions in Corollary 2.42 informally as follows:

$$Y(v_1, z_1)Y(v_2, z_2) \sim Y(v_2, z_2)Y(v_1, z_1) \tag{2.118}$$

and

$$Y(v_1, z_1)Y(v_2, z_2) \sim Y\big(Y(v_1, z_1 - z_2)v_2, z_2\big), \tag{2.119}$$

meaning that these expressions, defined in the domains indicated in Corollary 2.42 when the "matrix coefficients" of these expressions are taken as in this corollary, agree as operator-valued rational functions, up to analytic continuation.

Remark 2.44 Formulas (2.118) and (2.119) (or more precisely, (2.117)), express the meromorphic, or single-valued, version of "duality," in the language of conformal field theory. Formulas (2.119) (and (2.117)) express the existence and associativity of the single-valued, or meromorphic, operator product expansion. This is the statement that the product of two (vertex) operators can be expanded as a (suitable, convergent) infinite sum of vertex operators, and that this sum can be expressed in the form of an iterate of vertex operators, parametrized by the complex numbers $z_1 - z_2$ and z_2, in the format indicated; the infinite sum comes from expanding $Y(Y(v_1, z_1 - z_2)v_2, z_2)$ in powers of $z_1 - z_2$. A central goal of this work is to generalize (2.118) and (2.119), or more precisely, (2.117), to logarithmic intertwining operators in place of the operators $Y(\cdot, z)$. This will give the existence and also the associativity of the general, nonmeromorphic operator product expansion. This was done in the non-logarithmic setting in [72–74] and [53]. In the next section, we shall develop the concept of logarithmic intertwining operator.

Acknowledgements We are grateful to Bertram Kostant for asking whether the work of the first two authors described in [71] could (be generalized to) recover Kazhdan-Lusztig's braided tensor category structure [91–95]; the present work accomplishes this, as discussed below, and in fact, this question was one of our main initial motivations for embarking on this work. We would like to thank Sasha Kirillov Jr., Masahiko Miyamoto, Kiyokazu Nagatomo and Bin Zhang for their early interest in our work on logarithmic tensor product theory, and Bin Zhang in particular for inviting L.Z. to lecture on this ongoing work at Stony Brook. Y.-Z.H. would also like to thank Paul Fendley for raising his interest in logarithmic conformal field theory. We are grateful to the participants in our courses and seminars for their insightful comments on this work, especially Geoffrey Buhl, William Cook, Robert McRae, Thomas Robinson, Brian Vancil, and most especially, Shashank Kanade. We also thank Ingo Runkel for asking us whether our theory satisfies various conditions that he and his collaborators invoke in their work. (It does.) The authors gratefully acknowledge partial support from NSF grants DMS-0070800 and DMS-0401302. Y.-Z.H. is also grateful for partial support from NSF grant PHY-0901237 and for the hospitality of Institut des Hautes Études Scientifiques in the fall of 2007.

References

1. Abe, T.: A $\mathbb{Z}_2$-orbifold model of the symplectic fermionic vertex operator superalgebra. Math. Z. **255**, 755–792 (2007)
2. Adamović, D.: Rationality of Neveu-Schwarz vertex operator superalgebras. Int. Math. Res. Not. **1997**, 865–874 (1997)
3. Adamović, D.: Representations of the $N = 2$ superconformal vertex algebra. Int. Math. Res. Not. **1999**, 61–79 (1999)
4. Adamović, D.: Rationality of unitary $N = 2$ vertex operator superalgebras. math/9909055
5. Adamović, D., Milas, A.: Logarithmic intertwining operators and $\mathcal{W}(2, 2p - 1)$-algebras. J. Math. Phys. **48**, 073503 (2007)
6. Adamović, D., Milas, A.: On the triplet vertex algebra $\mathcal{W}(p)$. Adv. Math. **217**, 2664–2699 (2008)
7. Adamović, D., Milas, A.: The $N = 1$ triplet vertex operator superalgebras: twisted sector. SIGMA **4**, 087 (2008)
8. Adamović, D., Milas, A.: The $N = 1$ triplet vertex operator superalgebras. Commun. Math. Phys. **288**, 225–270 (2009)
9. Adamović, D., Milas, A.: Lattice construction of logarithmic modules for certain vertex operator algebras. Sel. Math. New Ser. **15**, 535–561 (2009)
10. Adamović, D., Milas, A.: On $\mathcal{W}$-algebras associated to $(2, p)$ minimal models and their representations. Int. Math. Res. Not. **2010**, 3896–3934 (2010)
11. Adamović, D., Milas, A.: The structure of Zhu's algebras for certain $\mathcal{W}$-algebras. Adv. Math. **227**, 2425–2456 (2011)
12. Bakalov, B., Kirillov, A. Jr.: Lectures on Tensor Categories and Modular Functors. University Lecture Series, vol. 21. Am. Math. Soc., Providence (2001)
13. Beilinson, A., Feigin, B., Mazur, B.: Introduction to algebraic field theory on curves. Preprint (provided by A. Beilinson 1996) (1991)
14. Belavin, A., Polyakov, A.M., Zamolodchikov, A.B.: Infinite conformal symmetries in two-dimensional quantum field theory. Nucl. Phys. B **241**, 333–380 (1984)
15. Borcherds, R.E.: Vertex algebras, Kac-Moody algebras, and the monster. Proc. Natl. Acad. Sci. USA **83**, 3068–3071 (1986)
16. Bushlanov, P.V., Feigin, B.L., Gaĭnutdinov, A.M., Tipunin, I.Yu.: Lusztig limit of quantum $s\ell(2)$ at root of unity and fusion of $(1, p)$ Virasoro logarithmic minimal models. Nucl. Phys. B **818**, 179–195 (2009)
17. Carqueville, N., Flohr, M.: Nonmeromorphic operator product expansion and C_2-cofiniteness for a family of $\mathcal{W}$-algebras. J. Phys. A **39**, 951–966 (2006)
18. Deligne, P.: Une description de catégorie tressée (inspiré par Drinfeld). Unpublished
19. Dong, C.: Vertex algebras associated with even lattices. J. Algebra **160**, 245–265 (1993)
20. Dong, C.: Representations of the moonshine module vertex operator algebra. In: Mathematical Aspects of Conformal and Topological Field Theories and Quantum Groups, South Hadley, MA, 1992. Contemp. Math., vol. 175, pp. 27–36. Am. Math. Soc., Providence (1994)
21. Dong, C., Lepowsky, J.: Generalized Vertex Algebras and Relative Vertex Operators. Progress in Math., vol. 112. Birkhäuser, Boston (1993)
22. Dong, C., Li, H., Mason, G.: Modular-invariance of trace functions in orbifold theory and generalized moonshine. Commun. Math. Phys. **214**, 1–56 (2000)
23. Drinfeld, V.G.: On quasitriangular quasi-Hopf algebras and a group closely connected with $\mathrm{Gal}(\bar{\mathbb{Q}}/\mathbb{Q})$. Algebra Anal. **2**, 149–181 (1990)

24. Eberle, H., Flohr, M.: Virasoro representations and fusion for general augmented minimal models. J. Phys. A **39**, 15245–15286 (2006)
25. Feigin, B.L., Gaĭnutdinov, A.M., Semikhatov, A.M., Tipunin, I.Y.: Logarithmic extensions of minimal models: characters and modular transformations. Nucl. Phys. B **757**, 303–343 (2006)
26. Feigin, B.L., Gaĭnutdinov, A.M., Semikhatov, A.M., Tipunin, I.Y.: Modular group representations and fusion in logarithmic conformal field theories and in the quantum group center. Commun. Math. Phys. **265**, 47–93 (2006)
27. Feigin, B.L., Gaĭnutdinov, A.M., Semikhatov, A.M., Tipunin, I.Yu.: The Kazhdan-Lusztig correspondence for the representation category of the triplet W-algebra in logarithmic conformal field theories. Theor. Math. Phys. **148**, 1210–1235 (2006)
28. Finkelberg, M.: Fusion categories. Ph.D. thesis, Harvard University (1993)
29. Finkelberg, M.: An equivalence of fusion categories. Geom. Funct. Anal. **6**, 249–267 (1996)
30. Flohr, M.: On modular invariant partition functions of conformal field theories with logarithmic operators. Int. J. Mod. Phys. A **11**, 4147–4172 (1996)
31. Flohr, M.: Bits and pieces in logarithmic conformal field theory. Int. J. Mod. Phys. A **18**, 4497–4591 (2003)
32. Flohr, M., Gaberdiel, M.R.: Logarithmic torus amplitudes. J. Phys. A **39**, 1955–1968 (2006)
33. Flohr, M., Knuth, H.: On Verlinde-like formulas in $c_{p,1}$ logarithmic conformal field theories. arXiv:0705.0545
34. Flohr, M., Grabow, C., Koehn, M.: Fermionic expressions for the characters of $c(p, 1)$ logarithmic conformal field theories. Nucl. Phys. B **768**, 263–276 (2007)
35. Frenkel, I.B., Zhu, Y.: Vertex operator algebras associated to representations of affine and Virasoro algebras. Duke Math. J. **66**, 123–168 (1992)
36. Frenkel, I.B., Lepowsky, J., Meurman, A.: Vertex operator calculus. In: Yau, S.-T. (ed.) Mathematical Aspects of String Theory, San Diego, 1986, pp. 150–188. World Scientific, Singapore (1987)
37. Frenkel, I.B., Lepowsky, J., Meurman, A.: Vertex Operator Algebras and the Monster. Pure and Appl. Math., vol. 134. Academic Press, Boston (1988)
38. Frenkel, I.B., Huang, Y.-Z., Lepowsky, J.: On axiomatic approaches to vertex operator algebras and modules. Mem. Am. Math. Soc. **104** (1993)
39. Fuchs, J.: On non-semisimple fusion rules and tensor categories. In: Huang, Y.-Z., Misra, K.C. (eds.) Lie Algebras, Vertex Operator Algebras and Their Applications, Proceedings of a Conference in Honor of J. Lepowsky and R. Wilson, 2005. Contemp. Math., vol. 442, pp. 315–337. Am. Math. Soc., Providence (2007)
40. Fuchs, J., Hwang, S., Semikhatov, A.M., Tipunin, I.Yu.: Nonsemisimple fusion algebras and the Verlinde formula. Commun. Math. Phys. **247**, 713–742 (2004)
41. Fuchs, J., Runkel, I., Schweigert, C.: Twenty five years of two-dimensional rational conformal field theory. J. Math. Phys. **51**, 015210 (2010)
42. Gaberdiel, M.: An algebraic approach to logarithmic conformal field theory. Int. J. Mod. Phys. A **18**, 4593–4638 (2003)
43. Gaberdiel, M., Kausch, H.G.: A rational logarithmic conformal field theory. Phys. Lett. B **386**, 131–137 (1996)
44. Gaberdiel, M., Kausch, H.G.: Indecomposable fusion products. Nucl. Phys. B **477**, 293–318 (1996)
45. Gaberdiel, M., Kausch, H.G.: A local logarithmic conformal field theory. Nucl. Phys. B **538**, 631–658 (1999)
46. Gaberdiel, M.R., Runkel, I.: The logarithmic triplet theory with boundary. J. Phys. A **39**, 14745–14780 (2006)
47. Gaberdiel, M.R., Runkel, I.: From boundary to bulk in logarithmic CFT. J. Phys. A **41**, 075402 (2008)
48. Gaberdiel, M.R., Runkel, I., Wood, S.: Fusion rules and boundary conditions in the $c = 0$ triplet model. J. Phys. A **42**, 325403 (2009)

49. Gaĭnutdinov, A.M., Tipunin, I.Y.: Radford, Drinfeld, and Cardy boundary states in $(1, p)$ logarithmic conformal field models. J. Phys. A **42**, 315207 (2009)
50. Gaĭnutdinov, A.M., Vasseur, R.: Lattice fusion rules and logarithmic operator product expansions. Nucl. Phys. B **868**, 223–270 (2013)
51. Gurarie, V.: Logarithmic operators in conformal field theory. Nucl. Phys. B **410**, 535–549 (1993)
52. Huang, Y.-Z.: On the geometric interpretation of vertex operator algebras. Ph.D. thesis, Rutgers University (1990)
53. Huang, Y.-Z.: A theory of tensor products for module categories for a vertex operator algebra, IV. J. Pure Appl. Algebra **100**, 173–216 (1995)
54. Huang, Y.-Z.: Virasoro vertex operator algebras, (nonmeromorphic) operator product expansion and the tensor product theory. J. Algebra **182**, 201–234 (1996)
55. Huang, Y.-Z.: A nonmeromorphic extension of the moonshine module vertex operator algebra. In: Moonshine, the Monster, and Related Topics, South Hadley, MA, 1994. Contemp. Math., vol. 193, pp. 123–148. Am. Math. Soc., Providence (1996)
56. Huang, Y.-Z.: Two-Dimensional Conformal Geometry and Vertex Operator Algebras. Progress in Math., vol. 148. Birkhäuser, Boston (1997)
57. Huang, Y.-Z.: Generalized rationality and a "Jacobi identity" for intertwining operator algebras. Sel. Math. New Ser. **6**, 225–267 (2000)
58. Huang, Y.-Z.: Conformal-field-theoretic analogues of codes and lattices. In: Sthanumoorthy, N., Misra, K.C. (eds.) Kac-Moody Lie Algebras and Related Topics, Proc. Ramanujan International Symposium on Kac-Moody Lie Algebras and Applications. Contemp. Math., vol. 343, pp. 131–145. Am. Math. Soc., Providence (2004)
59. Huang, Y.-Z.: Differential equations and intertwining operators. Commun. Contemp. Math. **7**, 375–400 (2005)
60. Huang, Y.-Z.: Differential equations, duality and modular invariance. Commun. Contemp. Math. **7**, 649–706 (2005)
61. Huang, Y.-Z.: Vertex operator algebras, the Verlinde conjecture, and modular tensor categories. Proc. Natl. Acad. Sci. USA **102**, 5352–5356 (2005)
62. Huang, Y.-Z.: Vertex operator algebras, fusion rules and modular transformations. In: Fuchs, J., Mickelsson, J., Rozenblioum, G., Stolin, A. (eds.) Non-commutative Geometry and Representation Theory in Mathematical Physics. Contemp. Math., vol. 391, pp. 135–148. Am. Math. Soc., Providence (2005)
63. Huang, Y.-Z.: Vertex operator algebras and the Verlinde conjecture. Commun. Contemp. Math. **10**, 103–154 (2008)
64. Huang, Y.-Z.: Rigidity and modularity of vertex tensor categories. Commun. Contemp. Math. **10**, 871–911 (2008)
65. Huang, Y.-Z.: Cofiniteness conditions, projective covers and the logarithmic tensor product theory. J. Pure Appl. Algebra **213**, 458–475 (2009)
66. Huang, Y.-Z.: Representations of vertex operator algebras and braided finite tensor categories. In: Bergvelt, M., Yamskulna, G., Zhao, W. (eds.) Vertex Operator Algebras and Related Topics, an International Conference in Honor of Geoffery Mason's 60th Birthday. Contemp. Math., vol. 497, pp. 97–111. Am. Math. Soc., Providence (2009)
67. Huang, Y.-Z.: Generalized twisted modules associated to general automorphisms of a vertex operator algebra. Commun. Math. Phys. **298**, 265–292 (2010)
68. Huang, Y.-Z., Lepowsky, J.: Toward a theory of tensor products for representations of a vertex operator algebra. In: Catto, S., Rocha, A. (eds.) Proc. 20th International Conference on Differential Geometric Methods in Theoretical Physics, New York, 1991, vol. 1, pp. 344–354. World Scientific, Singapore (1992)
69. Huang, Y.-Z., Lepowsky, J.: Vertex operator algebras and operads. In: Corwin, L., Gelfand, I., Lepowsky, J. (eds.): The Gelfand Mathematical Seminars, 1990–1992, pp. 145–161. Birkhäuser, Boston (1993)
70. Huang, Y.-Z., Lepowsky, J.: Operadic formulation of the notion of vertex operator algebra. In: Proc. 1992 Joint Summer Research Conference on Conformal Field Theory, Topological

Field Theory and Quantum Groups, Mount Holyoke, 1992. Contemp. Math., vol. 175, pp. 131–148. Am. Math. Soc., Providence (1994)
71. Huang, Y.-Z., Lepowsky, J.: Tensor products of modules for a vertex operator algebras and vertex tensor categories. In: Brylinski, R., Brylinski, J.-L., Guillemin, V., Kac, V. (eds.) Lie Theory and Geometry, in Honor of Bertram Kostant, pp. 349–383. Birkhäuser, Boston (1994)
72. Huang, Y.-Z., Lepowsky, J.: A theory of tensor products for module categories for a vertex operator algebra, I. Sel. Math. New Ser. **1**, 699–756 (1995)
73. Huang, Y.-Z., Lepowsky, J.: A theory of tensor products for module categories for a vertex operator algebra, II. Sel. Math. New Ser. **1**, 757–786 (1995)
74. Huang, Y.-Z., Lepowsky, J.: A theory of tensor products for module categories for a vertex operator algebra, III. J. Pure Appl. Algebra **100**, 141–171 (1995)
75. Huang, Y.-Z., Lepowsky, J.: Intertwining operator algebras and vertex tensor categories for affine Lie algebras. Duke Math. J. **99**, 113–134 (1999)
76. Huang, Y.-Z., Lepowsky, J.: A theory of tensor products for module categories for a vertex operator algebra, V
77. Huang, Y.-Z., Milas, A.: Intertwining operator superalgebras and vertex tensor categories for superconformal algebras, I. Commun. Contemp. Math. **4**, 327–355 (2002)
78. Huang, Y.-Z., Milas, A.: Intertwining operator superalgebras and vertex tensor categories for superconformal algebras, II. Trans. Am. Math. Soc. **354**, 363–385 (2002)
79. Huang, Y.-Z., Lepowsky, J., Li, H., Zhang, L.: On the concepts of intertwining operator and tensor product module in vertex operator algebra theory. J. Pure Appl. Algebra **204**, 507–535 (2006)
80. Huang, Y.-Z., Lepowsky, J., Zhang, L.: A logarithmic generalization of tensor product theory for modules for a vertex operator algebra. Int. J. Math. **17**, 975–1012 (2006)
81. Huang, Y.-Z., Lepowsky, J., Zhang, L.: Logarithmic tensor category theory, II: logarithmic formal calculus and properties of logarithmic intertwining operators. arXiv:1012.4196
82. Huang, Y.-Z., Lepowsky, J., Zhang, L.: Logarithmic tensor category theory, III: intertwining maps and tensor product bifunctors. arXiv:1012.4197
83. Huang, Y.-Z., Lepowsky, J., Zhang, L.: Logarithmic tensor category theory, IV: constructions of tensor product bifunctors and the compatibility conditions. arXiv:1012.4198
84. Huang, Y.-Z., Lepowsky, J., Zhang, L.: Logarithmic tensor category theory, V: convergence condition for intertwining maps and the corresponding compatibility condition. arXiv:1012.4199
85. Huang, Y.-Z., Lepowsky, J., Zhang, L.: Logarithmic tensor category theory, VI: expansion condition, associativity of logarithmic intertwining operators, and the associativity isomorphisms. arXiv:1012.4202
86. Huang, Y.-Z., Lepowsky, J., Zhang, L.: Logarithmic tensor category theory, VII: convergence and extension properties and applications to expansion for intertwining maps. arXiv:1110.1929
87. Huang, Y.-Z., Lepowsky, J., Zhang, L.: Logarithmic tensor category theory, VIII: braided tensor category structure on categories of generalized modules for a conformal vertex algebra. arXiv:1110.1931
88. Kac, V., Wang, W.: Vertex operator superalgebras and their representations. In: Mathematical Aspects of Conformal and Topological Field Theories and Quantum Groups, South Hadley, MA, 1992. Contemp. Math., vol. 175, pp. 161–191. Am. Math. Soc., Providence (1994)
89. Kausch, H.G.: Extended conformal algebras generated by multiplet of primary fields. Phys. Lett. B **259**, 448–455 (1991)
90. Kausch, H.G.: Symplectic fermions. Nucl. Phys. B **583**, 513–541 (2000)
91. Kazhdan, D., Lusztig, G.: Affine Lie algebras and quantum groups. Int. Math. Res. Not. **1991**, 21–29 (1991)
92. Kazhdan, D., Lusztig, G.: Tensor structures arising from affine Lie algebras, I. J. Am. Math. Soc. **6**, 905–947 (1993)

93. Kazhdan, D., Lusztig, G.: Tensor structures arising from affine Lie algebras, II. J. Am. Math. Soc. **6**, 949–1011 (1993)
94. Kazhdan, D., Lusztig, G.: Tensor structures arising from affine Lie algebras, III. J. Am. Math. Soc. **7**, 335–381 (1994)
95. Kazhdan, D., Lusztig, G.: Tensor structures arising from affine Lie algebras, IV. J. Am. Math. Soc. **7**, 383–453 (1994)
96. Knizhnik, V.G., Zamolodchikov, A.B.: Current algebra and Wess-Zumino models in two dimensions. Nucl. Phys. B **247**, 83–103 (1984)
97. Kondo, H., Saito, Y.: Indecomposable decomposition of tensor products of modules over the restricted quantum universal enveloping algebra associated to $\mathfrak{sl}_2$. J. Algebra **330**, 103–129 (2011)
98. Lepowsky, J.: From the representation theory of vertex operator algebras to modular tensor categories in conformal field theory, commentary on Y.-Z. Huang's PNAS article "Vertex operator algebras, the Verlinde conjecture and modular tensor categories,". Proc. Natl. Acad. Sci. USA **102**, 5304–5305 (2005)
99. Lepowsky, J., Li, H.: Introduction to Vertex Operator Algebras and Their Representations. Progress in Math., vol. 227. Birkhäuser, Boston (2003)
100. Li, H.: An analogue of the Hom functor and a generalized nuclear democracy theorem. Duke Math. J. **93**, 73–114 (1998)
101. Lubell, D.: Problem 10992, problems and solutions. Am. Math. Mon. **110**, 155 (2003)
102. Mac Lane, S.: Categories for the Working Mathematician. Graduate Texts in Mathematics, vol. 5. Springer, New York (1971)
103. Mathieu, P., Ridout, D.: Logarithmic $M(2, p)$ minimal models, their logarithmic couplings, and duality. Nucl. Phys. B **801**, 268–295 (2008)
104. Milas, A.: Weak modules and logarithmic intertwining operators for vertex operator algebras. In: Berman, S., Fendley, P., Huang, Y.-Z., Misra, K., Parshall, B. (eds.) Recent Developments in Infinite-Dimensional Lie Algebras and Conformal Field Theory. Contemp. Math., vol. 297, pp. 201–225. Am. Math. Soc., Providence (2002)
105. Milas, A.: Logarithmic intertwining operators and vertex operators. Commun. Math. Phys. **277**, 497–529 (2008)
106. Miyamoto, M.: Modular invariance of vertex operator algebras satisfying C_2-cofiniteness. Duke Math. J. **122**, 51–91 (2004)
107. Miyamoto, M.: Intertwining operators and modular invariance. math/0010180
108. Miyamoto, M.: Flatness of tensor products and semi-rigidity for C_2-cofinite vertex operator algebras I. arXiv:0906.1407
109. Miyamoto, M.: Flatness of tensor products and semi-rigidity for C_2-cofinite vertex operator algebras II. arXiv:0909.3665
110. Miyamoto, M.: A $\mathbb{Z}_3$-orbifold theory of lattice vertex operator algebra and $\mathbb{Z}_3$-orbifold constructions. arXiv:1003.0237
111. Moore, G., Seiberg, N.: Polynomial equations for rational conformal field theories. Phys. Lett. B **212**, 451–460 (1988)
112. Moore, G., Seiberg, N.: Classical and quantum conformal field theory. Commun. Math. Phys. **123**, 177–254 (1989)
113. Nagatomo, K., Tsuchiya, A.: Conformal field theories associated to regular chiral vertex operator algebras I: theories over the projective line. Duke Math. J. **128**, 393–471 (2005)
114. Nagatomo, K., Tsuchiya, A.: The triplet vertex operator algebra $W(p)$ and the restricted quantum group at root of unity. arXiv:0902.4607
115. Pearce, P.A., Rasmussen, J., Zuber, J.-B.: Logarithmic minimal models. J. Stat. Mech. Theory Exp. **2006**, P11017 (2006)
116. Pearce, P.A., Rasmussen, J., Ruelle, P.: Integrable boundary conditions and $\mathcal{W}$-extended fusion in the logarithmic minimal models $\mathcal{LM}(1, p)$. J. Phys. A **41**, 295201 (2008)
117. Pearce, P.A., Rasmussen, J., Ruelle, P.: Grothendieck ring and Verlinde formula for the $\mathcal{W}$-extended logarithmic minimal model $\mathcal{WLM}(1, p)$. J. Phys. A **43**, 045211 (2010)

118. Rahimi Tabar, M.R.: Disorder systems and logarithmic conformal field theory. Int. J. Mod. Phys. A **18**, 4703–4745 (2003)
119. Rasmussen, J.: $\mathcal{W}$-extended logarithmic minimal models. Nucl. Phys. B **807**, 495–533 (2009)
120. Rasmussen, J.: Polynomial fusion rings of $\mathcal{W}$-extended logarithmic minimal models. J. Math. Phys. **50**, 043512 (2009)
121. Rasmussen, J.: Fusion of irreducible modules in $\mathcal{WLM}(p, p')$. J. Phys. A **43**, 045210 (2010)
122. Rasmussen, J.: Fusion matrices, generalized Verlinde formulas, and partition functions in $\mathcal{WLM}(1, p)$. J. Phys. A **43**, 105201 (2010)
123. Rasmussen, J.: Graph fusion algebras of $\mathcal{WLM}(p, p')$. Nucl. Phys. B **830**, 493–541 (2010)
124. Rasmussen, J., Pearce, P.A.: Fusion algebras of logarithmic minimal models. J. Phys. **40**, 13711–13733 (2007)
125. Rasmussen, J., Pearce, P.A.: $\mathcal{W}$-extended fusion algebra of critical percolation. J. Phys. A **41**, 295208 (2008)
126. Read, N., Saleur, H.: Associative-algebraic approach to logarithmic conformal field theories. Nucl. Phys. B **777**, 316–351 (2007)
127. Reshetikhin, N., Turaev, V.: Ribbon graphs and their invariants derived from quantum groups. Commun. Math. Phys. **127**, 1–26 (1990)
128. Reshetikhin, N., Turaev, V.: Invariants of 3-manifolds via link polynomials and quantum groups. Invent. Math. **103**, 547–598 (1991)
129. Robinson, T.J.: Formal calculus, umbral calculus, and basic axiomatics of vertex algebras. Ph.D. thesis, Rutgers University (2009)
130. Robinson, T.J.: New perspectives on exponentiated derivations, the formal Taylor theorem, and Faà di Bruno's formula. In: Bergvelt, M., Yamskulna, G., Zhao, W. (eds.) Proc. Conference on Vertex Operator Algebras, Illinois State University, 2008. Contemp. Math., vol. 497, pp. 185–198. Am. Math. Soc., Providence (2009)
131. Robinson, T.J.: A recursion identity for formal iterated logarithms and iterated exponentials. arXiv:1012.0822
132. Rozansky, L., Saleur, H.: Quantum field theory for the multi-variable Alexander-Conway polynomial. Nucl. Phys. B **376**, 461–509 (1991)
133. Saleur, H., Schomerus, V.: On the $SU(2|1)$ WZW model and its statistical mechanics applications. Nucl. Phys. B **775**, 312–340 (2007)
134. Semikhatov, A.M.: A note on the logarithmic (p, p') fusion. arXiv:0710.5157
135. Tsuchiya, A., Ueno, K., Yamada, Y.: Conformal field theory on universal family of stable curves with gauge symmetries. In: Advanced Studies in Pure Math., vol. 19, pp. 459–566. Kinokuniya Company Ltd., Tokyo (1989)
136. Turaev, V.: Modular categories and 3-manifold invariants. Int. J. Mod. Phys. B **6**, 1807–1824 (1992)
137. Turaev, V.: Quantum Invariants of Knots and 3-Manifolds. de Gruyter Studies in Math., vol. 18. de Gruyter, Berlin (1994)
138. Verlinde, E.: Fusion rules and modular transformations in 2D conformal field theory. Nucl. Phys. B **300**, 360–376 (1988)
139. Wang, W.: Rationality of Virasoro vertex operator algebras. Duke Math. J. **71**, 197–211 (1993)
140. Wood, S.: Fusion rules of the $W_{p,q}$ triplet models. J. Phys. A **43**, 045212 (2010)
141. Zhang, L.: Vertex operator algebras and Kazhdan-Lusztig's tensor category. Ph.D. thesis, Rutgers University (2004)
142. Zhang, L.: Vertex tensor category structure on a category of Kazhdan-Lusztig. N.Y. J. Math. **14**, 261–284 (2008)
143. Zhu, Y.: Vertex operators, elliptic functions and modular forms. Ph.D. thesis, Yale University (1990)
144. Zhu, Y.: Modular invariance of characters of vertex operator algebras. J. Am. Math. Soc. **9**, 237–307 (1996)

C_2-Cofinite $\mathcal{W}$-Algebras and Their Logarithmic Representations

Dražen Adamović and Antun Milas

Abstract We discuss our recent results on the representation theory of $\mathcal{W}$-algebras relevant to Logarithmic Conformal Field Theory. First we explain some general constructions of $\mathcal{W}$-algebras coming from screening operators. Then we review the results on C_2-cofiniteness, the structure of Zhu's algebras, and the existence of logarithmic modules for triplet vertex algebras. We propose some conjectures and open problems which put the theory of triplet vertex algebras into a broader context. New realizations of logarithmic modules for $\mathcal{W}$-algebras defined via screenings are also presented.

1 Introduction: Irrational C_2-Cofinite Vertex Algebras

Vertex algebras are in many ways analogous to associative algebras, at least from the point of view of representation theory. Rational vertex operator algebras [2, 68] and regular vertex algebras have semisimple categories of modules and should be compared to (finite-dimensional) semisimple associative algebras. If we seek the same analogy with finite-dimensional non-semisimple associative algebras, we would eventually discover irrational C_2-cofinite vertex algebras (the C_2-condition guarantees that the vertex algebra has finitely many inequivalent irreducibles [68]). But oddly as it might seem, examples of such vertex algebras are rare and actually not much is known about them. For instance, it is not even known if there exists an irrational vertex algebra with finitely many indecomposable modules.

Motivated by important works of physicists [27–29, 41], in our recent papers [4, 6, 9, 11] (see also [1, 17]), among many other things, we constructed new families of irrational C_2-cofinite (i.e., *quasi-rational*) vertex algebras and superalgebras.

D. Adamović (✉)
Department of Mathematics, University of Zagreb, Zagreb, 10000, Croatia
e-mail: adamovic@math.hr

A. Milas
Department of Mathematics and Statistics, University at Albany (SUNY), Albany, NY 12222, USA
e-mail: amilas@math.albany.edu

C. Bai et al. (eds.), *Conformal Field Theories and Tensor Categories*,
Mathematical Lectures from Peking University, DOI 10.1007/978-3-642-39383-9_6,

The most surprising fact about quasi-rational vertex algebras is that all known examples are expected to be related to certain finite-dimensional quantum groups (Hopf algebras) via the conjectural Kazhdan-Lusztig correspondences [25, 28, 29]. It is also known that the module category of a C_2-cofinite vertex algebra has a natural finite tensor category structure [46, 51], although not necessarily rigid [62] (see also [39, 40] for related categorical issues).

2 Preliminaries

This paper deals mainly with the representation theory of certain vertex algebras. Because we are interested in their $\mathbb{Z}_{\geq 0}$-graded modules, the starting point is to recall the definition of Zhu's algebra for vertex operator (super)algebras following [52, 68].

Let $(V, Y, \mathbf{1}, \omega)$ be a vertex operator algebra. We shall always assume that V is of CFT type, meaning that it has $\mathbb{N}_{\geq 0}$ grading with the vacuum vector lying on the top component. Let $V = \coprod_{n \in \mathbb{Z}_{\geq 0}} V(n)$. For $a \in V(n)$, we shall write $\deg(a) = n$. As usual, vertex operator associated to $a \in V$ is denoted by $Y(a, x)$, with the mode expansion

$$Y(a, x) = \sum_{n \in \mathbb{Z}} a_n x^{-n-1}.$$

We define two bilinear maps: $* : V \times V \to V$, $\circ : V \times V \to V$ as follows. For homogeneous $a, b \in V$ let

$$a * b = \mathrm{Res}_x\, Y(a, x) \frac{(1+x)^{\deg(a)}}{x} b$$

$$a \circ b = \mathrm{Res}_x\, Y(a, x) \frac{(1+x)^{\deg(a)}}{x^2} b$$

Next, we extend $*$ and $\circ$ on $V \otimes V$ linearly, and denote by $O(V) \subset V$ the linear span of elements of the form $a \circ b$, and by $A(V)$ the quotient space $V/O(V)$. The space $A(V)$ has an associative algebra structure (with identity), with the multiplication induced by $*$. Algebra $A(V)$ is called the Zhu's algebra of V. The image of $v \in V$, under the natural map $V \mapsto A(V)$ will be denoted by $[v]$.

For a homogeneous $a \in V$ we define $o(a) = a_{\deg(a)-1}$. In the case when $V^{\bar{0}} = V$, V is a vertex operator algebra and we get the usual definition of Zhu's algebra for vertex operator algebras.

According to [68], there is an one-to-one correspondence between irreducible $A(V)$-modules and irreducible $\mathbb{Z}_{\geq 0}$-graded V-modules.

Moreover, if U is any $A(V)$-module. There is $\mathbb{Z}_{\geq 0}$-graded V-module $L(U)$ such that the top component $L(U)(0) \cong U$. V is called rational, if every $\mathbb{Z}_{\geq 0}$-graded module is completely reducible.

With V as above, we let $C_2(V) = \langle a_{-2} b : a, b \in V \rangle$, and $\mathcal{P}(V) = V/C_2(V)$. The quotient space $V/C_2(V)$ has an algebraic structure of a commutative Poisson algebra [68]. Explicitly, if we denote by $\bar{a}$ the image of a under the natural map $V \mapsto \mathcal{P}(V)$ the Poisson bracket is given by $\{\overline{a}, \overline{b}\} = \overline{a_0 b}$ and commutative product $\overline{a} \cdot \overline{b} = \overline{a_{-1} b}$. From the given definitions it is not hard to construct an increasing filtration of $A(V)$ such that $\mathrm{gr} A(V)$ maps onto $A(V)$.

3 Quantum $\mathcal{W}$-Algebras from Integral Lattices

$\mathcal{W}$-algebras are some of the most exciting objects in representation theory and have been extensively studied from many different point of views. There are several different types of $\mathcal{W}$-algebras in the literature, so to avoid any confusion we stress that (a) *finite* W-algebras are certain associative algebras associated to a complex semisimple Lie algebra $\mathfrak{g}$ and a nilpotent element $e \in \mathfrak{g}$ [20, 66], and can be viewed as deformations of Slodowy's slice, and (b) *affine* W-algebras are vertex algebras[1] obtained by Drinfeld-Sokolov reduction from affine vertex algebras [34]. The two algebras are related via a fundamental construction of Zhu (cf. [14, 21]). In this paper, (quantum) $\mathcal{W}$-algebras are vertex algebra generalizations of the affine W-algebras. More precisely

Definition 3.1 A $\mathcal{W}$-algebra V is a vertex algebra strongly generated by a finite set of primary vectors $u^1, \ldots, u^k$. Here strongly generated means that elements of the form

$$u^{i_1}_{-j_1} \cdots u^{i_m}_{-j_m} \mathbf{1}, \quad j_1, \ldots, j_m \geq 1 \tag{1}$$

form a spanning set of V. If $\deg(u_i) = r_i$ we say that V is of type $(2, r_1, \ldots, r_k)$.

Let us first outline the well-known construction of lattice vertex algebra V_L associated to a positive definite even lattice $(L, \langle\rangle)$. We denote by $\mathbb{C}[L]$ the group algebra of L. As a vector space

$$V_L = M(1) \otimes \mathbb{C}[L], \quad M(1) = S(\hat{\mathfrak{h}}_{<0}),$$

where $S(\hat{\mathfrak{h}}_{<0})$ is the usual Fock space. The vertex algebra V_L is known to be rational [22]. Denote by L° the dual lattice of L. For $\beta \in L^\circ$ we have "bosonic" vertex operators

$$Y(e^\beta, x) = \sum_{n \in \mathbb{Z}} e^\beta_n x^{-n-1},$$

introduced in [23, 37]. It is also known [22] that all irreducible V_L-modules are given by V_γ, $\gamma \in L^\circ / L$.

[1] n.b. For brevity, we shall often use "algebra" and "vertex algebra" when we mean "superalgebra" and "vertex superalgebra", respectively. From the context it should be clear whether the adjective "super" is needed.

Now, we specialize $L = \sqrt{p}Q$, where $p \geq 2$ and Q is a root lattice (of ADE type). We should say that this restriction is not that crucial right now, and in fact we can obtain interesting objects even if the lattice L is (say) hyperbolic. We equip V_L with a vertex algebra structure [16, 37] (by choosing an appropriate 2-cocycle). Let α_i denote the simple roots of Q. For the conformal vector we conveniently choose

$$\omega = \omega_{st} + \frac{p-1}{2\sqrt{p}} \sum_{\alpha \in \Delta_+} \alpha(-2)\mathbf{1},$$

where ω_{st} is the standard (quadratic) Virasoro generator [37, 54]. Then V_L is a conformal vertex algebra of central charge[2]

$$\mathrm{rank}(L) + 12(\rho, \rho)\left(2 - p - \frac{1}{p}\right).$$

Consider the operators

$$e_0^{\sqrt{p}\alpha_i}, e_0^{-\alpha_j/\sqrt{p}}, \quad 1 \leq i, j \leq \mathrm{rank}(L) \tag{2}$$

acting between V_L and V_L-modules. These are the so-called *screening operators*. More precisely,

Lemma 3.1 *For every i and j the operators $e_0^{\sqrt{p}\alpha_i}$ and $e_0^{-\alpha_j/\sqrt{p}}$ commute with each other, and they both commute with the Virasoro algebra.*

We shall refer to $e_0^{\sqrt{p}\alpha_i}$ and $e_0^{-\alpha_j/\sqrt{p}}$, as the *long* and *short* screening, respectively. It is well-known that the intersection of the kernels of residues of vertex operators is a vertex subalgebra (cf. [34]), so the next problem seems very natural to ask

Problem 1 What kind of vertex algebras can we construct from the kernels of screenings in (2)? What choices of (2) give rise to C_2-cofinite vertex (sub)algebras?

3.1 Affine $\mathcal{W}$-Algebras

The above construction with screening operators naturally leads to affine $\mathcal{W}$-algebras. The affine $\mathcal{W}$-algebra associated to $\hat{\mathfrak{g}}$ at level $k \neq -h^\vee$, denoted by $\mathcal{W}_k(\mathfrak{g})$ is defined as

$$H_k^*(\mathfrak{g}),$$

[2] Without the linear term the central charge would be $\mathrm{rank}(L)$.

where the cohomology is taken with respect to a quantized BRST complex for the Drinfeld-Sokolov Hamiltonian reduction [35]. As shown by Feigin and Frenkel (cf. [35] and [34] and citations therein) this cohomology is nontrivial only in degree zero. Moreover, it is known that $\mathcal{W}_k(\mathfrak{g})$ is a quantum $\mathcal{W}$-algebra (according to our definition) freely generated by rank($\mathfrak{g}$) primary fields. Although not evident from our discussion, the vacuum vertex algebra $V_k(\hat{\mathfrak{g}})$ coming from the affine Kac-Moody Lie algebra $\hat{\mathfrak{g}}$ enters in the definition of $H_k^0(\mathfrak{g})$ (see again [34]). It is possible to replace $V_k(\hat{\mathfrak{g}})$ with its irreducible quotient $L_k(\hat{\mathfrak{g}})$, but then the theory becomes much more complicated [14].

An important theorem of B. Feigin and E. Frenkel [34] says that if k is *generic* and $\mathfrak{g}$ is simply-laced, then there is an alternative description of $\mathcal{W}_k(\mathfrak{g})$. For this purpose, we let $\nu = k + h^\vee$, where k is generic. Then there are appropriately defined screenings

$$e_0^{-\alpha_i/\sqrt{\nu}} : M(1) \longrightarrow M(1, -\alpha_i/\sqrt{\nu}),$$

such that

$$\mathcal{W}_k(\mathfrak{g}) = \bigcap_{i=1}^{l} \mathrm{Ker}_{M(1)}\left(e_0^{-\alpha_i/\sqrt{\nu}}\right),$$

where $l = \mathrm{rank}(L)$. If we assume in addition that $\mathfrak{g}$ is simply laced (ADE type) then we also have the following important duality [34]

$$\mathcal{W}_k(\mathfrak{g}) = \bigcap_{i=1}^{l} \mathrm{Ker}_{M(1)}\left(e_0^{\sqrt{\nu}\alpha_i}\right).$$

Now, let us consider the case when $L = \sqrt{p}Q$, $p \in \mathbb{N}$, in connection with the problem we just raised. Having in mind the previous construction, it is natural to ask whether $p = k + h^\vee$ is also generic. For instance, it is known (cf. [36]) that $p = 1$ is generic. Next result seems to be known in the physics literature

Theorem 3.1 *Let $\mathfrak{g}$ be simply laced. Then $p = k + h^\vee \in \mathbb{N}_{\geq 2}$ is non-generic. More precisely,*

$$\bigcap_{i=1}^{l} \mathrm{Ker}_{M(1)}\, e_0^{-\alpha_i/\sqrt{p}}$$

is a vertex algebra containing $\mathcal{W}_k(\mathfrak{g})$ as a proper subalgebra.

Interestingly enough, for long screenings, we do expect "genericness" to hold:

Conjecture 3.1 *For $p \geq 2$ as above*

$$\mathcal{W}_k(\mathfrak{g}) = \bigcap_{i=1}^{l} \mathrm{Ker}_{M(1)}\, e_0^{\sqrt{p}\alpha_i}$$

This conjecture is known to be true in the rank one case, where the kernel of the long screening is precisely the Virasoro vertex algebra $L(c_{p,1},0)$ of central charge $1-\frac{6(p-1)^2}{p}$. But for the short screening we obtain the so-called singlet algebra $\overline{M(1)}$ of type $\mathcal{W}(2,2p-1)$, an extension of $L(c_{p,1},0)$ [3, 4] (the special case $p=2$ has been extensively studied in [1, 19, 65], etc.). Both vertex algebras are neither rational nor C_2-cofinite.

Let

$$h_{r,s}=\frac{(sp-r)^2-(p-1)^2}{4p}.$$

Theorem 3.2 [3] *Zhu's associative algebra $A(\overline{M(1)})$ is isomorphic to the commutative algebra $\mathbb{C}[x,y]/\langle P(x,y)\rangle$, where $\langle P(x,y)\rangle$ is the principal ideal generated by*

$$P(x,y)=y^2-C_p(x-h_{p,1})\prod_{i=1}^{p-1}(x-h_{i,1})^2 \quad (C_p\neq 0).$$

Now, by using results in Sect. 2, we see that irreducible $\overline{M(1)}$-modules are parameterized by zeros of a certain rational curve in $\mathbb{C}^2$. We expect that irreducible modules for vertex operator algebras from Theorem 3.1 also have interesting interpretation in the context of algebraic curves.

3.2 Further Extended Affine $\mathcal{W}$-Algebras

Instead of focusing on the charge zero subspace $M(1)$ (the Fock space), nothing prevented us from considering intersections of the kernels of screenings on the whole lattice vertex algebra V_L. Let us first examine the long screenings in this situation. Conjecturally, we expect to produce a certain vertex algebra denoted by $\mathcal{W}^{\diamond}(p)_Q:=\bigcap_{i=1}^{l}\mathrm{Ker}_{V_L}\, e_0^{\sqrt{p}\alpha_i}$, with a large ideal I such that $W^{\diamond}(p)_Q/I\cong\mathcal{W}_k(\mathfrak{g})$ (the structure of $\mathcal{W}^{\diamond}(p)$ was analyzed in [60] in connection to Feigin-Stoyanovsky's principal subspaces [24]).

Example 3.1 For $Q=A_1$, we have

$$W^{\diamond}(p)_Q=\langle e^{\sqrt{p}\alpha_1},\omega\rangle,$$

the smallest conformal vertex subalgebra of V_L containing the vector e^{α_1}. Here $\langle e^{\sqrt{p}\alpha_1}\rangle$ is the well-known FS principal subspace [24] (see also [60]).

3.3 Maximally Extended $\mathcal{W}$-Algebras: A Conjecture

Due to differences already observed in Theorem 3.1 and Conjecture 3.1, it is not surprising that the conformal vertex algebra

$$\mathcal{W}(p)_Q := \bigcap_{i=1}^{l} \mathrm{Ker}_{V_L} e_0^{-\alpha_j/\sqrt{p}} \tag{3}$$

will exhibit properties different to those observed for $\mathcal{W}^{\diamond}(p)_Q$.

We believe the following rather strong conjecture motivated by [25] holds.

Conjecture 3.2 *We have*

(1) *The vertex algebra $\mathcal{W}(p)_Q$ is irrational and C_2-cofinite.*
(2) *It is strongly generated by the generators of $\mathcal{W}_k(\mathfrak{g})$ and finitely many primary vectors.*
(3) $\mathrm{Soc}_{\mathcal{W}_k(\mathfrak{g})}(V_L) = \mathcal{W}(p)_Q$.
(4) *$\mathcal{W}(p)_Q$ admits logarithmic modules of $L(0)$-nilpotent rank at most* $\mathrm{rank}(L)+1$ *(for the explanation see Sect.* 5).
(5) $\dim A(\mathcal{W}(p)_Q) = \dim \mathcal{P}(\mathcal{W}(p)_Q)$.

Let us briefly comment on (5) first. For V a C_2-cofinite vertex operator algebra. M. Gaberdiel and T. Gannon in [42] initiated a relationship between $A(V)$ and $\mathcal{P}(V)$. They raised an interesting question: When does $\dim A(V) = \dim \mathcal{P}(V)$? For a large family of rational vertex operator algebras of affine type, the equality of dimensions holds (cf. [26, 30]). In [10], we studied this question for C_2-cofinite, irrational vertex operator superalgebras, and proved that (5) holds in the rank one case.

The first half of part (1) of the conjecture is known to be true in general, and this follows also from (4). We have already shown in [4] the conjecture to be true for $Q = A_1$. In this case we write $\mathcal{W}(p) = \mathcal{W}(p)_Q$ for brevity.

3.4 Triplet Vertex Algebra $\mathcal{W}(p)$

The next result was proven in [4] and [10].

Theorem 3.3 *The following holds:*

(1) *$\mathcal{W}(p)$ is C_2-cofinite and irrational.*
(2) *$\mathcal{W}(p)$ is strongly generated by ω and three primary vectors E, F and H of conformal weight $2p-1$.*
(3) *$\mathcal{W}(p)$ has exactly $2p$ irreducible modules, usually denoted by*

$$\Lambda(1), \ldots, \Lambda(p); \Pi(1), \ldots, \Pi(p).$$

(4)

$$\dim A(\mathcal{W}(p)) = \dim \mathcal{P}(\mathcal{W}(p)) = 6p - 1.$$

Let us here recall description of C_2-algebra $\mathcal{P}(\mathcal{W}(p))$. Generators of $\mathcal{P}(\mathcal{W}(p))$ are given by

$$\bar{\omega}, \bar{H}, \bar{E}, \bar{F},$$

and the relations are

$$\begin{aligned} \bar{E}^2 = \bar{F}^2 = \bar{H}\bar{F} = \bar{H}\bar{E} = 0, \\ \bar{H}^2 = -\bar{E}\bar{F} = \nu\bar{\omega}^{2p-1} \quad (\nu \neq 0), \\ \bar{\omega}^p \bar{H} = \bar{\omega}^p \bar{E} = \bar{\omega}^p \bar{F} = 0. \end{aligned}$$

The complete description of the structure of Zhu's algebra was obtained in [10], where we developed a new method for the determination of Zhu's algebra which was based on a construction of homomorphism of $\Phi : A(\overline{M(1)}) \to A(\mathcal{W}(p))$. Then we described the kernel of such homomorphism, and by using knowledge of Zhu's algebra $A(\overline{M(1)})$ for the singlet vertex algebra $\overline{M(1)}$ mentioned earlier, we get the following result:

Theorem 3.4 [4, 10] *Zhu's algebra $A(\mathcal{W}(p))$ decomposes as a direct sum*:

$$A(\mathcal{W}(p)) = \bigoplus_{i=2p}^{3p-1} \mathbb{M}_{h_{i,1}} \oplus \bigoplus_{i=1}^{p-1} \mathbb{I}_{h_{i,1}} \oplus \mathbb{C}_{h_{p,1}},$$

where $\mathbb{M}_{h_{i,1}}$ ideal isomorphic to matrix algebra $M_2(\mathbb{C})$, $\mathbb{I}_{h_{i,1}}$ is 2-dimensional ideal, $\mathbb{C}_{h_{p,1}}$ is 1-dimensional ideal. The structure of Zhu's algebra $A(\mathcal{W}(p))$ implies the existence of logarithmic modules.

The similar result was obtained in [10] for what we called the super-triplet vertex algebra $\mathcal{SW}(m)$. The advantage of the method used in [10] is that for the description of Zhu's algebra we don't use any result about the existence of logarithmic representations. In our approach, the existence of logarithmic representations is a consequence of the description of Zhu's algebra.

Corollary 3.1 [10] *For every $1 \le i \le p-1$, there exits a logarithmic, self-dual, $\mathbb{Z}_{\ge 0}$-graded $\mathcal{W}(p)$-module denoted by $\mathcal{P}_i^+$ such that the top component $\mathcal{P}_i^+(0)$ is two-dimensional and $L(0)$ acts on it (in some basis) as*

$$\begin{pmatrix} h_{i,1} & 1 \\ 0 & h_{i,1} \end{pmatrix}.$$

Remark 3.1 The vertex algebra $\mathcal{W}(p)$ also has $(p-1)$-logarithmic modules $\mathcal{P}_i^-$ which cannot be detected by $A(\mathcal{W}(p))$. These modules can be constructed explicitly as in [7] and [63]. On the other hand, one can apply the Huang-Lepowsky tensor product $\hat{\otimes}$ [51] and get:

$$\mathcal{P}_i^- := \mathcal{P}_{p-i}^+ \hat{\otimes} \Pi(1).$$

Remark 3.2 Almost everything in this section can be modified, along the lines of [5, 6], to $N = 1$ vertex superalgebras, by consideration of odd lattices and by tensoring V_L with the free fermion vertex superalgebra.

4 $\mathcal{W}$-Algebra Extensions of Minimal Models

If we consider $L = \sqrt{pp'}Q$, where p and p' are relatively prime and strictly bigger than one, there are additional degrees of freedom entering the construction of screening operators. These values allow us to construct more complicated vertex algebras, closely related to affine $\mathcal{W}$-minimal models.

For simplicity we only consider the case $Q = A_1$, well studied in the physics literature.

The setup is $L = \sqrt{pp'}\mathbb{Z}\alpha_1$, $\langle \alpha_1, \alpha_1 \rangle = 2$. To avoid (annoying) radicals, let $\alpha = \sqrt{pp'}\alpha_1$. Then

$$L = \mathbb{Z}\alpha, \qquad \langle \alpha, \alpha \rangle - 2pp'.$$

We construct V_L as before but now we choose

$$\omega_{p,p'} = \omega_{st} + \frac{p - p'}{2pp'}\alpha(-2),$$

such that the central charge is $1 - 6\frac{(p-p')^2}{pp'}$ (minimal central charges [65]). There are again two screening operators here [28, 29] (cf. [9, 11]):

$$\mathcal{Q} = e_0^{\alpha/p'} \quad \text{and} \quad \widetilde{\mathcal{Q}} = e_0^{-\alpha/p}$$

Although the rank is one, the replacement for $\mathcal{W}(p)_Q$ involves *both* screenings, namely

$$\mathcal{W}_{p,p'} := \mathrm{Ker}_{V_L} \mathcal{Q} \cap \mathrm{Ker}_{V_L} \widetilde{\mathcal{Q}}.$$

Compared to $\mathcal{W}(p)_Q$ this vertex algebra is more complicated and it is no longer simple [9]. The inner structure of V_L, and of $\mathcal{W}_{p,p'}$, as a Virasoro algebra module, can be visualized via the following diagram describing the semisimple filtration of V_L. Here all $\bullet$ symbols denote highest weight vectors for the Virasoro algebra and they generate the socle part of V_L. Similarly, all Δ symbols are representatives of the top part in the filtration, etc.

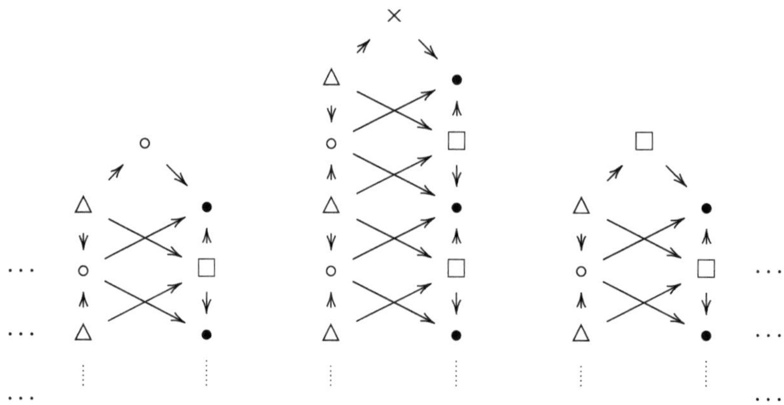

The $\mathcal{W}$-algebra $\mathcal{W}_{p,p'}$ is generated by all $\bullet$ (the socle part) and the vacuum vector $\times$. Clearly, the socle part forms a nontrivial ideal in $\mathcal{W}(p, p')$.

Conjecture 4.1 *Assume that* $(p, p') = 1$. *The vertex algebra* $\mathcal{W}_{p,p'}$ *is* C_2*-cofinite with* $2pp' + \frac{(p-1)(p'-1)}{2}$*-irreducible modules.*

4.1 The Triplet Vertex Algebra $\mathcal{W}_{p,2}$

There are not many rigorous results about the $\mathcal{W}$-algebras $\mathcal{W}_{p,p'}$, except for $p' = 2$ [8, 11]. We believe that some of the techniques introduced in [8, 11] are sufficient to prove the C_2-cofiniteness for all p and p'.

The triplet vertex algebra $\mathcal{W}_{p,2}$ can be realized as a subalgebra of V_L generated by ω and primary vectors

$$F = \mathcal{Q}e^{-3\alpha/2}, \qquad H = GF, \qquad E = G^2F,$$

where G is (new) screening operator defined by

$$G = \sum_{i=1}^{\infty} \frac{e_{-i}^{\alpha/2} e_i^{\alpha/2}}{i}.$$

Therefore, the triplet vertex algebra $\mathcal{W}_{p,2}$ is $\mathcal{W}$-algebra of type

$$\mathcal{W}(2, h, h, h) \quad \big(h = (2n+1)(pn + p - 1)\big).$$

The next result shows that Conjecture 4.1 holds for $p' = 2$.

Theorem 4.1 [9, 11] *We have*:

(1) *Every Virasoro minimal model for central charge* $c_{p,2}$ *is a module for* $\mathcal{W}_{p,2}$.

(2) $\mathcal{W}_{p,2}$ *has exactly* $4p + \frac{p-1}{2}$ *irreducible modules.*
(3) $\mathcal{W}_{p,2}$ *is* C_2*-cofinite.*
(4) $\mathcal{W}_{p,2}$ *is irrational and admits logarithmic modules.*

Let $p = 3$. Then $\mathcal{W}_{3,2}$ is called the triplet $c = 0$ vertex algebra. Let us recall Zhu's algebra for this vertex algebra.

Generators of $A(\mathcal{W}_{3,2})$: $[\omega]$, $[H]$, $[E]$, $[F]$.

Theorem 4.2 [10] *Zhu's algebra* $A(\mathcal{W}_{3,2})$ *decomposes as a direct sum*:

$$A(\mathcal{W}_{3,2}) = \bigoplus_{h \in S^{(2)}} \mathbb{M}_h \oplus \bigoplus_{h \in S^{(1)}} \mathbb{I}_h \oplus \mathbb{C}_{-1/24},$$

$$S^{(2)} = \left\{5, 7, \frac{10}{3}, \frac{33}{8}, \frac{21}{8}, \frac{35}{24}\right\}, \qquad S^{(1)} = \left\{0, 1, 2, \frac{1}{3}, \frac{1}{8}, \frac{5}{8}, \frac{-1}{24}\right\},$$

where $\mathbb{M}_h$ *is ideal isomorphic to* $M_2(\mathbb{C})$, $h \in S^{(2)}$, $\mathbb{I}_h$ *is* 2*-dimensional ideal,* $h \in S^{(1)}$, $h \neq 0$, $h \neq -1/24$, $\mathbb{C}_{-1/24}$ *is* 1*-dimensional ideal,* $\mathbb{I}_0$ *is* 3*-dimensional ideal.*

Remark 4.1 The previous theorem shows that in the category of $\mathcal{W}_{3,2}$-modules, the projective cover of trivial representation should have $L(0)$-nilpotent rank three. This result is used in the fusion rules analysis for the $c = 0$ triplet algebra (cf. [44, 45]).

In [11], we proved that in the category of $\mathcal{W}_{p,2}$-modules, the projective cover of every minimal model should have $L(0)$-nilpotent rank three. We expect the same result to hold for general minimal (p, p')-models.

5 Construction of Logarithmic Modules and Related Problems

In Sect. 3 we propose a large family of (conjecturally) C_2-cofinite vertex algebras coming from integral lattices. Now we examine indecomposable representations for these algebras.

5.1 Progenerator and Logarithmic Modules

A central question in representation theory of vertex algebra (or any algebraic structure) is to understand the structure of indecomposable modules. As it is well-known, for rational[3] vertex algebras it is sufficient to classify irreducible modules. In contrast, for irrational C_2-cofinite vertex algebras (with finitely many irreps) it is essential to analyze the projective covers P_i of irreducibles M_i, $i \in \mathrm{Irr}$ [46]. Provided

[3]Here for simplicity we assume strong rationality, meaning that for a given VOA every (weak) module is completely reducible.

that we have a good description of $P_i \to M_i$, we can then form a progenerator $P = \bigoplus_{i \in \mathrm{Irr}} P_i$, and compute

$$\mathcal{A} = \mathrm{End}_V(P)^{op},$$

which is known to be finite-dimensional. This associative algebra plays a major rule in representation theory, and the least it gives the Morita equivalence of abelian categories

$$\text{f.g } V\text{-Mod} \cong \text{f.d. } \mathcal{A}\text{-Mod}.$$

As we shall see later, the same algebra is also important for purposes of modular invariance. Because the category V-Mod has a natural braided tensor category structure [51], it is expected that one can do better and find a braided Hopf algebra $\mathcal{A}$ such that the above equivalence holds at the level of braided tensor categories (this is known in some cases [28, 29, 63]).

The main problem here is that there is no good construction of P_i even in the simplest case due to the fact that projective modules of irrational C_2-cofinite vertex algebras are often *logarithmic*, that is, non-diagonalizable with respect to the Virasoro operator $L(0)$. At the same time the C_2-cofinite vertex algebra is conformally embedded inside a rational lattice vertex algebra, which is known to have no logarithmic modules. Thus we cannot simply use the larger algebra to construct all relevant modules for the smaller algebra (except perhaps for the irreducibles [4]).

Thus, in order to maneuver ourselves into a situation in which $\mathrm{End}_V(P)$ can be studied, we first discuss construction of general logarithmic modules. There are other related problems such as construction of intertwining operators among irreducible modules [38, 50, 59], which we do not discuss here.

5.2 Screenings and Logarithmic Modules

Here we propose a very general construction of logarithmic modules [31, 47, 57–59] for vertex algebras coming from screenings operators as in Sect. 3. As we shall see, in some cases these modules are indeed projective covers. Our methods is based on screenings, local systems of vertex operators [54], together with deformation of the vertex algebra action [56] (cf. also [8]). Conjecturally, the method introduced here is sufficient to construct all projective covers for vertex algebras considered in Sect. 3.

Let V be a vertex algebra of CFT type and let $v \in V$ be a primary vector of conformal weight one, and

$$Y(v, x) = \sum_{n \in \mathbb{Z}} v_n x^{-n-1}.$$

As in [56] we let

$$\Delta(v,x) = x^{v_0} \exp\left(\sum_{n=1}^{\infty} \frac{v_n}{-n}(-x)^{-n}\right). \tag{4}$$

If v_0 acts semisimply on V and w is its eigenvector, the expression $x^{v_0}w$ is defined as $x^{\lambda}w$, where λ is the corresponding eigenvalue. But (4) is ambiguous if v_0 does not act semisimply, Still the next result [8] easily follows from [56].

Theorem 5.1 *Assume that V and v are as above. Let $\overline{V}$ be the vertex subalgebra of V such that $\overline{V} \subseteq \mathrm{Ker}_V\, v_0$. Assume that (M, Y_M) is a V-module. Define the pair $(\widetilde{M}, \widetilde{Y}_{\widetilde{M}})$ such that*

$$\widetilde{M} = M \quad \text{as a vector space,}$$
$$\widetilde{Y}_{\widetilde{M}}(a,x) = Y_M\big(\Delta(v,x)a, x\big) \quad \text{for } a \in \overline{V}.$$

Then $(\widetilde{M}, \widetilde{Y}_{\widetilde{M}})$ is a $\overline{V}$-module.

Corollary 5.1 *Assume that (M, Y_M) is a V-module such that $L(0)$ acts semisimply on M. Then $(\widetilde{M}, \widetilde{Y}_{\widetilde{M}})$ is a logarithmic $\overline{V}$-module if and only if v_0 does not act semisimply on M.*

By using this method logarithmic $\mathcal{W}(p)_Q$-modules (including projective covers) can be constructed by taking $v = e^{-\alpha_i/\sqrt{p}}$ [8]. But in general (cf. [8, 63]) one cannot construct all projective covers simply by taking v to be a primary vector inside the generalized vertex algebra V_{L°. Instead we require more complicated operators not present in the extended algebra V_{L° (for a recent application of this circle of ideas see [12]). Then, when combined with $\mathcal{W}(p)_Q$ (and not all of V_L!), these more complicated local operators $v^{[i]}(z)$ (here $[i]$ has no particular meaning; it merely indicates some sort of "power" construction) became mutually local with $\mathcal{W}(p)_Q$, which allows us to extend our $\mathcal{W}$-algebra with $v^{[i]}(z)$ by using Li's theory of local systems [54]. Then we cook up a Δ operator and consider the residue

$$v_0^{[i]} = \mathrm{Res}_{z_0}\, v^{[i]}(z),$$

which also annihilate $\mathcal{W}(p)_Q$, and again apply Theorem 5.1.

Already from this discussion we infer

Corollary 5.2 *The vertex algebra $\mathcal{W}(p)_Q$ is irrational.*

This result requires a single screening $e_0^{-\alpha_i/\sqrt{p}}$.

6 Some Logarithmic Modules for $\mathcal{W}(p)_Q$

In this section we shall describe a family of such logarithmic representations for $\mathcal{W}(p)_Q$ based on the second power of screening operators. We present a new locality result which enables us to use concepts developed in [8] and described above. To exemplify the construction we only consider $Q = A_1$, and focus on $\mathcal{W}(p)$, but everything in this section applies to α replaced by $\sqrt{p}\alpha_i$. Here $p > 1$. Define the following lattices

$$L = \mathbb{Z}\alpha, \qquad \widetilde{L} = \mathbb{Z}\frac{\alpha}{p},$$

where $\langle \alpha, \alpha \rangle = 2p$.

Then $V_{\widetilde{L}}$ has the structure of a generalized vertex operator algebra, and its subalgebra V_L is a vertex operator algebra with the Virasoro vector

$$\omega = \frac{1}{4p}\alpha(-1)^2 + \frac{p-1}{2p}\alpha(-2).$$

Let $a = e^{-\alpha/p}$. In the generalized vertex algebra $V_{\widetilde{L}}$ the following locality relation holds:

$$(z_1 - z_2)^{-2/p} Y(a, z_1) Y(a, z_2) - (z_2 - z_1)^{-2/p} Y(a, z_2) Y(a, z_1) = 0.$$

Define

$$\phi(t) = p t^{1/p} {}_2F_1\left(\begin{matrix} 1/p, 2/p \\ 1 + 1/p \end{matrix}; t\right) = \sum_{j=0}^{\infty} \frac{(-1)^j}{j + 1/p} t^{j+1/p} \binom{-2/p}{j},$$

$$G(z) = \operatorname{Res}_{z_1}\big(\phi(z/z_1) Y(a, z_1) Y(a, z) + \phi(z_1/z) Y(a, z) Y(a_1, z_1)\big).$$

Let $G(z) = \sum_{n\in\mathbb{Z}} G(n) z^{-n-1}$. Then, for $n \in \mathbb{Z}$, we get

$$G(n) = \sum_{j\geq 0} \frac{(-1)^j}{1/p + j} \binom{-2/p}{j} (a_{-1/p-j+n} a_{1/p+j} + a_{-1/p-j} a_{1/p+j+n}),$$

$$G(0) = 2 \sum_{j\geq 0} \frac{(-1)^j}{1/p + j} \binom{-2/p}{j} a_{-1/p-j} a_{1/p+j}.$$

We infer the following result:

Proposition 6.1 *We have*

(1)

$$\big[L(n), G(m)\big] = -m G(n+m) \quad (m, n \in \mathbb{Z}).$$

In particular, $G(0)$ is a screening operator.

(2) *The fields* $G(z)$ *and* $L(z)$ *are mutually local. More precisely:*

$$(z_1 - z_2)^3 \big[L(z_1), G(z_2)\big] = 0.$$

We also have:

$$L(z)_0 G(z) = G'(z), \qquad L(z)_1 G(z) = G(z), \qquad L(z)_n G(z) = 0 \quad \textit{for } n \geq 2.$$

(3) *Let* $\widetilde{L(n)} = L(n) + G(n)$. *Then operators* $\widetilde{L(n)}$ *define on*

$$M_2^+ = V_{L+((p+1)/(2p))\alpha} \oplus V_{L+((p-3)/(2p))\alpha},$$
$$M_2^- = V_{L+(1/(2p))\alpha} \oplus V_{L+(-3/(2p))\alpha}$$

the structure of the module for the Virasoro algebra.

Remark 6.1 One can also represent $G(z)$ by using contour integrals as in [63], and give a different proof of Proposition 6.1 by using methods developed in [64]. One defines

$$Q^{[2]}(z) = \int_\gamma dt Y(a, z) Y(a, tz) z$$

where γ is a certain contour. For $p \geq 2$, we can show that

$$\frac{\Gamma(2/p+1)\Gamma(-1/p)}{\Gamma(1/p+1)} G(z) = Q^{[2]}(z),$$

where $\Gamma(z)$ is the usual Γ-function.

Let $\widehat{a} = e^{\alpha - \alpha/p}$. Define

$$\begin{aligned} \overline{G}(z) &= \sum_{n \in \mathbb{Z}} \overline{G}(n) z^{-n-1} \\ &= \mathrm{Res}_{z_1}\big(\phi(z/z_1) Y(a, z_1) Y(\widehat{a}, z) + \phi(z_1/z) Y(\widehat{a}, z) Y(a, z_1)\big). \end{aligned}$$

Then

$$\overline{G}(n) = \sum_{j \geq 0} \frac{(-1)^j}{1/p + j} \binom{-2/p}{j} (\widehat{a}_{-1/p-j+n} a_{1/p+j} + a_{-1/p-j} \widehat{a}_{1/p+j+n}).$$

Let

$$\mu = \frac{p}{p-1}.$$

First we need the following result.

Lemma 6.1 *We have the following relations:*

(i) $[e_0^\alpha, G(n)] = -n\mu\overline{G}(n-1)$; *i.e.*, $[e_0^\alpha, G(z)] = \mu\overline{G}'(z)$;
(ii) $[e_0^\alpha, G(0)] = 0$,
(iii) $[L(n), \overline{G}(m)] = -(n+m+1)\overline{G}(n+m)$.

Proof Let us prove relation (i). First we notice that

$$e_0^\alpha a = \mu D\widehat{a}.$$

Then we have

$$\begin{aligned}
\left[e_0^\alpha, G(z)\right] &= \mu \operatorname{Res}_{z_1}\big(\phi(z/z_1)\partial_{z_1}Y(\widehat{a}, z_1)Y(a,z) + \phi(z_1/z)Y(a,z)\partial_{z_1}Y(\widehat{a}, z_1)\big) \\
&\quad + \mu \operatorname{Res}_{z_1}\big(\phi(z/z_1)Y(a, z_1)\partial_z Y(\widehat{a}, z) + \phi(z_1/z)\partial_z Y(\widehat{a}, z)Y(a, z_1)\big) \\
&= -\mu \operatorname{Res}_{z_1} z^{1/p} z_1^{1/p-1} \\
&\quad \times \big((z_1 - z)^{-2/p} Y(\widehat{a}, z_1)Y(a,z) - (z - z_1)^{-2/p} Y(a,z)Y(\widehat{a}, z_1)\big) \\
&\quad + \mu \operatorname{Res}_z z^{1/p-1} z_1^{1/p} \\
&\quad \times \big((z_1 - z)^{-2/p} Y(a, z_1)Y(\widehat{a}, z) - (z - z_1)^{-2/p} Y(\widehat{a}, z)Y(a, z_1)\big) \\
&\quad + \mu \operatorname{Res}_z\big(\overline{G}(z)\big) \\
&= \mu \operatorname{Res}_z\big(\overline{G}(z)\big).
\end{aligned}$$

This proves relation (i). The relation (ii) follows from (i). The proof of (iii) is similar to that of (ii). □

Recall that the triplet vertex algebra $\mathcal{W}(p)$ is realized as a subalgebra of V_L generated by the vectors

$$\omega, \qquad F = e^{-\alpha}, \qquad H = QF, \qquad E = Q^2 F,$$

where $Q = e_0^\alpha$.

The doublet vertex algebra $\mathcal{A}(p)$ is the subalgebra of $V_{\widetilde{L}}$ generated by

$$x^- = e^{-\alpha/2}, \qquad x^+ = Qa^{-\alpha/2}.$$

Clearly, $\mathcal{W}(p)$ is a subalgebra of $\mathcal{A}(p)$.

Proposition 6.2 *We have*

(i) *The fields* $G(z)$, $\overline{G}(z)$, $Y(x^-, z)$, $Y(x^+, z)$ *are mutually local.*
(ii)

$$\begin{aligned}
U &= \operatorname{span}_{\mathbb{C}}\{G(z); Y(v,z) | v \in \mathcal{W}(p)\}, \\
U^e &= \operatorname{span}_{\mathbb{C}}\{G(z), \overline{G}(z); Y(v,z) | v \in \mathcal{W}(p)\}
\end{aligned}$$

are local subspaces of fields acting on $M_2^{\pm}$.

Proof It is only non-trivial to prove that $G(z)$ and $Y(x^+, z)$ are local. We have

$$\begin{aligned}\left[Y\left(x^+, z_1\right), G(z_2)\right] &= \left[\mathcal{Q}, \left[Y\left(x^-, z_1\right), G(z_2)\right]\right] - \left[Y\left(x^-, z_1\right), \left[\mathcal{Q}, G(z_2)\right]\right] \\ &= -\left[Y\left(x^-, z_1\right), \left[\mathcal{Q}, G(z_2)\right]\right]. \end{aligned} \tag{5}$$

The proof easily follows if we invoke Lemma 6.1 and the fact that the fields $Y(a^-, z_1)$ and $e^{\alpha-\alpha/p}(z)$ are local. This proves (i). By using a standard result on locality of vertex operators [54, 55], we invoke that the field $G(z)$, $\overline{G}(z)$ are local with all fields $Y(v, z)$, $v \in \mathcal{A}(p)$. In particular, the sets U and U^e are local. □

Remark 6.2 We believe that this locality result is new. One can see that $G(z)$ is local only with $\mathcal{W}(p)$, but it is not local with all fields $Y(a, z)$, $a \in V_L$. In particular, $G(z)$ is not local with Heisenberg field $\alpha(z)$.

Let $\mathcal{V}$ (resp. $\mathcal{V}^e$) be the vertex algebra generated by local subspace U (resp. U^e). It is clear that

$$v \mapsto Y(v, z) \quad \left(v \in \mathcal{W}(p)\right)$$

is a injective homomorphism of vertex algebras. So $\mathcal{W}(p)$ can be considered as a subalgebra of $\mathcal{V}$.

Theorem 6.1 *We have*:

(i) $\mathcal{W}(p).G(z) = \mathrm{span}_{\mathbb{C}}\{Y(v, z)_n G(z) | v \in \mathcal{W}(p), n \in \mathbb{Z}\} \cong \Pi(p-1)$.
(ii) $\mathcal{V} \cong \mathcal{W}(p) \oplus \Pi(p-1)$.
(iii) $\mathcal{V}^e = \mathcal{W}(p) \oplus E$, *where* $E = \mathcal{W}(p).\overline{G}(z)$.
(iv) *There is a non-split extension*

$$0 \to \Pi(p-1) \to E \to \Lambda(1) \to 0.$$

Proof It is clear that $\mathcal{V} = \mathcal{W}(p) \oplus \mathcal{W}(p).G(z)$. So it remains to identify cyclic $\mathcal{W}(p)$-module $\mathcal{W}(p).G(z)$. The locality relations

$$(z_1 - z_2)^{2p-1}\left[Y(x, z_1), G(z_2)\right] = 0 \quad \left(x \in \{E, F, H\}\right),$$

imply that $\mathcal{W}(p).G(z)$ is a $\mathbb{Z}_{\geq 0}$-graded $\mathcal{W}(p)$-module with lowest weight 1. Top component is 2-dimensional and spanned by $G(z)$ and $[\mathcal{Q}, G(z)]$. By using representation-theoretic results from [4] we see that this module is isomorphic to $\Pi(p-1)$, and that $E/\Pi(p-1) \cong \Lambda(1)$. The proof follows. □

Remark 6.3 We know that there is also a non-split extension

$$0 \to \Pi(p-1) \to V_{L+\alpha-\alpha/p} \to \Lambda(1) \to 0.$$

But, $V_{L+\alpha-\alpha/p} \ncong E$.

The operators $\widetilde{G(z)}_n, n \in \mathbb{Z}$, define on $\mathcal{V}$ the structure of a module for the Heisenberg algebra such that $\widetilde{G(z)}_0$ acts trivially. Therefore the field

$$\Delta\bigl(\widetilde{G(z)}, z_1\bigr) = z_1^{\widetilde{G(z)}_0} \exp\biggl(\sum_{n=1} \frac{\widetilde{G(z)}_n}{-n} z_1^{-n}\biggr)$$

is well defined on $\mathcal{V}$. As in [8] we have the following result:

Theorem 6.2 *Assume that $(M, Y_M(\cdot, z_1))$ is a weak $\mathcal{V}$-module. Define the pair $(\widetilde{M}, \widetilde{Y}_{\widetilde{M}}(\cdot, z_1))$ such that*

$$\widetilde{M} = M \quad \text{as a vector space,}$$
$$\widetilde{Y}_{\widetilde{M}}\bigl(v(z), z_1\bigr) = Y_M\bigl(\Delta\bigl(\widetilde{G(z)}, z_1\bigr)v(z), z_1\bigr).$$

Then $(\widetilde{M}, \widetilde{Y}_{\widetilde{M}}(\cdot, z_1))$ is a weak $\mathcal{V}$-module. In particular, $(\widetilde{M}, \widetilde{Y}_{\widetilde{M}}(\cdot, z_1))$ is a $\mathcal{W}(p)$-module.

Recall that $M_2^{\pm}$ are modules for the vertex algebra $\mathcal{V}$ with the vertex operator map

$$Y\bigl(v(z), z_0\bigr) = v(z_0), \quad v(z) \in \mathcal{V}.$$

Applying the above construction we get a (new) explicit realization of logarithmic modules for $\mathcal{W}(p)$.

Theorem 6.3 *$(\widetilde{M_2^{\pm}}, \widetilde{Y})$ is a $\mathcal{W}(p)$-module such that*

$$\widetilde{Y}(v, z) = Y(v, z) + \sum_{n=1}^{\infty} \frac{G(z)_n Y(v, z)}{-n} (-z)^{-n}, \quad v \in \mathcal{W}(p).$$

In particular,

$$\widetilde{Y}(\omega, z) = \widetilde{L}(z).$$

The operator $\widetilde{L(0)}$ acts on $\widetilde{M_2^{\pm}}$ as

$$\widetilde{L(0)} = L(0) + G(0)$$

and it has nilpotent rank 2.

Remark 6.4 By applying the methods developed in [8] we see that $\widetilde{M_2^{\pm}}$ are self-dual, logarithmic modules of semisimple rank three. Moreover, $\widetilde{M_2^{+}}$ (resp. $\widetilde{M_2^{-}}$) is projective cover of $\Lambda(2)$ (resp. $\Pi(p-2)$). The same modules have been constructed in [63] by using a slightly different method.

7 Conclusion

We hope that we have conveyed the main ingredients behind the plethora of $\mathcal{W}$-algebras connected to Logarithmic Conformal Field Theory. There are still numerous problems to be resolved at the structural level (e.g. C_2-cofiniteness), but we hope that the present techniques in vertex algebra theory—with further constructions as in Sect. 6—are sufficient to resolve the main conjectures in the paper, including construction of projective covers. Eventually this development on the vertex algebra side will play an important role in finding a precise relationship with the finite-dimensional quantum groups at root of unity proposed in [25, 27–29].

There are several aspects of C_2-cofinite W-algebras that we did not discuss in this paper. Here we briefly outline on these developments.

(1) There is an important (simple-current) extension of the triplet vertex algebra $\mathcal{W}(p)$, called the doublet $\mathcal{A}(p)$. If p is even, that $\mathcal{A}(p)$ carries the structure of a vertex algebra (or vertex superalgebra). Its representation theory has been developed in [13]. This extension can be constructed in the higher rank as well. Also, a large portion of the present work extends to $N = 1$ vertex operator superalgebras.
(2) We expect to see rich combinatorics underlying $\mathcal{W}(p)_Q$, including properties of graded dimensions of modules and of some distinguished subspaces examined in [60] (see also [33]). Another important facet of the theory was initiated in [9, 11] in connection to constant term identities of Morris-Macdonald type (see also [18]). These identities are expected to play a role in the theory of higher Zhu's algebras.
(3) Modular invariance and one-point functions on the torus are important ingredients in CFT [68]. In [7] (cf. also [32]) we have shown that the space of one-point functions for $\mathcal{W}(p)$ is $3p - 1$ dimensional. But in view of [61], it is not completely obvious how to describe the space of one-point functions explicitly via certain pseudotraces. For $\mathcal{W}_{p,p'}$ we still do not know precisely even its dimension, although there is an obvious guess by looking at the properties of irreducible characters [27–29]. One-point functions for the C_2-cofinite vertex algebra SF^+ coming from *symplectic fermions* [1] have been recently studied in [15]. Some general results about "logarithmic modular forms" are obtained in [53].
(4) There is ongoing effort in the direction of constructing the *full* rational conformal field theory [48, 49]. Although it is not clear how to generalize the notion of full field algebra to general C_2-cofinite vertex algebras, some progress has been achieved recently on the construction of the bulk space, in the case of the triplet vertex algebra $\mathcal{W}(p)$ and $\mathcal{W}_{2,3}$ [43–45] (cf. also [67] for general p and p').

Acknowledgements This note is based on lectures given by the authors at the conference on "Tensor Categories and Conformal Field Theory", June 2011, Beijing. Thus, except for Sect. 5 and some constructions in Sect. 2, all the material is based on earlier works by the authors (we should say that some constructions in Sect. 2 were independently introduced in [25]). We are indebted to the organizers for invitation to this wonderful conference. We also thank A. Semikhatov, A. Gaĭnutdinov, A. Tsuchiya, I. Runkel, Y. Arike, J. Lepowsky, L. Kong and Y.-Z. Huang for the interesting discussion during and after the conference. We are also grateful to Jinwei Yang for helping us around in Beijing.

References

1. Abe, T.: A Z_2-orbifold model of the symplectic fermionic vertex operator superalgebra. Math. Z. **255**, 755–792 (2007)
2. Abe, T., Buhl, G., Dong, C.: Rationality, regularity and C_2-cofiniteness. Trans. Am. Math. Soc. **356**, 3391–3402 (2004)
3. Adamović, D.: Classification of irreducible modules of certain subalgebras of free boson vertex algebra. J. Algebra **270**, 115–132 (2003)
4. Adamovic, D., Milas, A.: On the triplet vertex algebra $W(p)$. Adv. Math. **217**, 2664–2699 (2008)
5. Adamovic, D., Milas, A.: The $N = 1$ triplet vertex operator superalgebras: twisted sector. SIGMA **4**, 087 (2008)
6. Adamovic, D., Milas, A.: The $N = 1$ triplet vertex operator superalgebras. Commun. Math. Phys. **288**, 225–270 (2009)
7. Adamovic, D., Milas, A.: An analogue of modular BPZ equation in logarithmic conformal field theory. Contemp. Math. **497**, 1–17 (2009)
8. Adamovic, D., Milas, A.: Lattice construction of logarithmic modules for certain vertex algebras. Sel. Math. New Ser. **15**(4), 535–561 (2009)
9. Adamovic, D., Milas, A.: On W-algebras associated to $(2, p)$ minimal models and their representations. Int. Math. Res. Not. **2010**, 3896–3934 (2010)
10. Adamovic, D., Milas, A.: The structure of Zhu's algebras for certain W-algebras. Adv. Math. **227**, 2425–2456 (2011)
11. Adamovic, D., Milas, A.: On W-algebra extensions of $(2, p)$ minimal models: $p > 3$. J. Algebra **344**, 313–332 (2011)
12. Adamovic, D., Milas, A.: An explicit realization of logarithmic modules for the vertex operator algebra $\mathcal{W}_{p,p'}$. J. Math. Phys. **53**, 073511 (2012)
13. Adamovic, D., Milas, A.: The doublet vertex operator algebra $\mathcal{A}(p)$ and $\mathcal{A}_{2,p}$. Contemp. Math. **602** (2013)
14. Arakawa, T.: Representation theory of W-algebras. Invent. Math. **169**, 219–320 (2007). arXiv:math/0506056
15. Arike, Y., Nagatomo, A.: Some remarks on pseudo-trace functions for orbifold models associated with symplectic fermions. Preprint. arXiv:1104.0068
16. Borcherds, R.E.: Vertex algebras, Kac–Moody algebras, and the monster. Proc. Natl. Acad. Sci. USA **83**, 3068–3071 (1986)
17. Carqueville, N., Flohr, M.: Nonmeromorphic operator product expansion and C_2-cofiniteness for a family of W-algebras. J. Phys. A, Math. Gen. **39**, 951 (2006)
18. Chappell, T., Lascoux, A., Warnaar, S., Zudilin, W.: Logarithmic and complex constant term identities. arXiv:1112.3130
19. Creutzig, T., Ridout, D.: Relating the archetypes of logarithmic conformal field theory. arXiv:1107.2135
20. de Boer, J., Tjin, T.: Quantization and representation theory of finite W-algebras. Commun. Math. Phys. **158**, 485–516 (1993)
21. De Sole, A., Kac, V.: Finite vs. affine W-algebras. Jpn. J. Math. **1**, 137–261 (2006). arXiv:math-ph/0511055
22. Dong, C.: Vertex algebras associated with even lattices. J. Algebra **160**, 245–265 (1993)
23. Dong, C., Lepowsky, J.: Generalized Vertex Algebras and Relative Vertex Operators. Progress in Mathematics. Birkhäuser, Boston (1993)
24. Feigin, B., Stoyanovsky, A.: Quasi-particles models for the representations of Lie algebras and geometry of flag manifold. arXiv:hep-th/9308079

25. Feigin, B., Tipunin, I.: Logarithmic CFTs connected with simple Lie algebras. arXiv: 1002.5047
26. Feigin, B., Feigin, E., Littelmann, P.: Zhu's algebras, C_2 algebras and abelian radicals. J. Algebra **329**(1), 130–146 (2011). arXiv:0907.3962
27. Feigin, B.L., Gaĭnutdinov, A.M., Semikhatov, A.M., Tipunin, I.Y.: The Kazhdan-Lusztig correspondence for the representation category of the triplet W-algebra in logarithmic conformal field theories. Teor. Mat. Fiz. **148**(3), 398–427 (2006) (Russian)
28. Feigin, B.L., Gaĭnutdinov, A.M., Semikhatov, A.M., Tipunin, I.Y.: Logarithmic extensions of minimal models: characters and modular transformations. Nucl. Phys. B **757**, 303–343 (2006)
29. Feigin, B.L., Gaĭnutdinov, A.M., Semikhatov, A.M., Tipunin, I.Y.: Modular group representations and fusion in logarithmic conformal field theories and in the quantum group center. Commun. Math. Phys. **265**, 47–93 (2006)
30. Feigin, E., Littelmann, P.: Zhu's algebra and the C_2-algebra in the symplectic and the orthogonal cases. J. Phys. A, Math. Theor. **43**, 135206 (2010). arXiv:0911.2957v1
31. Fjelstad, J., Fuchs, J., Hwang, S., Semikhatov, A.M., Tipunin, I.Y.: Logarithmic conformal field theories via logarithmic deformations. Nucl. Phys. B **633**, 379–413 (2002)
32. Flohr, M.: On modular invariant partition functions of conformal field theories with logarithmic operators. Int. J. Mod. Phys. A **11**(22), 4147–4172 (1996)
33. Flohr, M., Grabow, A., Koehn, R.: Fermionic expressions for the characters of $c(p,1)$ logarithmic conformal field theories. Nucl. Phys. B **768**, 263–276 (2007)
34. Frenkel, E., Ben-Zvi, D.: Vertex Algebras and Algebraic Curves. Mathematical Surveys and Monographs, vol. 88. Am. Math. Soc., Providence (2001)
35. Frenkel, E., Kac, V., Wakimoto, M.: Characters and fusion rules for W-algebras via quantized Drinfeld-Sokolov reduction. Commun. Math. Phys. **147**, 295–328 (1992)
36. Frenkel, E., Kac, V., Radul, A., Wang, W.: $W_{1+\infty}$ and $W(gl_N)$ with central charge N. Commun. Math. Phys. **170**, 337–357 (1995)
37. Frenkel, I.B., Lepowsky, J., Meurman, A.: Vertex Operator Algebras and the Monster. Pure and Appl. Math., vol. 134. Academic Press, New York (1988)
38. Frenkel, I.B., Huang, Y.-Z., Lepowsky, J.: On axiomatic approaches to vertex operator algebras and modules. Mem. Am. Math. Soc. **104** (1993)
39. Fuchs, J.: On non-semisimple fusion rules and tensor categories. Contemp. Math. **442**, 315–337 (2007)
40. Fuchs, J., Schweigert, C.: Hopf algebras and finite tensor categories in conformal field theory. Rev. Unión Mat. Argent. **51**, 43–90 (2010)
41. Fuchs, J., Hwang, S., Semikhatov, A.M., Tipunin, I.Y.: Nonsemisimple fusion algebras and the Verlinde formula. Commun. Math. Phys. **247**(3), 713–742 (2004)
42. Gaberdiel, M., Gannon, T.: Zhu's algebra, the C_2-algebra, and twisted modules. Contemp. Math. **497**, 65–78 (2009)
43. Gaberdiel, M., Runkel, I.: From boundary to bulk in logarithmic CFT. J. Phys. A **41**, 075402 (2008)
44. Gaberdiel, M., Runkel, I., Wood, S.: Fusion rules and boundary conditions in the $c = 0$ triplet model. J. Phys. A, Math. Theor. **42**, 325–403 (2009). arXiv:0905.0916
45. Gaberdiel, M., Runkel, I., Wood, S.: A modular invariant bulk theory for the $c = 0$ triplet model. J. Phys. A, Math. Theor. **44**, 015204 (2011). arXiv:1008.0082v1
46. Huang, Y.-Z.: Cofiniteness conditions, projective covers and the logarithmic tensor product theory. J. Pure Appl. Algebra **213**, 458–475 (2009). arXiv:0712.4109
47. Huang, Y.-Z.: Generalized twisted modules associated to general automorphisms of a vertex operator algebra. Preprint. arXiv:0905.0514
48. Huang, Y.-Z., Kong, L.: Full field algebras. Commun. Math. Phys. **272**, 345–396 (2007)
49. Huang, Y.-Z., Kong, L.: Modular invariance for conformal full field algebras. Trans. Am. Math. Soc. **362**, 3027–3067 (2010)
50. Huang, Y.-Z., Li, H., Lepowsky, J., Zhang, L.: On the concepts of intertwining operator and tensor product module in vertex operator algebra theory. J. Pure Appl. Algebra **204**, 507–535 (2006)

51. Huang, Y.-Z., Lepowsky, J., Zhang, L.: Logarithmic tensor product theory for generalized modules for a conformal vertex algebra. arXiv:0710.2687 (also Parts I–VIII: arXiv:1012.4193, arXiv:1012.4196, arXiv:1012.4197, arXiv:1012.4198, arXiv:1012.4199, arXiv:1012.4202, arXiv:1110.1929, arXiv:1110.1931)
52. Kac, V., Wang, W.: Superconformal vertex operator superalgebras and their representations. Contemp. Math. **175**, 161–191 (1994)
53. Knopp, M., Mason, G.: Logarithmic vector-valued modular forms. arXiv:0910.3976
54. Lepowsky, J., Li, H.: Introduction to Vertex Operator Algebras and Their Representations. Birkhäuser, Boston (2003)
55. Li, H.: Local systems of vertex operators, vertex superalgebras and modules. J. Pure Appl. Algebra **109**, 143–195 (1996)
56. Li, H.: The physics superselection principle in vertex operator algebra theory. J. Algebra **196**, 436–457 (1997)
57. Milas, A.: Weak modules and logarithmic intertwining operators for vertex operator algebras. Contemp. Math. **297**, 201–225 (2002)
58. Milas, A.: Fusion rings associated to degenerate minimal models. J. Algebra **254**, 300–335 (2002)
59. Milas, A.: Logarithmic intertwining operators and vertex operators. Commun. Math. Phys. **277**, 497–529 (2008)
60. Milas, A., Penn, M.: Lattice vertex algebras and combinatorial bases: general case and W-algebras. N.Y. J. Math. **18**, 621–650 (2012)
61. Miyamoto, M.: Modular invariance of vertex operator algebras satisfying C_2-cofiniteness. Duke Math. J. **122**, 51–91 (2004)
62. Miyamoto, M.: Flatness of tensor products and semi-rigidity for C_2-cofinite vertex operator algebras I. arXiv:0906.1407
63. Nagatomo, K., Tsuchiya, A.: The triplet vertex operator algebra $W(p)$ and the restricted quantum group at root of unity. In: Exploring New Structures and Natural Constructions in Mathematical Physics. Adv. Stud. Pure Math., vol. 61, pp. 1–49. Math. Soc. Japan, Tokyo (2011). arXiv:0902.4607
64. Tsuchiya, A., Kanie, Y.: Fock space representations of the Virasoro algebra—intertwining operators. Publ. Res. Inst. Math. Sci. **22**, 259–327 (1986)
65. Wang, W.: Classification of irreducible modules of W_3 algebra with $c = -2$. Commun. Math. Phys. **195**, 113–128 (1998)
66. Wang, W.: Nilpotent orbits and finite W-algebras. Fields Inst. Commun. **59**, 71–105 (2011)
67. Wood, S.: Fusion rules of the $W_{p,q}$ triplet models. J. Phys. A **43**(4), 045212 (2010)
68. Zhu, Y.: Modular invariance of characters of vertex operator algebras. J. Am. Math. Soc. **9**, 237–302 (1996)

C_1-Cofiniteness and Fusion Products for Vertex Operator Algebras

Masahiko Miyamoto

Abstract Let V be a vertex operator algebra. We prove that if U and W are C_1-cofinite $\mathbb{N}$-gradable V-modules, then a fusion product $U \boxtimes W$ is also a C_1-cofinite $\mathbb{N}$-gradable V-module, where the fusion product is defined by (logarithmic) intertwining operators.

1 Introduction

The tensor product theory is a powerful tool in the theory of representations. Unfortunately, in the theory of vertex operator algebras (shortly VOA), a tensor product (we call "fusion product") for some modules may not exist in the category of modules of vertex operator algebras. In order to avoid such an ambiguity, we will introduce a new approach to treat fusion products. Let us explain it briefly. The details are given in Sect. 3. Let $V = \bigoplus_{n=K}^{\infty} V_n$ be a vertex operator algebra (shortly VOA) and $\mathrm{mod}_{\mathbb{N}}(V)$ denote the set of $\mathbb{N}$-gradable (weak) V-modules, where a (weak) V-module W is called $\mathbb{N}$-gradable if $W = \bigoplus_{m=0}^{\infty} W_{(m)}$ such that

$$v_k w \in W_{(m+\mathrm{wt}(v)-k-1)}$$

for any homogeneous element $v \in V_{\mathrm{wt}(v)}$, $k \in \mathbb{Z}$ and $w \in W_{(m)}$. It is well-known that

$$g(V) := V \otimes_{\mathbb{C}} \mathbb{C}[x, x^{-1}] \Big/ \left(L(-1) \otimes 1 - 1 \otimes \frac{d}{dx} \right) V \otimes_{\mathbb{C}} \mathbb{C}[x, x^{-1}]$$

has a Lie algebra structure and all (weak) V-modules are $g(V)$-modules (see [1]). For $U, W \in \mathrm{mod}_{\mathbb{N}}(V)$, we introduce a $g(V)$-module $U \boxtimes W$ (or its isomorphism class) as a projective limit of a direct set of V-modules (by viewing them as $g(V)$-modules). So, a $g(V)$-module $U \boxtimes W$ always exists. The key point is that a fusion product for $U, W \in \mathrm{mod}_{\mathbb{N}}(V)$ exists if and only if $U \boxtimes W$ is a V-module (and so it is a fusion product).

The main purpose of this paper is to explain the fusion products by emphasizing the importance of C_1-cofiniteness. The importance of the C_1-cofiniteness conditions on modules was firstly noticed by Huang in [2], where he has proved that

M. Miyamoto
Institute of Mathematics, University of Tsukuba, Tsukuba 305, Japan

C. Bai et al. (eds.), *Conformal Field Theories and Tensor Categories*,
Mathematical Lectures from Peking University, DOI 10.1007/978-3-642-39383-9_7,

intertwining operators of a C_1-cofinite $\mathbb{N}$-gradable module from a C_1-cofinite $\mathbb{N}$-gradable module to an $\mathbb{N}$-gradable module satisfy a differential equation. He has also shown the associativity of intertwining operators among C_1-cofinite $\mathbb{N}$-gradable modules by using the space of solutions of this differential equation. We will prove Key Theorem by using his idea. In order to follow his arguments, we give a slightly different definition of C_m-cofiniteness for modules.

Definition 1 Set $m = 1, 2, \dots$. A V-module U is said to be "C_m-cofinite as a V-module" if $C_m(U) := \mathrm{Span}_{\mathbb{C}}\{v_{-m}u \mid u \in U, v \in V, \mathrm{wt}(v) > 1-m\}$ has a finite codimension in U.

This is slightly different from the old one. For example, any VOA V is always C_1-cofinite as a V-module in our definition. Since $(L(-1)v)_{-m} = mv_{-m-1}$ and $\mathrm{wt}(L(-1)v) = \mathrm{wt}(v) + 1$, C_m-cofiniteness implies C_{m-1}-cofiniteness for $m = 2, 3, \dots$.

We will prove the following theorem.

Key Theorem *Let V be a VOA. For each $m = 1, 2, \dots$ and C_m-cofinite $\mathbb{N}$-gradable V-modules U and W, there is an integer $f_m(U, W)$ such that if T is an $\mathbb{N}$-gradable V-module and there is a surjective (logarithmic) intertwining operator in $I\binom{T}{U\ W}$, then $\dim T/C_m(T) < f_m(U, W)$. In particular, T is also C_m-cofinite as a V-module.*

Hereafter $I\binom{T}{U\ W}$ denotes the space of (logarithmic) intertwining operators of type $\binom{T}{U\ W}$. For $\phi \in \mathrm{Hom}(T, S)$ and $\mathcal{Y} \in I\binom{T}{U\ W}$, we can define an intertwining operator $\phi \circ \mathcal{Y}$ of type $\binom{S}{U\ W}$ by

$$(\phi \circ \mathcal{Y})(u, z)w := \phi\big(\mathcal{Y}(u, z)w\big) \quad \text{for } u \in U, w \in W.$$

$\mathcal{Y} \in I\binom{T}{U\ W}$ is called "surjective" if for any proper injection $\epsilon : E \to T$ and any intertwining operator $\mathcal{J} \in I\binom{E}{U\ W}$, we have $\epsilon \circ \mathcal{J} \neq \mathcal{Y}$. As an application, we will prove the following theorem.

Main Theorem *Let $m = 1, 2, \dots$ and let V be a VOA. If U and W are $\mathbb{N}$-gradable C_m-cofinite V-modules, then a fusion product $U \boxtimes W$ is also a C_m-cofinite $\mathbb{N}$-gradable V-module.*

As an application of Main Theorem, we have:

Corollary 2 *Let V be a simple VOA with $V \cong V'$, where V' denotes the restricted dual of V. If there is a V-module W such that W and its restricted dual W' are both C_2-cofinite, then V is C_2-cofinite.*

2 Proof of Key Theorem

We first assume that U and W are indecomposable. Sincc U and W are C_m-cofinite, there is $N \in \mathbb{N}$ such that $U = C_m(U) + E$ and $W = C_m(W) + F$, where $E = \bigoplus_{k=0}^{N} U_{r_U+k}$ and $F = \bigoplus_{k=0}^{N} W_{r_W+k}$ and r_U and r_W denote the lowest weights of V-modules U and W, respectively. We fix bases $\{p^i \mid i \in I\}$ of E and $\{q^j \mid j \in J\}$ of F consisting of homogeneous elements, respectively.

Let $\mathcal{Y} \in I\binom{T}{U\ W}$ be a surjective intertwining operator and let T' denote the restricted dual of T. For each $\theta \in T'$, $u \in U$, $w \in W$, we define a bilinear form

$$\langle \theta, \mathcal{Y}(u,z)w \rangle \in \mathbb{C}\{z\}[\log z]$$

by $\theta(\mathcal{Y}(u,z)w)$. Applying the idea in [2] to $\theta \in \mathrm{Annih}(C_m(T)) \cap T'$, we have the following lemma.

Lemma 3 *For $p \in U$, $q \in W$ and $\theta \in \mathrm{Annih}(C_m(T)) \cap T'$,*

$$F(\theta, p, q; z) := \langle \theta, \mathcal{Y}(p,z)q \rangle$$

is a linear combination of $\{F(\theta, p^i, q^j; z) \mid i \in I, j \in J\}$ with coefficients in $\mathbb{C}[z, z^{-1}]$. We are able to choose these coefficients independently from the choice of θ. Moreover, there is an integer $f_m(U, W)$ given by U and W only such that $\dim(C/C_m(T)) < f_m(U, W)$.

Proof We will prove Lemma 3 for $m = 1$. For $m \geq 2$, the proofs are similar. We will prove the first assertion in Lemma 3 by induction on $\mathrm{wt}(p) + \mathrm{wt}(q)$. Clearly, we may assume that $\mathrm{wt}(p) > N + r_U$ or $\mathrm{wt}(q) > N + r_W$. If $\mathrm{wt}(p) > N + r_U$, then $p = \sum_k v^k_{-1} a^k$ with $v^k \in V$ and $a^k \in U$. We note that this expression does not depend on the choice of θ. Since p is a linear sum, we are able to treat each term separately, that is, we may assume $p = v_{-1}a$ with $v \in V$ and $a \in U$. Then for $\theta \in \mathrm{Annih}(C_1(T)) \cap T'$, we have:

$$\begin{aligned} \langle \theta, \mathcal{Y}(p,z)q \rangle &= \langle \theta, \mathcal{Y}(v_{-1}a, z)q \rangle \\ &= \langle \theta, Y^-(v,z)\mathcal{Y}(a,z)q + \mathcal{Y}(a,z)Y^+(v,z)q \rangle \\ &= \langle \theta, \mathcal{Y}(a,z)Y^+(v,z)q \rangle, \end{aligned} \tag{2.1}$$

where $Y^-(v,z) = \sum_{h<0} v_h z^{-h-1}$ and $Y^+(v,z) = \sum_{h \geq 0} v_h z^{-h-1}$. This is a reduction on the sum of weights because $\mathrm{wt}(v_h q) < \mathrm{wt}(v) + \mathrm{wt}(q)$ for $h \geq 0$, that is, all terms of $Y^+(v,z)q$ have less weights than $\mathrm{wt}(v) + \mathrm{wt}(q)$. An important thing is that the processes of these reductions are irrelevant with the choice of θ.

Similarly, if $\mathrm{wt}(q) > N + r_W$, then we may assume $q = v_{-1}b$ with $v \in V$ and $b \in W$ and we have:

$$\begin{aligned} \langle \theta, \mathcal{Y}(p,z)q \rangle &= \langle \theta, \mathcal{Y}(p,z)v_{-1}b \rangle \\ &= \langle \theta, v_{-1}\mathcal{Y}(p,z)b \rangle + \left\langle \theta, \sum_{i=0}^{\infty} \binom{-1}{i} z^{-1-i} \mathcal{Y}(v_i p, z)b \right\rangle \\ &= \sum_{i=0}^{\infty} \binom{-1}{i} z^{-1-i} \langle \theta, \mathcal{Y}(v_i p, z)b \rangle. \end{aligned} \tag{2.2}$$

We also note that this calculation is independent of the choice of θ and this is also a reduction on the weights because $\mathrm{wt}(v_i p) + \mathrm{wt}(b) < \mathrm{wt}(v_{-1}b) + \mathrm{wt}(p)$ for $i \geq 0$. Therefore, $\langle \theta, \mathcal{Y}(p, z)q \rangle$ is a linear combination of $\{\langle \theta, \mathcal{Y}(p^i, z)q^j \rangle \mid i \in I, j \in J\}$ with coefficients in $\mathbb{C}[z, z^{-1}]$ and these coefficients are independent of the choice of θ.

We next prove the second assertion in Lemma 3. We consider an $|I| \times |J|$-dimensional vector $A = (\langle \theta, \mathcal{Y}(p^i, z)q^j \rangle)_{i,j}$. Then there is a square matrix $B \in M_{|I|\times|J|,|I|\times|J|}(\mathbb{C}[z, z^{-1}])$ which does not depend on the choice of θ such that

$$\frac{d}{dz}A = \left(\langle \theta, \mathcal{Y}\left(L(-1)p^i, z\right)q^j \rangle\right)_{i,j} = BA. \tag{2.3}$$

The space of solutions of the above differential equation (2.3) has a finite dimension and so the space of the choice of θ in A is also of finite dimension. Since $\mathcal{Y}$ is surjective, $\dim T/C_1(T)$ is bounded by a number $f_1(U, W)$ which does depend on U and W only. In particular, T is C_1-cofinite.

We next assume that $U = \bigoplus U^{(r)}$ and $W = \bigoplus W^{(k)}$ with indecomposable V-modules $U^{(r)}$ and $W^{(k)}$. Clearly, $U^{(r)}$ and $W^{(k)}$ are also C_1-cofinite as V-modules. Since $\mathrm{wt}(v_{-1}u) > \mathrm{wt}(u)$ for $v \in V$, $u \in U^{(r)}$ with $\mathrm{wt}(v) > 0$, we have $U^{(r)}/C_1(U^{(r)}) \neq 0$ for every r and so U is a finite direct sum of indecomposable modules. Similarly, so is W. For $\mathcal{Y} \in I\binom{T}{U\ W}$ and for each (r, k), we define $\mathcal{Y}^{(r,k)} \in I\binom{T^{(r,k)}}{U^{(r)}\ W^{(k)}}$ by restrictions, that is,

$$\mathcal{Y}^{(r,k)}\left(u^{(r)}, z\right)w^{(k)} := \mathcal{Y}\left(u^{(r)}, z\right)w^{(k)} \quad \text{for } u^{(r)} \in U^{(r)}, w^{(k)} \in W^{(k)}$$

and $T^{(r,k)}$ is the subspace of T spanned by all coefficients of $\mathcal{Y}^{(r,k)}(u^{(r)}, z)w^{(k)}$. As we showed, there are integers $f_1(U^{(r)}, W^{(k)})$ such that $\dim T^{(r,k)}/C_1(T^{(r,k)}) \leq f_1(U^{(r)}, W^{(k)})$. Since $T = \sum_{r,k} T^{(r,k)}$ and $C_1(T^{(r,k)}) \subseteq C_1(T)$, we have

$$\dim T/C_1(T) \leq \sum_{r,k} \dim T^{(r,k)}/C_1\left(T^{(r,k)}\right) \leq \sum_{r,k} f_1\left(U^{(r)}, W^{(k)}\right)$$

as we desired.

This completes the proof of Key Theorem. □

3 On Fusion Products

In this section, we would like to explain our approach to the fusion product of two modules. The fusion product of modules in the theory of vertex operator algebra are firstly defined by Huang and Lepowsky in several ways (see [3]). We go back to the original concept of tensor products, that is, as stated in the introduction of [3], it should be a universal one in the following sense, that is, if U and W are V-modules, then a fusion product is a pair $(U \boxtimes W, \mathcal{Y}^{U\boxtimes W})$ of a V-module $U \boxtimes W$ and an intertwining operator $\mathcal{Y}^{U\boxtimes W} \in I\binom{U\boxtimes W}{U\ W}$ such that for any V-module T and any intertwining operator $\mathcal{Y} \in I\binom{T}{U\ W}$, there is a homomorphism $\phi : U \boxtimes W \to T$ such that $\phi \circ \mathcal{Y}^{U\boxtimes W} = \mathcal{Y}$, that is,

$$\phi\big(\mathcal{Y}^{U\boxtimes W}(u,z)w\big) = \mathcal{Y}(u,z)w$$

for any $u \in U$ and $w \in W$. Unfortunately, in the theory of vertex operator algebra, unlike the categories of vector spaces, a fusion product module may not exist.

Our idea in this paper is that we first construct a $g(V)$-module $U \boxtimes W$. Furthermore, as we will see, $U \boxtimes W$ satisfies all conditions (Commutativity, etc.) to be a V-module except lower truncation property. Associativity is also true if it is well-defined, that is, if the lower truncation property holds on $U \boxtimes W$.

Let us construct it. We fix $U, W \in \mathrm{mod}_{\mathbb{N}}(V)$ and consider the set of surjective intertwining operators $\mathcal{Y}$ of U from W:

$$\mathcal{F}(U,W) = \left\{(F,\mathcal{Y}) \;\middle|\; F \in \mathrm{mod}_{\mathbb{N}}(V), \mathcal{Y} \in I\begin{pmatrix} F \\ U\ W \end{pmatrix} \text{ is surjective}\right\}.$$

Here the set of intertwining operators includes not only formal $\mathbb{C}$-power series but also all intertwining operators of logarithmic forms.

We define $(F^1, \mathcal{Y}_1) \cong (F^2, \mathcal{Y}_2)$ if there is an isomorphism $f : F^1 \to F^2$ such that $f \circ \mathcal{Y}_1 = \mathcal{Y}_2$, that is, $f(\mathcal{Y}_1(u,z)w) = \mathcal{Y}_2(u,z)w$ for any $u \in U$, $w \in W$. We also define a partial order $\leq$ in $\mathcal{F}(U,W)/\cong$ as follows: For $(F^1, \mathcal{Y}_1), (F^2, \mathcal{Y}_2) \in \mathcal{F}(W,U)$,

$$\mathcal{Y}_1 \leq \mathcal{Y}_2 \quad \Leftrightarrow \quad {}^{\exists} f \in \mathrm{Hom}_V\big(F^2, F^1\big) \text{ such that } f \circ \mathcal{Y}_2 = \mathcal{Y}_1.$$

We note that since $\mathcal{Y}_1$ and $\mathcal{Y}_2$ are surjective, f is uniquely determined. Clearly, if $\mathcal{Y}_1 \leq \mathcal{Y}_2$ and $\mathcal{Y}_2 \leq \mathcal{Y}_1$, then we have $(F^1, \mathcal{Y}_1) \cong (F^2, \mathcal{Y}_2)$.

Lemma 4 *$\mathcal{F}(U,W)/\cong$ is a (right) directed set.*

Proof For $\mathcal{Y}_1 \in I\binom{F^1}{U\ W}$ and $\mathcal{Y}_2 \in \mathcal{I}\binom{F^2}{U\ W}$, we define $\mathcal{Y}$ by

$$\mathcal{Y}(u,z)w = \big(\mathcal{Y}_1(u,z)w, \mathcal{Y}_2(u,z)w\big) \in \big(F^1 \times F^2\big)\{z\}[\log z] \quad \text{for } u \in U, w \in W.$$

Clearly, $\mathcal{Y} \in I\binom{F^1\times F^2}{U\ W}$. Let $F \subseteq F^1 \times F^2$ be the subspace spanned by all coefficients of $\mathcal{Y}(u,z)w$ with $u \in U$, $w \in W$, then $(F, \mathcal{Y}) \in \mathcal{F}(U,W)$. Moreover, by the projections $\pi_i : F^1 \times F^2 \to F^i$, we have $\pi_1 \circ \mathcal{Y} = \mathcal{Y}_1$ and $\pi_2 \circ \mathcal{Y} = \mathcal{Y}_2$, that is, we have $(F^1, \mathcal{Y}_1) \leq (F, \mathcal{Y})$ and $(F^2, \mathcal{Y}_2) \leq (F, \mathcal{Y})$ as we desired. □

Since $\mathcal{F}(U,W)/\cong$ is a direct set, we can consider a projective limit of $\mathcal{F}(U,W)/\cong$ and we denote it (or a representative of its isomorphism class) by $(U \boxtimes W, \mathcal{Y}^{U\boxtimes W})$ (or simply $U \boxtimes W$). Since T is a $g(V)$-module for any $(T, \mathcal{Y}) \in \mathcal{F}(U,W)$, a projective limit $U \boxtimes W$ is also a $g(V)$-module. In order to see the actions of $g(V)$ on $U \boxtimes W$, let us show a direct construction of projective limit. Let $\{(F^i, \mathcal{Y}_i) \mid i \in I\}$ be the set of all representatives of $\mathcal{F}(U,W)/\cong$ and set $\mathcal{Y}_i(u,z) = \sum_{j=0}^{K} \sum_{m\in\mathbb{C}} u^{\mathcal{Y}_i}_{(j,m)} z^{-m-1} \log^j z$ for $u \in U$. We consider the product $(\prod_{i\in I} F^i, \prod_{i\in I} \mathcal{Y}_i)$ and take subspaces F_r of $\prod_{i\in I} F^i$ spanned by all coefficients $\prod_{i\in I} u^{\mathcal{Y}_i}_{j,\mathrm{wt}(w)-1-r+\mathrm{wt}(u)} w$ of weights $r \in \mathbb{C}$ for homogeneous elements $u \in U$, $w \in W$ and $j \in \mathbb{N}$. We then set

$$F = \coprod_{r\in\mathbb{C}} F_r \quad \left(\subseteq \prod_{i\in I} F^i\right). \tag{3.1}$$

We note that F is $\mathbb{C}$-gradable by the definition.

For $v^1, v^2 \in V$, since the commutator formula

$$[v^1_n, v^2_m] = \sum_{i=0}^{\infty} \binom{n}{i} (v^1_i v^2)_{n+m-i} \tag{3.2}$$

holds on every F^i, we have the commutator formula (3.2) on $\prod_{i\in I} F^i$ and also on F. $L(-1)$-derivative property is also true on F since it is true on every F^i. For Associativity, since

$$(v^1_n v^2)_m = \sum_{i=0}^{\infty} \binom{n}{i} (-1)^i \{v^1_{n-i} v^2_{m+i} - (-1)^n v^2_{n+m-i} v^1_i\} \tag{3.3}$$

is true for each F^i, it is also true on F if the right side of (3.3) is well-defined on F. Therefore, if V acts on F truncationally, then F becomes a (weak) V-module.

Let us show $F \cong U \boxtimes W$ as $g(V)$-modules. Let $\pi_i : F \subseteq \prod_{h\in I} F^h \to F^i$ be a projection. By the definition, $\pi_i \circ (\prod_{h\in I} \mathcal{Y}_h) = \mathcal{Y}_i$. We note that if $(F^j, \mathcal{Y}_j) \leq (F^i, \mathcal{Y}_i)$ for $i, j \in I$, that is, if there is a surjective homomorphism $\phi_{i,j} : F^i \to F^j$ such that $\phi_{i,j} \circ \mathcal{Y}_i = \mathcal{Y}_j$, then $\pi_j \circ (\prod_{h\in I} \mathcal{Y}_h) = \mathcal{Y}_j = \phi_{i,j} \circ (\mathcal{Y}_i) = \phi_{i,j} \circ (\pi_i \circ (\prod_{h\in I} \mathcal{Y}_h))$. Therefore, for any $\alpha \in F$, $\pi_j(\alpha) = \phi_{i,j}(\pi_i(\alpha))$ and so we have the following commutative diagram:

$$\begin{array}{ccc} \pi_i : \left(F, \prod_{h\in I} \mathcal{Y}_h\right) & \longrightarrow & (F^i, \mathcal{Y}_i) \\ \| \text{ Identity} & & \downarrow \phi_{i,j} \\ \pi_j : \left(F, \prod_{h\in I} \mathcal{Y}_h\right) & \longrightarrow & (F^j, \mathcal{Y}_j). \end{array}$$

We note that since $\mathcal{Y}_i$ are surjective, $\phi_{i,j}$ are uniquely determined and $\phi_{jk}(\phi_{ij}(\mathcal{Y}_i(u, z)w)) = \mathcal{Y}_k(u, z)w = \phi_{ik}(\mathcal{Y}_i(u, z)w)$. Therefore, there is a surjective homomorphism $\phi : F \to U \boxtimes W$ such that $\phi \circ (\prod_{h\in I} \mathcal{Y}_h) = \mathcal{Y}^{U\boxtimes W}$. On the other hand, since $U \boxtimes W$ is a projective limit of $(F^i, \mathcal{Y}_i)_{i\in I}$, there are $\varphi_i : U \boxtimes W \to F^i$ such that $\varphi_i \circ (\mathcal{Y}^{U\boxtimes W}) = \mathcal{Y}_i$. Therefore, $\varphi_i \circ (\phi \circ (\prod_{h\in I} \mathcal{Y}_h)) = \mathcal{Y}_i = \pi_i \circ (\prod_{h\in I} \mathcal{Y}_h)$ and so $\varphi_i \phi = \pi_i$. Since $\bigcap_{h\in I} \mathrm{Ker}\, \pi_h = 0$ by the definition, $\mathrm{Ker}\, \phi = 0$ and so $(F, \prod_{h\in I} \mathcal{Y}^h)$ is isomorphic to a projective limit $(U \boxtimes W, \mathcal{Y}^{U\boxtimes W})$ of $\mathcal{F}(U, W)/\cong$.

Definition 5 We will call $U \boxtimes W$ a "fusion product" of U and W (even if it is not a V-module).

We have to note that since $\mathcal{Y}^{U\boxtimes W}$ is a projective limit, the powers of $\log z$ in $\mathcal{Y}^{U\boxtimes W}(w, z)u$ may not have an upper bound even if $U \boxtimes W$ is a (weak) V-module.

However, since $\mathcal{Y}^{U\boxtimes W}$ satisfies $L(-1)$-derivative property, Commutativity and Associativity and etc., we still treat it as an intertwining operator.

4 Proof of Main Theorem

Before we start the proof of Main Theorem, we prove the following lemma.

Lemma 6 *If an* $\mathbb{N}$*-gradable module* $U=\bigoplus_{m=0}^{\infty}U_{(m)}$ *is* C_1*-cofinite as a* V*-module, then* $\dim U_{(m)}<\infty$ *for any* $m\in\{0,1,\ldots\}$.

Proof We will prove it by induction on m. For $m=0$, since $U_{(0)}\cap C_1(U)=\{0\}$ by the definition of C_1-cofiniteness, we have $\dim U_{(0)}\le\dim(U/C_1(U))<\infty$. We next assume $\dim U_{(k)}<\infty$ for $k=0,\ldots,m-1$. Then

$$(V_i)_{-1}U_{(m-i)}:=\mathrm{Span}_{\mathbb{C}}\{v_{-1}u\mid v\in V_i,\, u\in U_{(m-i)}\}$$

is also of finite dimension for $i=1,2,\ldots,m$ since $\dim V_i<\infty$. Furthermore, since $U_{(m)}/\sum_{i=1}^{m}(V_i)_{-1}U_{(m-i)}\subseteq U/C_1(U)$, we have $\dim U_{(m)}<\infty$ as we desired. □

Let U be a C_1-cofinite $\mathbb{N}$-gradable module. Since $\dim U_{(m)}<\infty$ and $L(0)$ acts on $U_{(m)}$, $U_{(m)}$ is a finite direct sum of generalized eigenspaces of $L(0)$. Furthermore, since U is C_1-cofinite as a V-module, there is an integer r such that $U_{(m)}\subseteq C_1(U)$ for $m\ge r$. Hence there is a finite set $\{a_i\in\mathbb{C}\mid i\in I\}$ such that

$$U=\bigoplus_{i\in I}\bigoplus_{m=0}^{\infty}U_{a_i+m},$$

where U_{a_i+m} is a generalized eigenspace of U for $L(0)$ with eigenvalue a_i+m. We note $a_i\not\equiv a_j \pmod{\mathbb{Z}}$ for $i\ne j$. In particular, we have

$$\mathrm{wt}(U)\subseteq\bigcup_{i\in I}(a_i+\mathbb{N}),$$

where $\mathrm{wt}(U)$ denotes the set of all weights of elements in U. We may assume $U_{(r)}=\bigoplus_{i\in I}U_{a_i+r}$ by rearranging the $\mathbb{N}$-grading. We note that if $\phi:P\to Q$ is surjective and $P,Q\in\mathrm{mod}_{\mathbb{N}}(V)$, then $\mathrm{wt}(Q)\subseteq\mathrm{wt}(P)$.

Let us start the proof of Main Theorem.

By the same arguments in the proof of Key Theorem, we know that

$$\left(\bigoplus_r U^{(r)}\right)\boxtimes\left(\bigoplus_k W^{(k)}\right)\cong\bigoplus_{r,k}\left(U^{(r)}\boxtimes W^{(k)}\right)$$

and so it is sufficient to prove Main Theorem for indecomposable V-modules U and W.

As we showed in the proof of Lemma 6, for $T=\bigoplus_{i=0}^{\infty}T_{(i)}\in\mathrm{mod}_{\mathbb{N}}(V)$, we have $C_m(T)\cap T_{(0)}=\{0\}$ by the definition of $C_m(T)$. We also note that if $A=\bigoplus_{i=0}^{\infty}A_{(i)}$

and $B = \bigoplus_{j=0}^{\infty} B_{(j)}$ are $\mathbb{N}$-graded V-modules and $\phi : A \to B$ is a surjective V-homomorphism, then $\phi(C_m(A)) = C_m(B)$ since $v_{-m}\phi(a) = \phi(v_{-m}a)$ for $a \in A$ and $v \in V$. In particular, $\dim(B/C_m(B)) \le \dim(A/C_m(A))$.

On the other hand, by Key Theorem, there is a number $f_m(U, W) \in \mathbb{N}$ such that $\dim T/C_m(T) < f_m(U, W)$ for any $(T, \mathcal{Y}) \in \mathcal{F}(U, W)$. Therefore, there is $(S, \mathcal{J}) \in \mathcal{F}(U, W)$ such that for any $(T, \mathcal{Y}) \in \mathcal{F}(U, W)$ we have $\dim(T/C_m(T)) \le \dim(S/C_m(S))$. As we have shown, there is a finite set $\{a_i \in \mathbb{C} \mid i \in I\}$ such that

$$\mathrm{wt}(S) \subseteq \bigcup_{i \in I}(a_i + \mathbb{N}).$$

We fix $(S, \mathcal{J})$ and $\{a_i \mid i \in I\}$ for a while.

Lemma 7 *For any* $(T, \mathcal{Y}) \in \mathcal{F}(U, W)$, *we have* $\mathrm{wt}(T) \subseteq \bigcup_{i \in I}(a_i + \mathbb{N})$.

Proof For $(T, \mathcal{Y}) \in \mathcal{F}(U, W)$, there is $(P, \mathcal{I}) \in \mathcal{F}(U, W)$ such that $(P, \mathcal{I}) > (T, \mathcal{Y})$ and $(P, \mathcal{I}) > (S, \mathcal{J})$. Since $(P, \mathcal{I}) > (T, \mathcal{Y})$, we have $\mathrm{wt}(T) \subseteq \mathrm{wt}(P)$ and so it is sufficient to prove Lemma 7 for $(P, \mathcal{I})$. Therefore, we may assume $(T, \mathcal{Y}) > (S, \mathcal{J})$. Let $\phi : T \to S$ be a surjection. In this case, since $\phi(C_m(T)) \subseteq C_m(S)$ and $\dim(T/C_m(T)) \le \dim(S/C_m(S))$, we have $\mathrm{Ker}(\phi) \subseteq C_m(T)$ and $\mathrm{Ker}(\phi) \cap T_{(0)} = \{0\}$. Therefore, we have

$$\mathrm{wt}(T) \subseteq \mathrm{wt}(T_0) + \mathbb{N} \subseteq \mathrm{wt}(S) + \mathbb{N} \subseteq \bigcup_{i \in I}(a_i + \mathbb{N})$$

as we desired. □

We come back to the proof of Main Theorem. Since

$$(U \boxtimes W)_r \subseteq \prod_{(T,\mathcal{Y}) \in \mathcal{F}(U,W)} T_r$$

by the construction (3.1),

$$\mathrm{wt}(U \boxtimes W) \subseteq \bigcup_{i \in I}(a_i + \mathbb{N}).$$

Namely, the weights of elements in $U \boxtimes W$ is bounded below and so $v_n w = 0$ for a sufficiently large n for $w \in U \boxtimes W$ and $v \in V$. Therefore, $U \boxtimes W$ is a (weak) $\mathbb{N}$-gradable V-module. It is also C_1-cofinite as a V-module by Key Theorem.

The remaining thing is to show that $\mathcal{Y}^{U \boxtimes W}$ is a (logarithmic) intertwining operator. By the construction of $\mathcal{Y}^{U \boxtimes W}$, it has a form:

$$\mathcal{Y}^{U \boxtimes W}(u, z)w = \sum_{i=0}^{\infty} \sum_{n \in \mathbb{C}} u_{(i,n)} z^{-n-1} \log^i z. \tag{4.1}$$

We have to prove that powers of $\log z$ in (4.1) are bounded. Since the homogeneous subspaces of U, W and $U \boxtimes W$ are of finite dimension by Lemma 6, $L(0)^{\mathrm{nil}} = L(0) - \mathrm{wt}$ acts nilpotently on every homogeneous spaces $U_{(n)}$, $W_{(n)}$ and $(U \boxtimes W)_{(n)}$. Furthermore, since $L(0)^{\mathrm{nil}}$ commutes with all actions v_k for $v \in V$

and $k \in \mathbb{Z}$ and $U \boxtimes W$ is C_1-cofinite as a V-module, there is an integer N such that $(L(0)^{\mathrm{nil}})^N(U \boxtimes W) = 0$. Similarly, we may assume $(L(0)^{\mathrm{nil}})^N W = 0$ and $(L(0)^{\mathrm{nil}})^N U = 0$ by taking N large enough. From the $L(-1)$-derivative property for $\mathcal{Y}^{U\boxtimes W}$, we have

$$\begin{aligned}(i+1)u_{(i+1,n)}w = &-L(0)^{\mathrm{nil}}(u_{(i,n)}w) + \left(L(0)^{\mathrm{nil}}u\right)_{(i,n)}w \\ &+ u_{(i,n)}\left(L(0)^{\mathrm{nil}}w\right),\end{aligned} \tag{4.2}$$

Therefore, $u_{(k,n)}w = 0$ for $k \geq 3N$, $u \in U$ and $w \in W$, which implies that $\mathcal{Y}^{U\boxtimes W}$ is a (logarithmic) intertwining operator.

This completes the proof of Main Theorem. □

5 Discussion

At last, we would like to note one more thing. Although we have treated all logarithmic intertwining operators, if we restrict intertwining operators into formal $\mathbb{C}$-power series, that is, if we consider

$$\mathcal{F}_{fp}(U,W) = \left\{(T,\mathcal{Y}) \in \mathcal{F}(U,W) \mid \mathcal{Y}(u,z)w \text{ is a formal } \mathbb{C}\text{-power series} \right.$$
$$\left. \text{for any } u \in U \text{ and } w \in W\right\}$$

then $\mathcal{F}_{fp}(U, W)$ is still a direct set and so we are able to consider its projective limit, that is, another fusion product $U \boxtimes_{fp} W$ in this sense.

References

1. Borcherds, R.E.: Vertex algebras, Kac-Moody algebras, and the monster. Proc. Natl. Acad. Sci. USA **83**, 3068–3071 (1986)
2. Huang, Y.-Z.: Differential equations and intertwining operators. Commun. Contemp. Math. **7**, 375–400 (2005)
3. Huang, Y.-Z., Lepowsky, J.: A theory of tensor products for module categories for a vertex algebra, I. Sel. Math. New Ser. **1**, 699–756 (1995)

MIX
Papier aus verantwortungsvollen Quellen
Paper from responsible sources
FSC® C105338

If you have any concerns about our products,
you can contact us on
ProductSafety@springernature.com

In case Publisher is established outside the EU,
the EU authorized representative is:
Springer Nature Customer Service Center GmbH
Europaplatz 3, 69115 Heidelberg, Germany

Printed by Libri Plureos GmbH
in Hamburg, Germany